P9-ELX-255

COGENERATION PLANNER'S HANDBOOK

COGENERATION PLANNER'S HANDBOOK

By
Joseph A. Orlando, Ph.D., P.E.

Published by
THE FAIRMONT PRESS, INC.
700 Indian Trail
Lilburn, GA 30247

Library of Congress Cataloging-in-Publication Data

Orlando, J. A.
Cogeneration planner's handbook / by Joseph A. Orlando.
p. cm.
Includes bibliographical references and index.
ISBN 0-88173-111-0
1. Cogeneration of electric power and heat--Evaluation. I. Title.
TK1041.075 1991 333.79'3--dc20 90-40857
CIP

Published by The Fairmont Press, Inc.
700 Indian Trail
Lilburn, GA 30247

Printed in the United States of America

10 9 8 7 6 5 4 3 2 1

ISBN 0-88173-111-0 FP

ISBN 0-13-372921-4 PH

Distributed by Prentice-Hall, Inc.
A division of Simon & Schuster
Englewood Cliffs, NJ 07632

Prentice-Hall International (UK) Limited, London
Prentice-Hall of Australia Pty. Limited, Sydney
Prentice-Hall Canada Inc., Toronto
Prentice-Hall Hispanoamericana, S.A., Mexico
Prentice-Hall of India Private Limited, New Delhi
Prentice-Hall of Japan, Inc., Tokyo
Simon & Schuster Asia Pte. Ltd., Singapore
Editora Prentice-Hall do Brasil, Ltda., Rio de Janeiro

Table of Contents

Chapter 1

INTRODUCTION

With the conclusion of the twentieth century and despite the many uncertainties about our energy future, it is highly likely that both the role and importance of electric power will continue to change and grow. These changes may have an impact which surpasses all that has happened to date. It was during the early years of this century that the concepts and inventions discovered by Humprey Davy, Nikola Tesla and Thomas Edison were forged together creating the ability for the controlled production and use of electric power. Then the many small industrial "power houses" which produced steam for mechanical drives as well as for process and heating discovered the advantages of the electric motor as a source of mechanical power. Wires replaced steam pipes and belted pulley systems, and motors replaced the steam drive. These small power houses quickly adopted cogeneration in order to produce the more useful electric power and steam, which was still required for heating and process use.

Subsequently, Edison developed his concept of an electric system that included centralized power generation delivering power to many electric customers. As these electric systems developed and competed with one another, it was Samuel Insul who put the last component of the modern electric system in place by developing the concept of electric power as a natural monopoly subject to control through regulation rather than by the market place. With the institution of state and, subsequently, federal regulation, the industry matured and developed inter-utility grids and regional power pools. Over the period 1940 through 1970, the electric utility industry delivered an increasingly useful and reliable product at a decreasing cost. Today, the basic electric industry developed in the first half of this century has emerged as the supplier of most of our non-transportation power needs.

During that same time period, through a combination of federal regulatory and market conditions, fuel costs were also maintained at low levels. Most end users found that the most cost effective energy supply was power purchased from an electric utility and either natural gas or oil for their packaged on-site boilers.

While cogeneration could be an efficient energy alternative, its application was limited primarily to those industrial facilities, such as petroleum refineries, pulp and paper mills or chemical plants, where a unique combination of energy requirements and the availability of by-product fuels and on-site engineering made cogeneration cost effective. Within the utility industry, a number of electric utilities developed a significant amount of cogeneration capacity, distributing the cogenerated power through the electric grid and selling steam through a district heating system. However, as shown in Figure 1-1, the amount of power produced by cogenerators became a smaller and smaller component of the total generation.

It was not until the late 1960s and early 1970s that any serious effort was made to revitalize interest in cogeneration. The first attempt was led by gas utilities as they attempted to expand their markets and, more importantly, develop non-seasonal gas loads. Cogeneration systems with high annual load factors provided this opportunity. After some initial success, resistance from competitive electric utilities and the gas shortages of the early 1970s resulted in a falling off of interest in Total Energy, as the concept was then called.

Almost immediately thereafter, the perceived need to conserve more critical resources such as oil and natural gas led to a renewal of interest in cogeneration as an energy and resource conservation measure. A properly designed and operated cogeneration system would require less energy to meet an end user's energy requirements than would be required using the more common approach of separate power and fuel systems. As a result of these concerns, the National Energy Acts of 1978 included provisions intended to encourage cogeneration development. Specifically, the Public Utilities Regulatory Policies Act (PURPA) removed barriers that developed in the 1960s by allowing cogenerators to interconnect with the

FIGURE 1-1
COGENERATION TREND

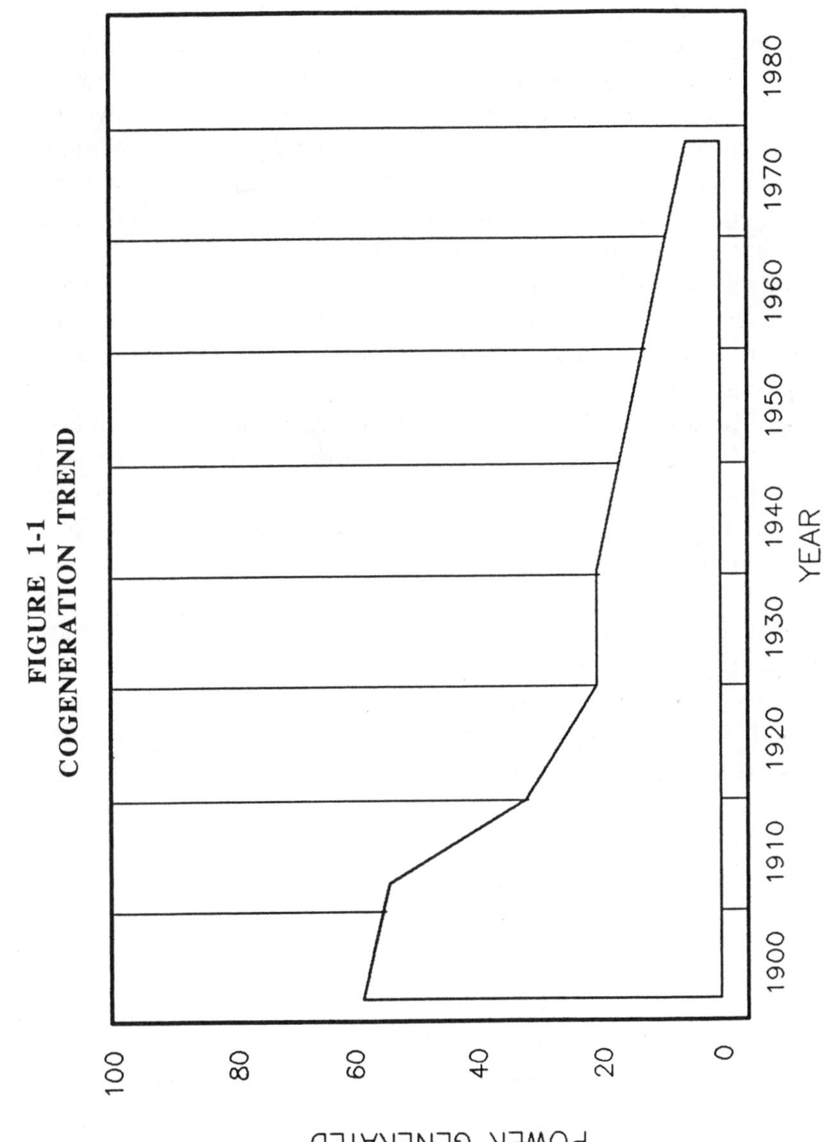

electric utility grid, both to purchase and to sell power. The cogenerators' power purchases were to be at non-discriminatory rates, while power sales by cogenerators were to be at rates that were based on the costs that the utility would incur to produce the power. The two most important provisions of PURPA applied to power sales to utilities. These incentives were exemption from rate regulation by the Federal Energy Regulatory Commission (FERC), the body charged with approving rates for inter-utility or wholesale power transactions, and the pricing of that power at the purchasing utility's avoided cost. These benefits were only available to those cogenerators who met federally specified operating and efficiency requirements, and to Small Power Producers using renewable sources of energy.

The option to sell power to an electric utility proved to be the basis of the resurgence of interest in cogeneration. By some counts, the cogeneration industry has brought over 15,000 megawatts of new power generation capacity on-line since the passage of PURPA in 1978, with another 26,000 megawatts in active development. Small Power Producers have developed another 5,000 megawatts. While the future is uncertain, it is anticipated that the cogeneration industry will provide between 20,000 megawatts and 40,000 megawatts of new capacity in the next decade. Today, cogenerators, Small Power Producers, and Independent Power (non-regulated central power plants) are considered vital to the nation's ability to meet future power requirements.

This chapter provides a basic introduction to the concept of cogeneration and the basis for its economic viability.

THE BASICS

Cogeneration is defined as the sequential use of a primary energy stream to produce two useful energy forms–thermal and power. It is an energy conversion process and is, in itself, independent of 1) the form of the output–power can be either electricity or shaft horsepower, thermal energy can be either heating or cooling; 2) the disposition of the power–it can be sold to a utility or used at the site; 3) ownership–a cogeneration system can be owned by a regulated utility, an energy end user or an independent developer; or 4) size–the cogeneration system can be viable in modules as small as a few kilowatts and as large as several hundred megawatts.

Figure 1-2 illustrates a typical cogeneration system. In this illustration fuel, either natural gas or oil, is burned in a combustion turbine producing shaft horsepower which powers an electric generator. The electricity can be used on-site in the facility, sold to a utility or other organization, or a combination of both. The turbine exhaust gases, which may be at temperatures of 1,000°F or more, flow through a boiler where they produce steam which is used on-site for process requirements. This type of cogeneration system where the fuel is first used to produce power, with the byproduct heat recovered and usefully applied, is defined as a *topping cycle*. In contrast, a *bottoming cycle* cogeneration system is one in which the fuel is first used to produce heat for some thermal process with the process exhaust gases then used to produce electric power.

Figure 1-3 illustrates the more common or conventional utility approach. A central power plant converts fuel, which can be coal, oil, natural gas or enriched uranium, to power. That power is then delivered to the energy end user using a transmission and distribution system. The heat lost from the power plant boilers or rejected from the plant's condensers, which is usually at a few hundred degrees or less, is usually dumped into the atmosphere through a cooling tower or into a nearby lake or river. The end user's steam requirements are usually met by burning oil or natural gas in a packaged on-site boiler.

Both systems, conventional utilities and cogeneration, deliver the same amount of power and steam to the end user. The difference between the two, and the potential for conservation, is the amount of fuel that must be input to the system.

As shown in Figure 1-2, a 20-megawatt combustion turbine is capable of operating at an efficiency of almost 33%. In addition, over 40% of the input can be recovered and used to produce steam and the overall system efficiency can approach 74%. In comparison, the typical central power plant operates at an efficiency of 33%, with the transmission and distribution system being between 90% and 95% efficient. The overall efficiency of the electric component of the conventional utility system is approximately 30%. The on-site boilers that produce steam can operate at rather high efficiencies,

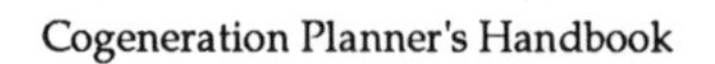

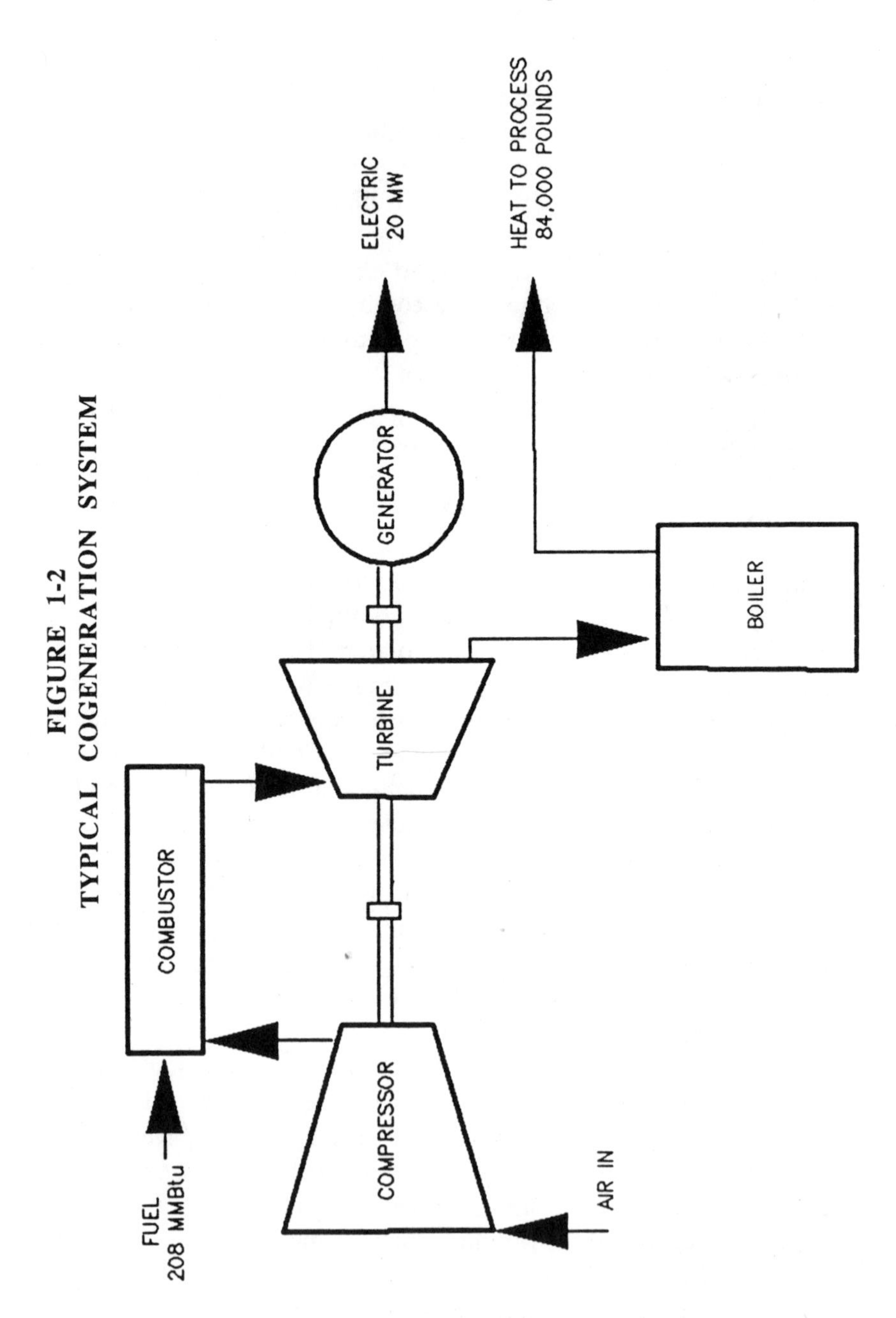

FIGURE 1-2
TYPICAL COGENERATION SYSTEM

FIGURE 1-3
TYPICAL CONVENTIONAL UTILITY SYSTEM

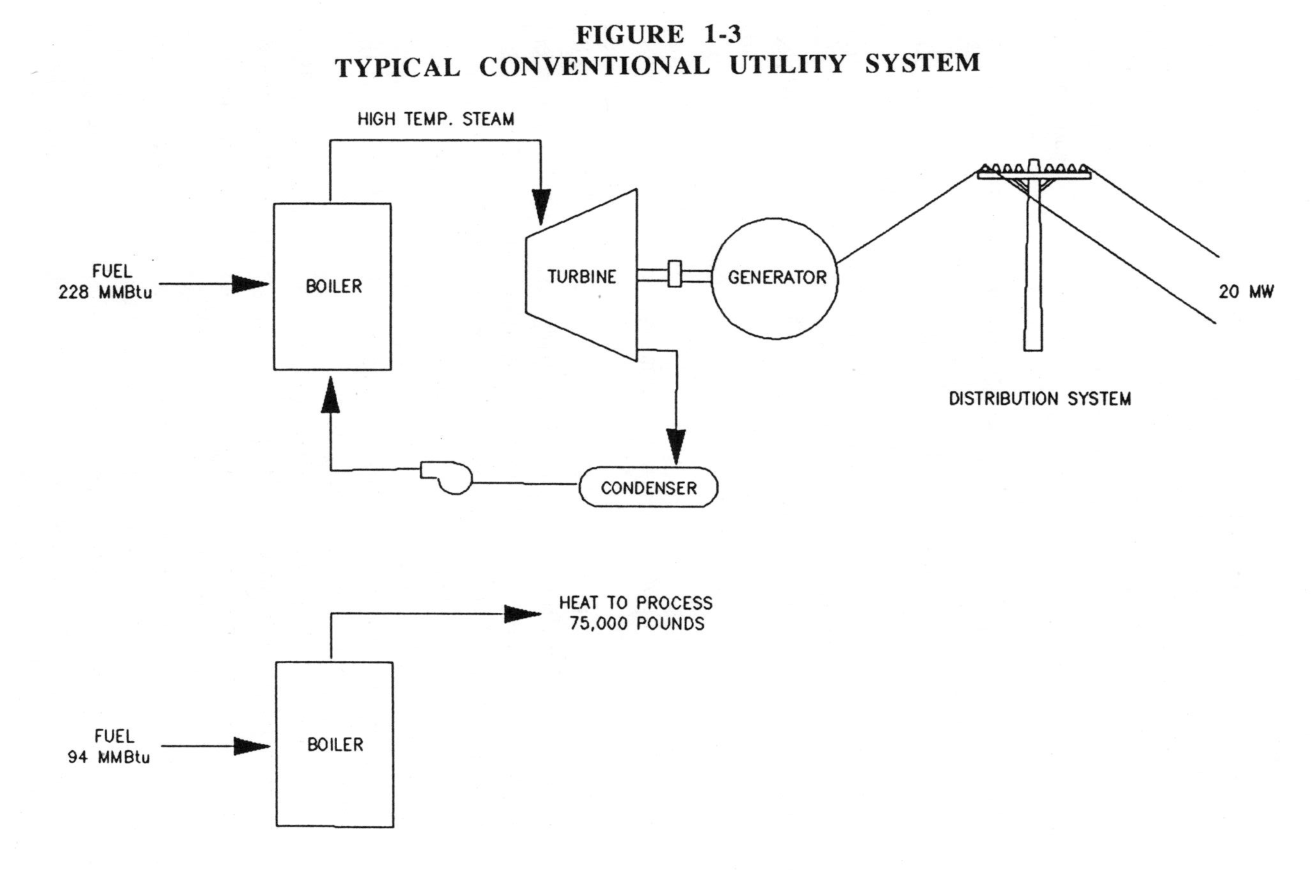

ranging from between 70% to 85%. Depending on an end user's mix of power and thermal requirements, the overall efficiency of the typical energy customer can vary between 35% for end users with modest heating requirements and 65% for end users with modest power requirements.

The magnitude of these efficiency differences can be illustrated with reference to Figures 1-2 and 1-3. In this case, it is assumed that the end user requires 20 megawatt hours of power and 75,000 pounds of steam. At a 30% efficiency, the conventional electric utility will require 228 million British Thermal Units (Btus) to produce the required power, while the boiler which operates at an 80% efficiency will require almost 94 million Btus (MMBtu). The total energy requirement is 322 MMBtu. In comparison, the combustion turbine, operating at an efficiency of 32.8%, will require a total fuel input of 208 MMBtu. However, it will also produce 84 MMBtu of heat which can be used to produce almost 84,000 pounds of steam, thus concurrently meeting the end user's thermal requirements. The total fuel input to the cogeneration system is 208 MMBtu or over 35% less than the energy input to the conventional system.

The cost impact of this efficiency gain can be illustrated using Table 1-1 which presents the variable operating costs for a 390-megawatt, coal-fired, base-loaded central power plant. The most significant variable cost component is fuel, which totaled $32.3 million for 1988 ($.0124 per kilowatt hour) or 76% of the total cost. The utility's cost of fossil fuel for its entire system was over $200 million or $.0196 per kilowatt hour. With revenues averaging $.087 per kilowatt hour, the cost of fuel was but a small fraction of the retail customer's overall cost of power.

This central power plant had a heat rate of slightly more than 9,570 Btus per kilowatt hour (equivalent to an efficiency of 35.6%) with a total annual input of just less than one million tons of coal (24.3 million MMBtu). Approximately 64.4% of the fuel input (over 16.5 million MMBtu), equivalent to $20.8 million, was rejected to the environment while another 2.6%, or $840,000, was lost in transmission and distribution. Only 33% of the input energy was delivered to the end user, while a total of 67% of the fuel input, costing $21.6 million, was lost or rejected.

Table 1-1
Central Power Plant Variable Operating Costs

Expense	Annual Cost ($)	Percent of Total (%)
OPERATIONS		
Supervision & Eng	440,000	1.03
Fuel	32,341,000	76.15
Steam Expenses	411,000	0.97
Electric Expenses	155,000	0.36
Misc. Steam	2,194,000	5.17
MAINTENANCE		
Supervision & Eng	581,000	1.37
Structures	828,000	1.95
Boiler Plant	3,647,000	8.59
Electrical Plant	1,623,000	3.83
Misc. Steam Plant	246,000	0.58
TOTAL	42,466,000	100.00

Applying these same ratios to the utility's total fossil requirements indicates that these central power plants reject annually the equivalent of almost $129 million in fuel costs. It will not be possible to effectively use all this energy, however, even a small increase in the efficiency with which source energy is utilized can result in significant cost savings.

PURPA

In 1978, the Public Utility Regulatory Policies Act was enacted, leveling major barriers to the development of cogeneration facilities. The most significant of PURPA's provisions were the right to interconnect with an electric utility for the purchase and/or sale of electric power at non-discriminatory rates, and exemption from federal regulation of rates for sales to utilities. These provisions, when coupled with 1981 tax incentives for cogenerators, provided the basis for the first wave of post PURPA development.

Interconnection for Power Sales and/or Purchases

The right of a cogenerator to interconnect to the utility carried with it the ability to sell to and/or purchase power from that utility. Along with the right to interconnect, the cogenerator also has the obligation to pay for any incremental interconnection costs incurred by the utility. In addition, the legislation and subsequent rules developed by FERC established the basis for the pricing of those transactions.

Under PURPA, the rate that the cogenerator receives for power (either capacity, energy or both) that is sold to an electric utility is equal to the cost that the utility avoids by the purchase of that power. The rather simple concept of *avoided cost* includes several important provisions.

1. The rate that the cogenerator receives for power sales to the utility is based on the utility's costs and not the cogenerator's costs.

2. The rate that the cogenerator receives for power sales can include two components; a capacity component and an energy component, and the rates for each component must be based respectively, on the utility's costs for capacity and energy that are avoided through the purchase of cogenerated power.

3. The rate that the cogenerator receives for power is tied to a specific utility source of power, either a utility-owned power plant or power purchased from another utility, *and*, because the rate is tied to a specific source of power, the rate paid for energy and capacity from that same source.

4. The cogenerator could sell energy to a utility at the energy-only cost even if that utility does not need additional capacity.

The result of these provisions is that the cogenerator who meets FERC requirements is entitled to sell capacity and energy to an electric utility, replacing an investment in capacity that the utility would have been required to make. If the cogenerator does displace the need for the utility-owned capacity, it is entitled to receive a rate for the power that transfers the avoided power costs to the cogenerator. By linking the avoided cost to a specific source of utility-supplied power, the rate is established with consistent capacity and energy components. If the cogenerator eliminates the need for a more costly baseloaded power plant burning a less expensive fuel such as coal, the cogenerator is then entitled to a high capacity payment and a low energy charge. Conversely, if the cogenerator replaces a less expensive peaking power plant that burns a more costly fuel, then the cogenerator receives a lower capacity payment but a higher energy payment. Thus, independent cogenerators can become part of the national electric grid.

Because the purchases from a cogenerator would cost no more than the cost of utility-produced power, the electric utility's retail customers would pay nothing additional for power produced by cogenerators. In addition, because the utility is only required to pay for actual purchases, these same customers are insulated from the financial risk associated with the construction of that capacity.

The second benefit of interconnection is the ability to purchase power from the electric utility, either to supplement or to replace cogenerated power. The availability of utility-supplied power at non-discriminatory, *partial service* rates allows the cogenerator to design and operate a system that is based on the end user's own thermal and electrical requirements, thus maximizing the overall energy use efficiency. The availability of supplemental power at non-discriminatory rates allows a cogenerator to operate the on-site capacity to satisfy the base or minimum electric or thermal requirement. Requirements that exceed the on-site capacity are met with power purchased from the electric utility. A cogeneration facility sized to the continuous or base loads of an end user will minimize the amount of rejected energy.

The cogenerator also has the right to purchase power from the utility during scheduled or unscheduled outages of the cogeneration system. Again, PURPA requires these costs to be non-discriminatory and cost based, thus insuring that neither the utility's full service customers nor the partial service cogenerators bear any special cost burdens.

Exemption from Federal Rate Regulation

The second major provision of PURPA is an exemption of those cogenerators who sell power for resale by an electric utility from rate regulation by the FERC. An immediate result of this provision was the development of a number of large cogeneration systems selling power to an electric utility under a long-term contract, with the steam sold to a local industrial host. This revenue structure provides the security of a long-term market for cogenerated power at the utility's avoided cost. In addition, it allows non-utility developers to enter the power generation market without fear that their profits from traditional non-generation business activities would be subject to FERC scrutiny or control.

While PURPA provides an exemption from FERC regulation for wholesale transactions, it does not address the regulation of any retail transactions between cogenerators and end users, as these are subject to state regulation.

Qualifying Facility Status

The PURPA benefits were intended to encourage increased efficiency and, to insure this efficiency gain, FERC established specific operating and efficiency requirements as a condition for the availability of PURPA benefits. While anyone can develop a cogeneration system, only those cogenerators who are considered a *Qualifying Facility (QF)* by FERC are entitled to PURPA benefits. FERC has established two performance tests for QF status for cogeneration topping cycles.

The first criterion is output. The FERC requirement is that a minimum of 5% of the cogeneration system's annual output must be in the form of useful thermal energy. This thermal energy can take the form of steam, hot water or chilled water as might be produced from an absorption chiller.

The second criterion is operating efficiency and FERC established two tests based on the following efficiency formula.

(1) FERC EFF = (Power Out + [Useful Thermal/2])/Fuel Input

where

FERC EFF	=	The computed efficiency for purposes of the FERC efficiency test
Power Out	=	The useful power produced by the cogeneration system
Useful Thermal	=	The useful thermal energy supplied by the cogeneration system. The useful thermal energy may not be greater than the recovered thermal energy.
Fuel Input	=	The fuel input to the cogeneration system measured in the fuel's Lower Heating Value

The efficiency calculation considers electrical energy to be twice as valuable as thermal energy, and weighs it accordingly.

The FERC rule requires that a cogeneration system that is fired with oil and/or natural gas and whose thermal output is 15% or more of the total energy output must operate at an annual efficiency, as defined by Equation 1, of 42.5% or more. The efficiency can be computed for any consecutive 12-month period and need not be based on a conventional calendar year. Cogeneration systems with a total useful thermal output of less than 15%, but at least 5%, must achieve an annual efficiency of at least 45.0%.

Finally, there is an ownership test which is applicable to the determination of QF status. Simply an electric utility, or an electric utility affiliate, cannot own more than 50% of a QF.

PURPA Results

Since the passage and implementation of PURPA, cogeneration developers have completed in a total of 15,000 megawatts of capacity, most of which sells cogenerated power to an electric utility. In fact, many of these plants produce only the amount of thermal energy required to maintain QF status, and their economics are highly dependent on power sale revenues with only modest levels of steam sale revenues.

The growth of the PURPA industry and its ability to contribute to the electric grid's needs for power is so large that, in many states, the potential independent generation capacity exceeds current demand. The result has been the evolution of capacity bidding procedures, wherein a utility in need of additional capacity conducts an auction to determine both who will provide that capacity and what the utility will pay for the capacity and energy. While many factors are considered in these auctions, cost has been a dominant factor and most of the capacity has been awarded to independent suppliers with the lowest overall cost.

The PURPA industry was originally based on two components: cogenerators and Small Power Producers. More recently, a third component, the *Independent Power Producer (IPP)*, has evolved. An IPP is a non-utility owned source of generation which sells all of its output to an electric utility. Because the IPP does not engage in retail transactions, it is subject to rate regulation by the FERC. Additionally, most IPPs would be subject to financial controls imposed on electric utility holding companies by the Public Utility Holding Company Act of 1935 (PUHCA).

A characteristic of IPPs is that, in and of themselves, they are generally perceived to be lacking in any monopoly power, or even strong market power. Secondly, they are not required to be either cogenerators or small power producers and as such they do not necessarily operate as efficiently as QFs. PUHCA restrictions on electric utility holding companies are likely to inhibit the full development of this concept.

In summary, PURPA enabled the development of an independent power industry which has brought about remarkable changes to the concept of a regulated electric utility. Insul's doctrine, that generation was a natural monopoly with the requirement that only electric utilities should own generation, has been shattered. The disruptions caused by the high, and sometimes imprudent, costs of nuclear power plants have provided utilities and regulators with an added incentive to identify risk-free mechanisms for financing future capacity additions. The performance of the independent power industry over the past decade has demonstrated that the competitive market place can provide one alternative.

REVIEW

Cogeneration is but one option for utility service. It is based on the concept that end users who require two energy forms–power and thermal–may be more cost effectively served using an on-site efficient utility plant.

In general, cogenerated thermal energy, whether it is steam, hot water or chilled water, is used on-site reducing the amount of fuel that is required in a conventional boiler. There are, however, several practical alternatives for the use of cogenerated power. First, it may be used on-site by the end user reducing the amount of power that is purchased from an electric utility. This type of cogeneration facility is referred to as an *internal use* project and is common for plants ranging from tens of kilowatts to tens of megawatts. A second use for cogenerated power is its sale to an electric utility, and this type facility is referred to as a *sellback* project. The typical sellback plant ranges in size from tens of megawatts to hundreds of megawatts. Finally, a single cogeneration facility can produce power both for sale to a utility and for use on-site. Both internal use and sellback cogeneration projects are based on the same concept, the efficient generation of power and thermal energy, and the technologies discussed in this book are applicable to both.

The analysis of the viability of cogeneration at any specific site will be dependent on the energy use and cost characteristics of that site. Whether designed as a sellback or internal use project, a cogeneration system's economic viability will be strongly influenced by the ability to utilize recoverable heat and, therefore, each site's thermal requirements must be examined in detail. An internal use cogeneration project will require a detailed analysis of the site's electric use patterns and purchased power rates. While energy use patterns and retail rates are less of a concern with sellback projects, an electric utility's need for additional generation capacity and the potential cost of that capacity is critical.

This text is divided into two parts. The first half deals with technologies including prime movers, heat recovery, cooling, electric generators, interconnection with the electric grid, and thermal distribution systems. A discussion of integrated, factory assembled or packaged cogeneration systems is also included. The second half describes the overall development process and analytic techniques with particular attention to the level and cost of the analyses required at each step of development. Collection of site data and equipment performance modelling is reviewed, with a subsequent discussion of project financing and third-party ownership. Finally, several analytic case studies are included.

Chapter 2

PRIME MOVERS

Cogeneration consists of a concept whereby both power and useful thermal energy are sequentially produced from a single energy source. While the applications of that concept can be quite diverse, utilizing many different technologies, the heart of any cogeneration system is a prime mover which converts thermal or chemical energy into power. The most commonly employed, commercially available cogeneration prime movers include internal combustion reciprocating engines, combustion turbines and steam turbines. Other technologies, less frequently found in cogeneration systems, include reciprocating steam engines, Stirling engines and fuel cells. These prime movers offer potential cogenerators a broad range of sizes (ranging from a few hundred watts through to several hundred megawatts) and performance characteristics. Prior to any discussion of prime movers, it is useful to clarify a few basic definitions. In this discussion, the term engine is broadly applied to devices for converting energy from one form to another rather than to a specific type of prime mover.

BASIC TERMS AND DEFINITIONS

The most basic considerations in a discussion of a cogeneration system's performance are the capacity and efficiency. The prime mover's *rating* is the power output of the engine under specified ambient and operating conditions. Several different ratings may be provided for the same prime mover depending on the anticipated operating mode. The *prime power or continuous duty* rating is the rating which the manufacturer assigns to the prime mover when it is operated as a primary source of power. Typically this rating is applied to a prime mover that is continuously operated 24 hours a day at a steady output level. The *standby or intermittent duty* rating is applied when the engine is operated for 24 hours a day for a limited number of days. As the name implies, this type of operation is associated with standby or emergency use when the primary source of power is unavailable. Finally, the *overload or short term* rating is the maximum output for short periods of time, and typically for one or two hours in any single day. These ratings are all listed at specific temperatures and altitudes and they each specify the amount of power that the engine can deliver without excessive maintenance costs or shortened engine life.

The second important consideration is *efficiency* and, as was the case with the rating, it is necessary to consider different efficiencies. The *mechanical efficiency* is the prime mover's mechanical output divided by the fuel energy input. It may be expressed as a percentage, where the output and input are converted to consistent units, or as is frequently the case with engines, it may be expressed as a heat rate. The engine *heat rate* is a measure of the system's electrical efficiency and is defined as the amount of fuel energy input required to produce a kilowatt hour of power. *Steam rate* , a special measure applied to steam turbines, is defined as the amount of steam, usually measured in pounds, required to produce a kilowatt hour.

While efficiency may seem a rather straightforward concept (simply equal to output divided by input), the measurement of input energy requires special attention. For purposes of engine performance, the energy content of a fuel is its heating or calorific value, a parameter which can be readily determined.

When a hydrocarbon fuel such as oil or natural gas is burnt, one of the products of combustion is water vapor which is created as steam. Part of the chemical energy of the fuel must be used to create the water vapor and that heat is stored in the vapor as the latent heat of vaporization. If the water is exhausted from the engine as a vapor along with the carbon monoxide and carbon dioxide, the heat of vaporization is also exhausted. If the water vapor is condensed within the engine and the heat of vaporization used to produce power, then that energy is usefully applied.

The total heat that is released during combustion is defined as a fuel's *Higher Heating Value (HHV)*. The amount of heat that is available after the water's heat of vaporization is deducted, is the fuel's *Lower Heating Value (LHV)*. Since the heat of vaporization is typically exhausted, most engine manufacturers specify the engine's performance based on LHV. The ratio of the LHV to HHV for natural gas is approximately 0.90; light oils have a ratio of 0.93 and heavy oils have a ratio of 0.96. Incorrect use of these terms can result in a 10% underestimate of the amount of fuel that is required to operate an engine.

Cogeneration systems are concerned with the production of two useful forms of energy–power and heat–and there are terms which are specifically applicable to the thermal side of a prime mover. *Recoverable heat* is the amount of useful thermal energy that can be recovered from a prime mover. Unlike heat rate, the amount of recoverable heat is not a specific number based on the engine's characteristics, but rather is based on the thermal requirements. The amount of recoverable heat will depend on the difference between the temperature at which heat is required and the temperature at which it is available from the prime mover. Thus, if an application requires heat at 1,000°F and the maximum temperature available from the engine is 800°F, unless another source of energy is available to boost the temperature from 800°F to 1,000°F, the amount of useful recoverable heat is zero.

The last term considered here is the equipment's *scheduled availability*, which is the fraction of the time that an engine can be expected to be available for service after the time required for scheduled maintenance procedures is subtracted.

RECIPROCATING ENGINES

Internal combustion, reciprocating engines (see Figure 2-1) have proven to be cost effective in cogeneration applications. While these engines have been used for direct drive applications including mechanical equipment, air compressors, mechanical chillers and heat pumps, their use in cogeneration systems is primarily for electric power generation. Reciprocating engines are extremely efficient in smaller sizes and are available in a large number of different sizes.

FIGURE 2-1
RECIPROCATING ENGINE

With typical availabilities of 7,600 to 8,400 hours per year, they are popular in applications of up to two or three megawatts.

Concept

The basic concept underlying the operation of an internal combustion engine is to rapidly burn fuel in a cylinder where the expanding gases operate against a piston, causing it to move in a reciprocal motion. This motion is then converted into a rotational motion by an off-center linkage to a shaft, with the shaft then driving a generator. While the basic operation may be simple, these engines can be categorized into a number of types depending on various criteria.

The most common reciprocating engine is the *spark ignited, Otto Cycle,* where the air and fuel mixture are compressed by the piston and then ignited by an externally supplied spark. As the compressed air/fuel mix burns, it expands and does work by acting against the piston. The Otto Cycle is capable of operating on a broad range of fuels including gasoline, natural gas, propane and sewage plant or landfill produced methane.

The alternative thermodynamic cycle, the *Diesel Cycle,* also ignites the fuel in a cylinder; however, no external spark is required. In this cycle, the air is compressed by the piston until the air temperature is higher than the fuel's ignition temperature. At that point in the cycle, the fuel is injected into the cylinder where it spontaneously ignites, burns and expands against the piston. The flow of fuel to the cylinder is controlled so that the burning gases maintain constant pressure against the piston. Diesel engines generally operate on liquid fuels ranging from No. 2 distillate oil through to No. 6 residual oil.

Both the engine capacity and efficiency are directly related to the engine's *compression ratio,* which is defined as the ratio of the volume of the uncompressed fuel/air mixture to the minimum volume of the cylinder. Higher compression ratios increase efficiency, and engine designs seek to achieve high compression ratios. However, as was pointed out above, high compression ratios can ignite the engine fuel and, therefore, the compression ratio of an Otto Cycle engine must be kept low enough to avoid spontaneous ignition (dieseling) of the air/fuel mix. Among the negative consequences of preignition are loss of efficiency and capacity, and increased maintenance costs. Methane can be compressed to a ratio of 15 to 1; however, most spark-ignited, natural gas engines operate with compression ratios ranging from 9:1 to 12:1. While the Otto Cycle reciprocating engine has a higher theoretical efficiency than does the Diesel Cycle engine, limitations on the compression ratio result in better performance from the Diesel Cycle.

Engines can also be categorized as to fuel capability. *Gaseous fuel* engines can burn a broad variety of fuels including natural gas, propane, butane, methane, sewage treatment and landfill gases. They can operate on fuels with a calorific value as low as 200 Btu per cubic foot, although these low quality fuels require derating of the engine. When gases with a low or varying heating value are available, they are often mixed with pipeline quality natural gas to insure a constant quality gas to the engine. *Liquid fuel* engines can operate on a variety of oils ranging from light distillates through to No. 6 or residual oils. Heavier oils require additional fuel conditioning and increase maintenance costs; however, they are usually available at significant discounts as compared to natural gas and light oils. *Dual fuel* reciprocating engines can operate on both liquid and gaseous fuels. In general, these engines require a small amount of liquid fuel even when operating in the gaseous mode. This pilot charge, which can be 5% to 10% of the fuel requirement, acts as an ignition source for the gaseous fuel. When required, these engines can operate on 100% liquid fuel.

A second criterion for categorizing an engine is its type of *aspiration*; either naturally aspirated or supercharged. *Naturally aspirated* engines supply the air/fuel to the piston cylinder at atmospheric pressure and, therefore, only require very low pressure gaseous fuels. Typically, the low pressure natural gas that is common to residential gas distribution systems is at an adequate pressure. *Supercharged* or *turbocharged* engines deliver a higher pressure air/fuel mix to the piston cylinder. The increase in the quantity of air that is delivered to the cylinder permits combustion of a greater quantity of fuel and, therefore, increases the output of the engine. Supercharged models of an engine block will have a rating which is 30% to 40% greater than the rating for the same block with natural aspiration, and are less expensive on a dollars per kilowatt basis than are naturally aspirated engines. Additionally, the process of supercharging increases the mixing of the air/fuel mix thereby improving efficiency.

There are alternative approaches to supercharging. One of the most common consists of using the engine exhaust gases to drive a turbine, which in turn compresses the intake air. The fuel is then mixed with the higher pressure air. This *turbocharger* approach has several disadvantages. First, it is necessary to provide gaseous fuels at 20 to 40 psi and, in many applications, it will be necessary to install a fuel gas compressor or booster to achieve these pressures. The energy required for this compressor reduces the engine output. Second, the turbocharger increases the air temperature and, in order to avoid preignition in the cylinder, cooling of the air leaving the turbocharger is required. The rating of turbocharged engines is based on the temperature of the water available for cooling the air exiting the turbocharger.

Another approach, which has recently been applied, is to supercharge the air/fuel mixture after the carburetor, cooling it to avoid preignition. The primary benefit of this approach is that it eliminates the need for a fuel pressure booster. One engine manufacturer provides engines with a small booster sized to compress only the fuel required to start up the engine.

Engine speed is also an important consideration in the selection of reciprocating engines. The engine capacity is a direct function of engine speed, and increasing the engine speed will decrease unit costs. Higher speed engines, ranging from 900 rpm to 1,800 rpm, can attain efficiencies approaching 35%, and are generally restricted to lighter fuels. Slow speed engines (below 300 rpm) are rather expensive; however, they are capable of efficiencies approaching 50% and can fire heavy oils.

One final categorization of engines is by *operating cycle,* either two-stroke or four-stroke. The *four-stroke* engine is most common, with four piston strokes per cycle (see Figure 2-2). It consists of an intake stroke during which either the air or the air/fuel mix is drawn into the cylinder; a compression stroke during which the air or the mix is compressed by the piston; a power stroke during which the burning fuel works against the piston; and an exhaust stroke during which the combustion products are exhausted from the engine. The

FIGURE 2-2
FOUR-STROKE ENGINE CYCLE

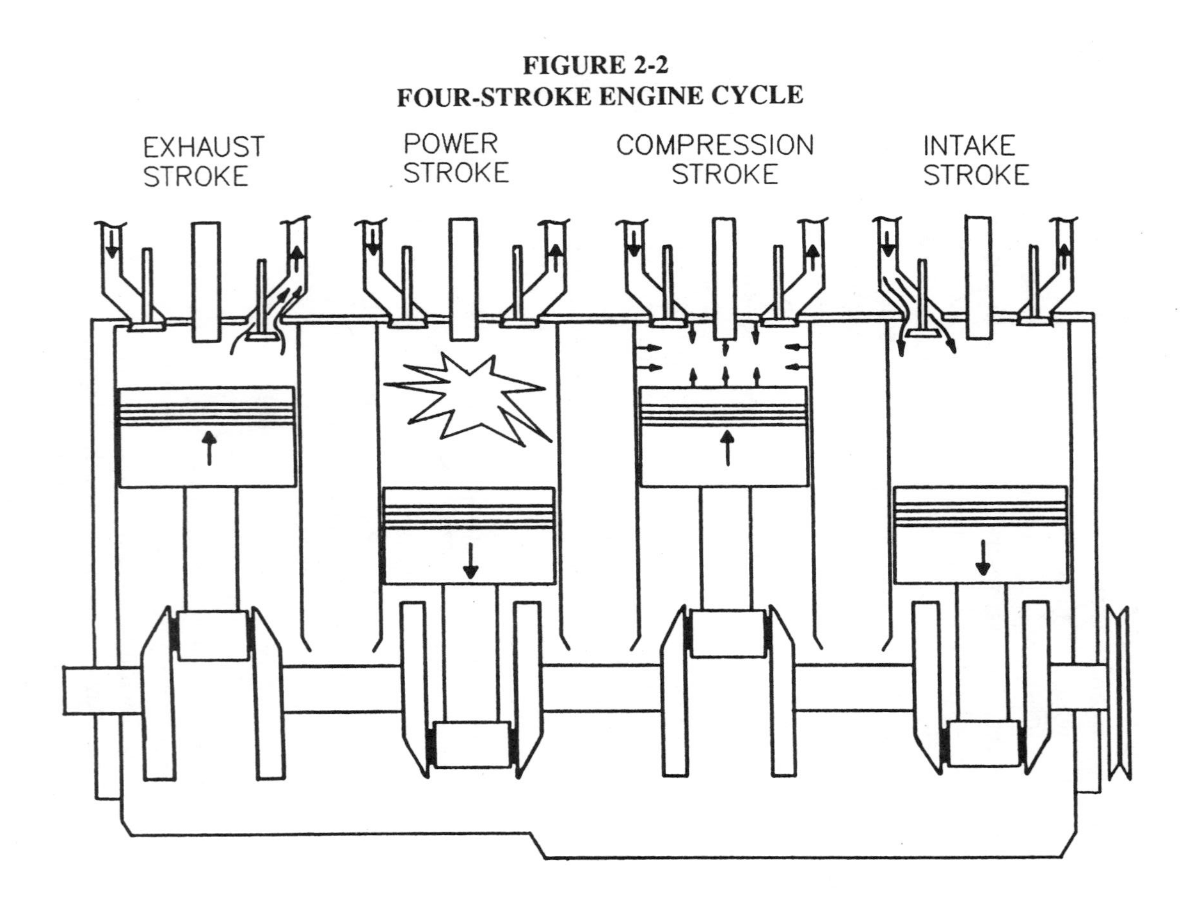

four-stroke, or four-cycle engine, requires two revolutions of the crankshaft for every power stroke of the piston.

In comparison, in the *two-stroke* engine each revolution of the crankshaft is accompanied by a power stroke of the piston. The initial stroke performs two functions; it draws in the air or the air/fuel mixture while simultaneously compressing it. The second, or power stroke, allows for the expansion of the burning fuel, delivering power to the crankshaft, and exhausts the combustion products.

Because the two-stroke engine delivers power from each cylinder for each revolution of the crankshaft, it can have twice the capacity of a similar block when operated in a four-stroke cycle. The two-stroke engine can be considerably less expensive per kilowatt than can the four-stroke engine, although the two-stroke engine's part-load efficiency drops off significantly.

Equipment Availability and Performance Characteristics

Reciprocating engines can be classified as being automative derivative, off-road vehicle derivative, marine derivative or industrial. Automotive derivative engines have been limited to applications of 100 kilowatts or less,while off-road and industrial engines have been used in cogeneration applications ranging from less than 100 kilowatts to two megawatts. Slower speed marine derivative engines have been employed in applications ranging from a few megawatts to 15 megawatts.

Engines are rated based on the type of duty to which they will be applied and the environmental conditions under which they are operated. Alternative rating procedures have been developed, each based on a unique set of "standard conditions." The NEMA rating is specified at a temperature of 80°F and an elevation of 1,000 feet, while the SAE rating is based on 85°F and a 500-foot elevation and the DEMA rating is based on 90°F and a 1,500-foot elevation. Another standard that is frequently used is 85°F and an air pressure of 29.5 inches of mercury. Each manufacturer's specification sheet should be reviewed since other conditions are sometimes used.

When used in a particular application, these ratings must also be corrected for altitude and temperature. While each engine's performance will vary, the ASHRAE *1988 Equipment Handbook* suggests an altitude derating of 3% per 1,000 feet for naturally aspirated engines and 2% per 1,000 feet for turbocharged engines. The Handbook also suggests a temperature correction of 1% per 10°F increase in inlet air temperature.

Reciprocating engine efficiency is more significantly affected by engine speed, compression ratio and type of aspiration, than by operating output. Figure 2-3 illustrates part-load performance for a typical reciprocating engine.

FIGURE 2-3
PART-LOAD PERFORMANCE

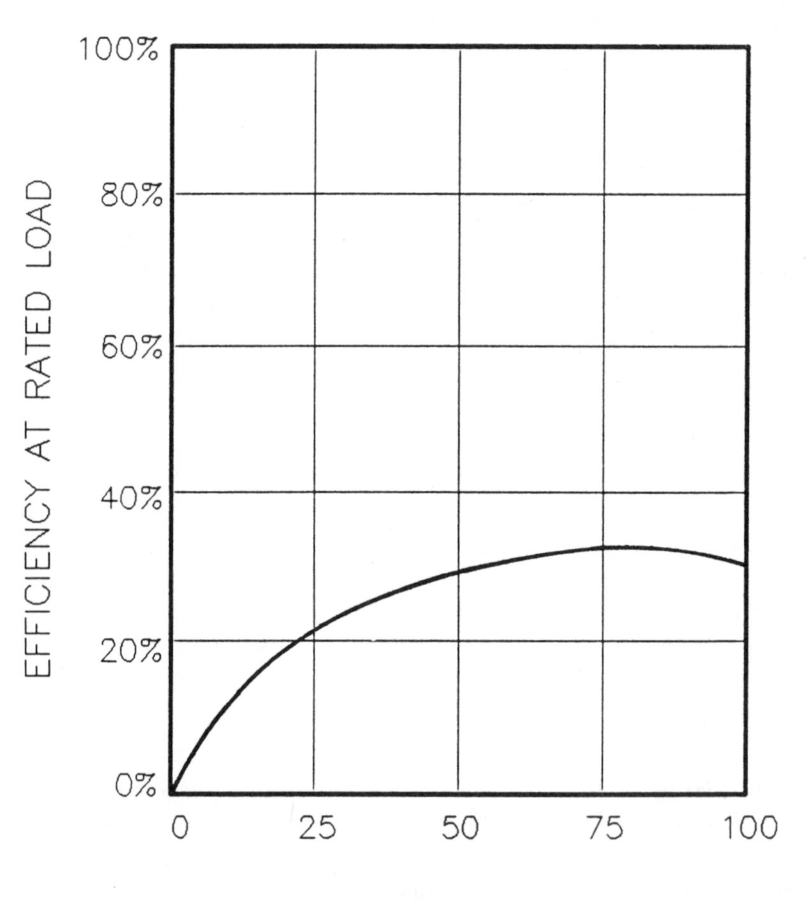

Maintenance Characteristics

Reciprocating engines are capable of unattended operation for prolonged lengths of time. However, as with any piece of operating equipment, a preventive maintenance program should include periodic, preferably daily, physical inspections for signs of failure. The objectives of the visual inspection are to identify either coolant or oil leaks, unusual odors or sounds and equipment which may be operating abnormally. In commercial and institutional facilities, this daily inspection need not be conducted by engine mechanics but rather by maintenance personnel. In larger facilities, such as hospitals and industrial plants, plant operators generally have all the skill required for daily inspections.

Other maintenance activities can be scheduled on an elapsed engine time basis and, for this reason, most quotations for engine maintenance contracts are on an engine hour basis. Air and oil filters, belts and linkages should be checked at intervals ranging from once per week to once every 250 hours. This procedure can usually be performed by facility maintenance personnel and need not involve a maintenance contractor. A more extensive preventive maintenance procedure, including changes of lubricating oil and all filters, belt replacement, valve checks, check of controls, sensors and set points, is usually required at intervals of 750 to 1,000 hours. It is generally good practice to have a lube oil sample analyzed to identify the presence of water, chemicals (including acids), and foreign materials, such as metal shavings. Depending on site operating conditions, this procedure may be scheduled at shorter or more extended intervals.

High-speed engines firing natural gas require minor overhauls that are scheduled at intervals of 12,000 to 18,000 hours, while major overhauls are scheduled at intervals of 36,000 hours or more. Engines firing a light distillate fuel will require similar maintenance intervals; however, engines operating on heavier oils will require significantly shorter intervals. Large, marine engines burning residual oils may require a major overhaul on an annual basis.

The monthly procedure for high speed, off-road or industrial engines can be completed within a few hours,while the minor and major overhauls may require anywhere from a few days to one week. Large marine engines may require as much as 4 weeks for a major overhaul. All maintenance for non-automotive derivative engines is usually performed on-site. In contrast, because automotive derivative engines are somewhat smaller and lighter, some cogeneration system designs are based on an engine replacement rather than an on-site major overhaul. This approach can significantly reduce the time and cost of a maintenance program.

Maintenance contracts are generally available from the equipment vendor although, in some cases, the equipment vendor may simply subcontract to a local service organization. The cost of this contract will vary depending on numerous factors including engine speed, type of aspiration, engine operating temperatures, type of heat recovery, scheduled operating hours, engine environment, any guarantees about engine performance including guaranteed heat rates, distance from the service organization to the engine location, any system availability guarantees, and any service response time guarantees. Finally, in negotiating a service contract, it is important to determine whether consumables including lubricating oil and coolant are included in the quoted prices.

Maintenance costs for smaller engines of 100 kilowatts or less, operating on natural gas or a light oil, will vary between $1 and $1.50 per engine hour. Engines ranging from 100 kilowatts to approximately 1,000 kilowatts will experience maintenance costs of approximately $.0075 per kilowatt hour to $.0150 per kilowatt hour. Larger engines will have maintenance costs that are slightly less. The use of heavier fuels or operation in a dirty environment can increase the above costs by 50% to 100%. The above costs are based on contracted service. They can be reduced if some or all of the maintenance functions are taken on by in-house personnel.

Heat Recovery Characteristics

Reciprocating engines can have high mechanical efficiencies, with over 30% of the input energy being converted to power. The remaining energy is converted to heat and must be removed from the engine. While heat rejection is mandatory for the operation of an engine, the application of that heat to a useful function is required for the economic viability of a cogeneration system.

Figure 2-4 is an energy balance for a 1,000 kilowatt, naturally aspirated reciprocating engine. Approximately 32% of the input is

FIGURE 2-4
ENGINE ENERGY BALANCE
Naturally Aspirated

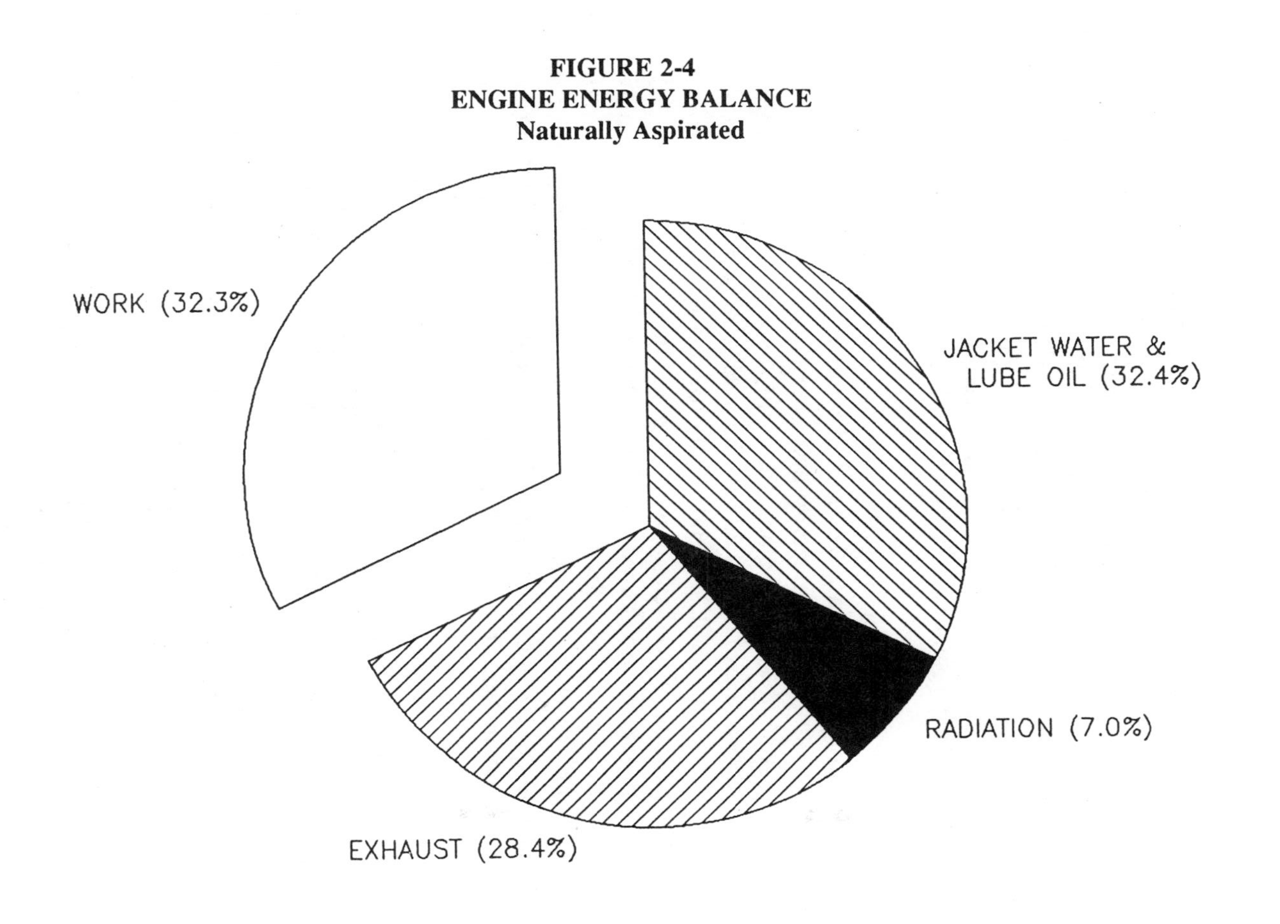

converted to shaft horsepower, and 68% must be rejected to avoid excessive engine overheating. This thermal energy can be rejected in the following manner: heat radiation from the engine itself (approximately 7%), heat carried out by the lubricating oil (approximately 3%), exhaust gases (approximately 28%), and the jacket water (approximately 30%). A turbocharged engine, (see Figure 2-5) also requires that heat be rejected from the turbocharger.

The energy balance will vary from engine to engine, and will vary as a function of engine loading. Figure 2-6 shows the variation in the energy balance as a function of loading for another typical engine.

The most commonly used sources of thermal energy from a reciprocating engine are the jacket coolant and the exhaust gases. In practice, almost all the energy that is rejected to the jacket coolant, which cools the engine block, the heads and the exhaust manifold, can be recovered. The quality of the available heat (temperature or steam pressure) will depend on the design of the engine cooling system, either forced circulation or ebullient.

A forced circulation (see Figure 2-7) cooling system removes heat from the engine by increasing the temperature of the engine coolant. A jacket water pump is required to supply coolant to the engine, where the temperature increases and removes the heat of combustion from the engine. The coolant is then routed to a heat exchanger where the engine heat is rejected. The temperature of the coolant exiting the engine will depend on the engine design; however, temperatures as high as 260°F have been used. The supply temperature to the engine will be determined by the allowable rise in coolant temperature within the engine. Lower temperatures within the engine reduce thermal stresses; however, they require larger coolant flow rates, pumps and heat exchangers, thus increasing system cost. Higher temperatures result in reduced capital costs; however, they may result in higher maintenance costs and shortened engine life.

Figure 2-7 includes two heat rejection heat exchangers. One allows for the jacket heat to be rejected to a process use, thus reducing the cost of conventional boiler fuels. The heat exchanger also provides isolation between the process and the engine loop, and in cases where the engine is used for potable water, a double

FIGURE 2-5
ENGINE ENERGY BALANCE
Turbocharged

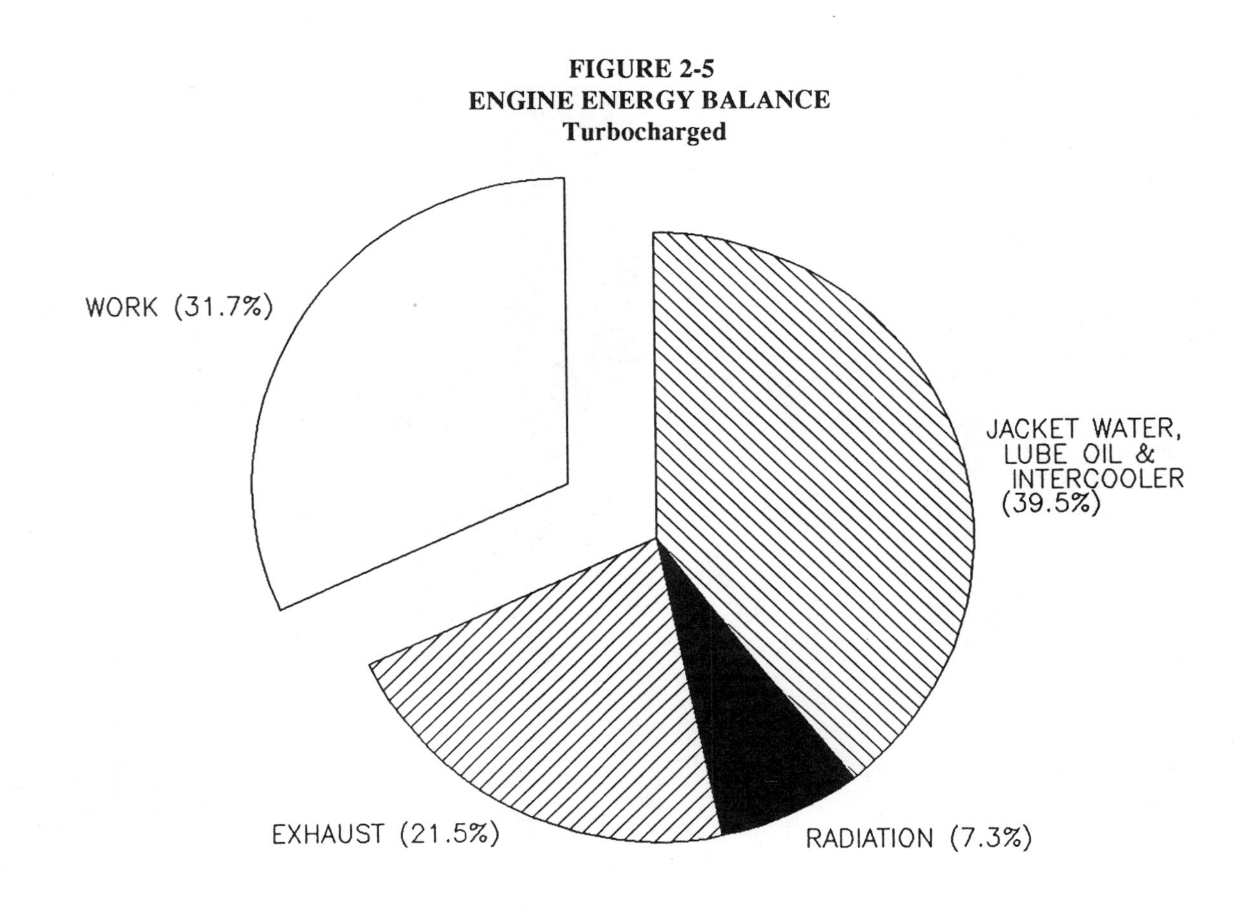

FIGURE 2-6
PART LOAD ENERGY BALANCE

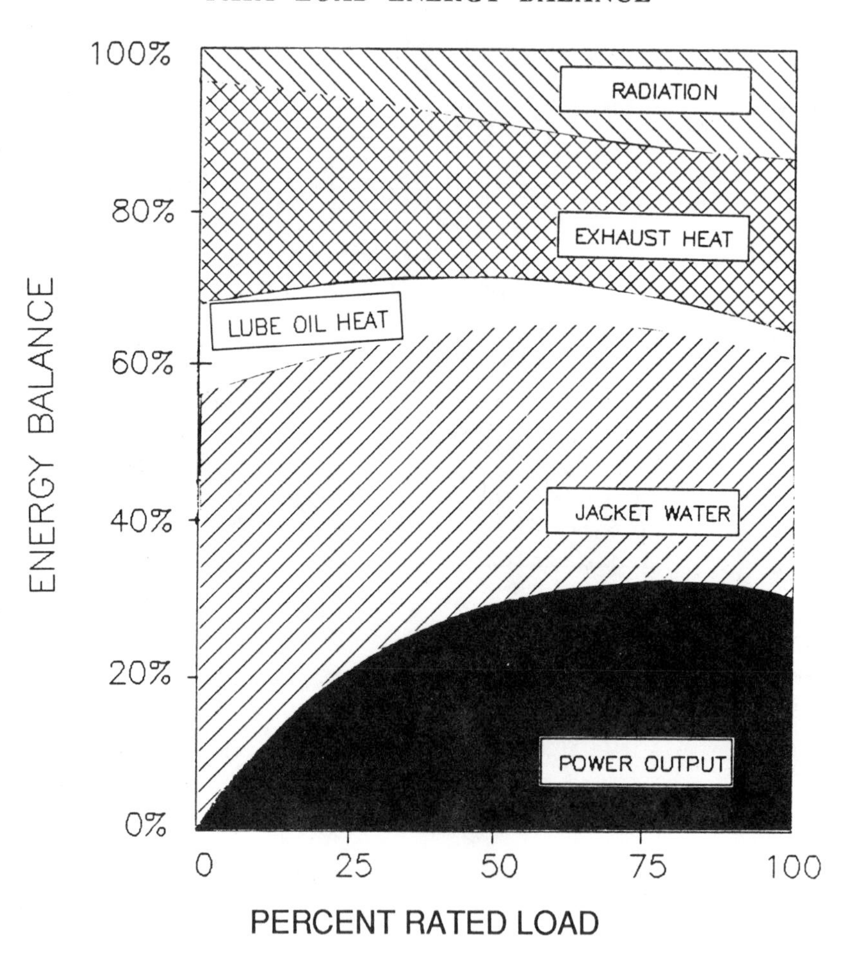

wall heat exchanger will be required. The second heat exchanger allows for rejection of the jacket heat to the atmosphere or to a source of cooling water. While this type of operation does not provide the economic benefit of cogeneration, it does allow continued operation of the engine during periods when the site thermal load is either inadequate or unavailable. This heat exchanger provides both physical and economic protection to the cogeneration system.

FIGURE 2-7
HOT WATER COOLED ENGINE

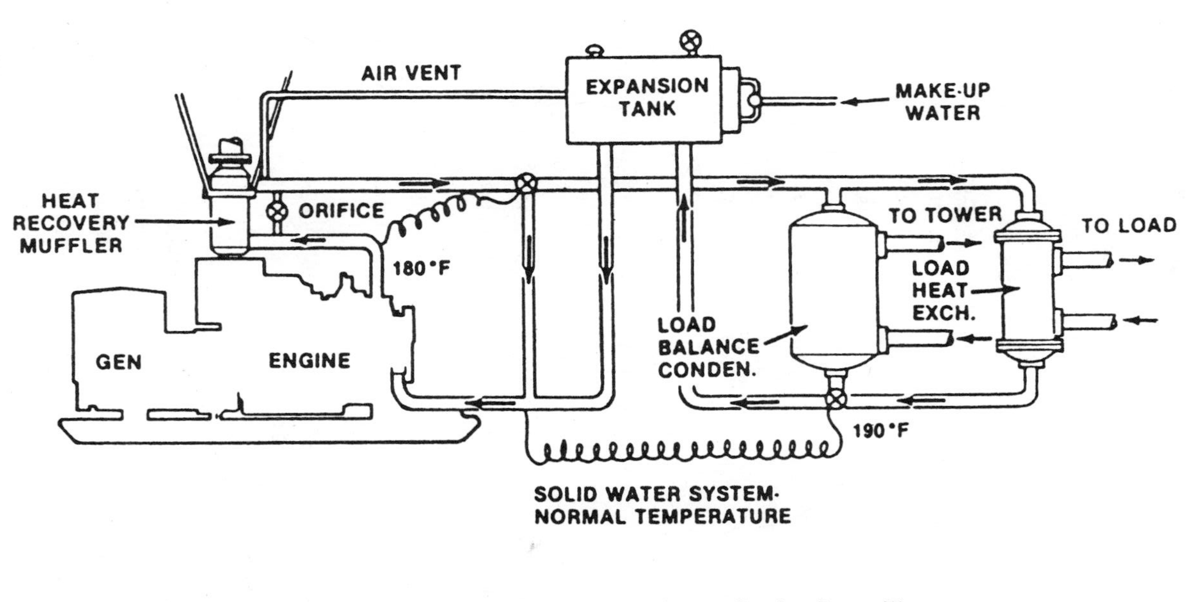

SOURCE: *Spark Ignited Application and Installation Guide,* Caterpillar Engine Division, Caterpillar Tractor Co., May 1986.

In some cases, it is necessary to supply low pressure steam to the process and there are two potential mechanisms for using jacket heat to produce steam. One is based on a forced circulation cooling design, where the coolant is routed through a heat exchanger to produce hot water which is then flashed to steam. The second approach, which is more commonly used, is to *ebulliently cool* the engine (see Figure 2-8). An ebulliently cooled engine rejects heat by changing the coolant phase. The coolant enters the engine as a pressurized liquid where it is heated by the engine. Since the coolant is at its boiling point, the heat does not increase the coolant temperature but, rather, causes the coolant to change phase. The coolant, which is then a steam/water mix, is less dense than the liquid coolant and gravity causes the mixture to rise to the top of the engine. On exiting the engine, the steam/water mixture is routed to a steam separator where the steam is made available for process and the water is returned to the jacket.

Pressure is maintained in the engine by locating the steam separator well above the engine (see Figure 2-9). If this pressure is lost, the steam bubbles within the engine will expand causing a loss of coolant flow and overheating of the engine block. Loss of pressure can result in severe damage to the engine.

Ebulliently cooled engines can provide low pressure steam for process and, because the coolant temperature is constant, thermal stresses within the engine are minimized. This approach to engine cooling also eliminates the need for an external coolant pump. Not all engines, and particularly engines of 100 kilowatts or less, are capable of ebullient cooling and the manufacturer should be consulted prior to any attempt to apply ebullient cooling. Finally, some manufacturers derate the engine slightly for ebullient cooling.

The second significant source of engine heat is the exhaust which can reach temperatures as high as 1,300°F. However, unlike the jacket water heat, it is not economically possible to recover all the exhaust heat. In order to recover this heat, it is necessary to include an air-to-water heat exchanger in the exhaust system. If the temperature of the gases leaving the heat exchanger falls below 300°F to 350°F, these gases may condense in the engine stack and the exhaust system design must eliminate any potential for the condensate to flow back into the engine. Condensation will also require the use of more costly corrosion resistant metals. In addition,

FIGURE 2-8
EBULLIENT COOLING HEAT RECOVERY SYSTEM

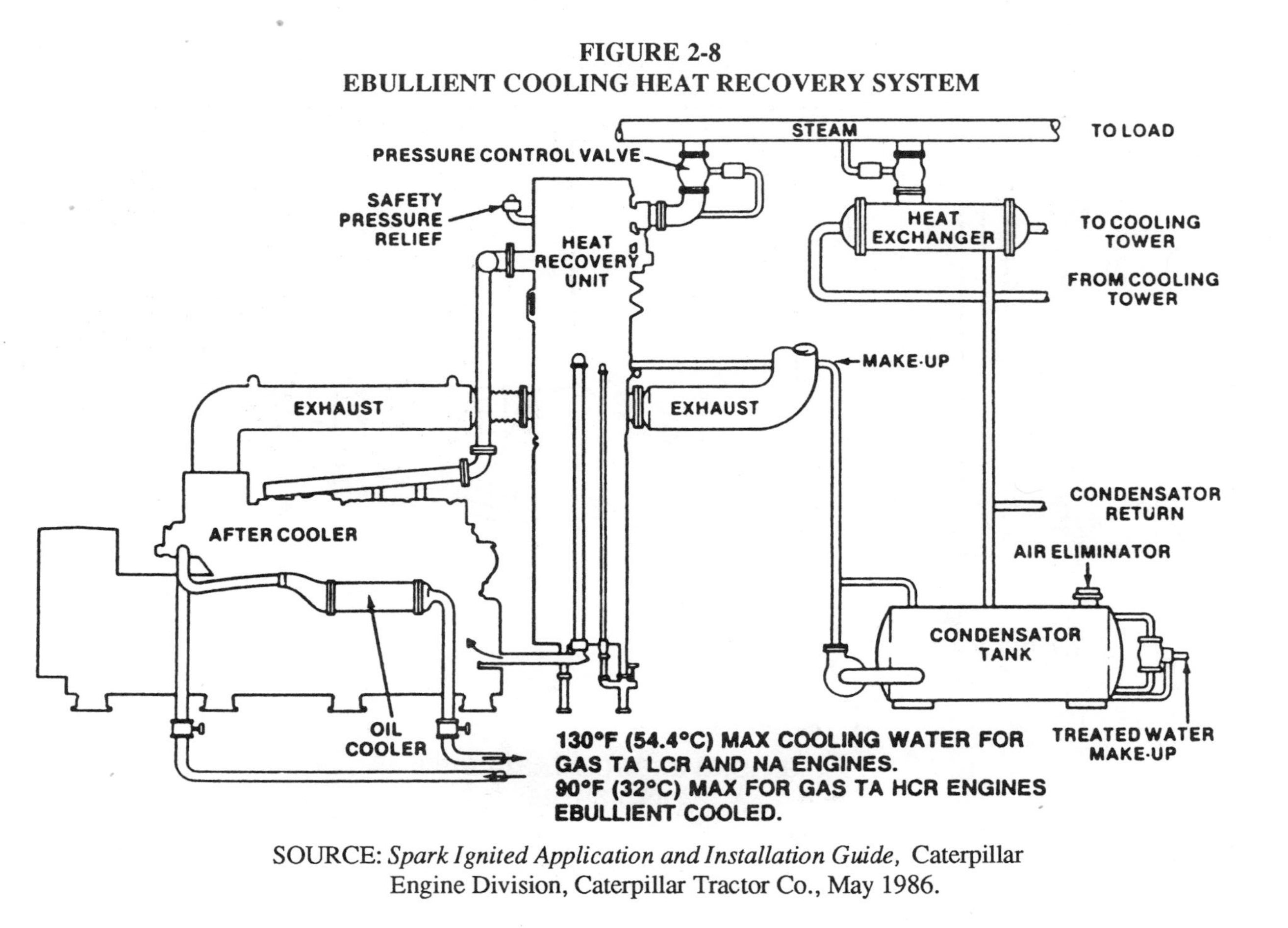

SOURCE: *Spark Ignited Application and Installation Guide*, Caterpillar Engine Division, Caterpillar Tractor Co., May 1986.

FIGURE 2-9
EBULLIENT COOLED ENGINE

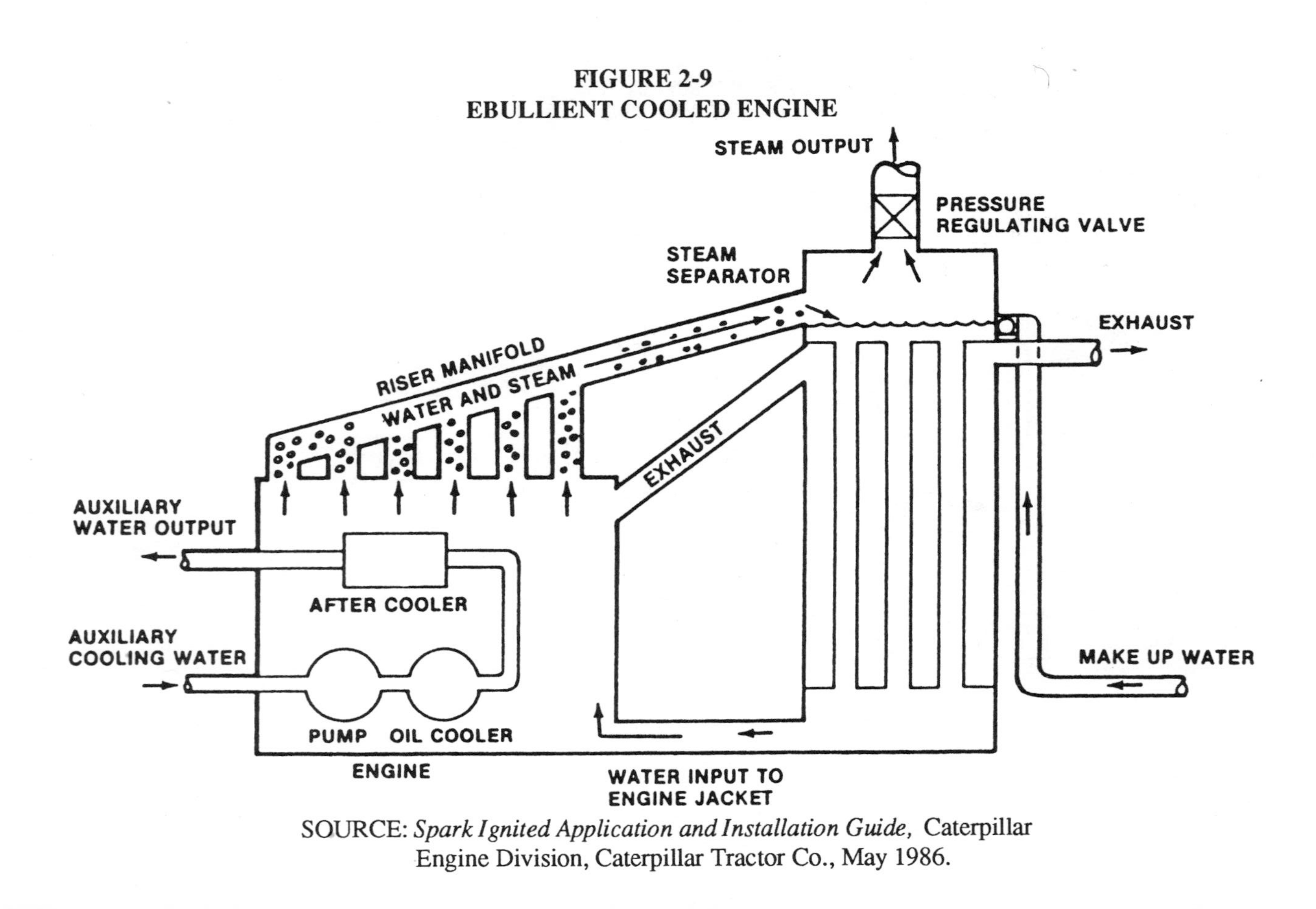

SOURCE: *Spark Ignited Application and Installation Guide*, Caterpillar Engine Division, Caterpillar Tractor Co., May 1986.

low stack temperatures may result in poor dispersion of exhaust gases. As a rule of thumb, only 50% of the exhaust heat is recoverable; however, this fraction will depend on the engine operation and the exhaust temperature.

Heat recovery is possible from the lube oil and the turbocharger aftercooler. However, these heat sources are at relatively low temperatures, usually less than 160°F, and are therefore of limited economic use. Heat recovery from these sources is usually limited to domestic water, dying and washing processes and boiler feedwater preheating. High temperatures in the lube oil loop can lead to breakdown of the oil and excessive engine wear.

Finally, it should be noted that all engines will radiate heat from higher temperature sources such as the turbocharger and manifold. One design consideration is adequate ventilation to remove this heat and avoid overheating the engine. In some cases, where the air being exhausted from the engine room can be applied to a useful purpose, radiated engine heat can be recovered.

Environmental Considerations

As with all prime movers, reciprocating engine emissions are being significantly limited. Emissions of nitrogenous oxides (NO_X), carbon monoxide (CO) and unburned hydrocarbons are of concern with all fuels, while engines operating on oil must also meet requirements for sulfurous oxides (SO_X) controls.

There are two approaches to meeting NO_X and CO emission requirements. One approach is to pass the exhaust through a catalyst or series of catalysts which selectively remove specific pollutants. In some cases, the unburned hydrocarbons can also be removed as part of the catalytic reaction. NO_X removal can be controlled using Selective Catalytic Reduction (SCR), and this technique is discussed in more detail in the review of turbine technology.

A second approach is to limit emissions with controlled combustion, either through a lean mixture of fuel and air or through staged ignition. This latter process sometimes involves the use of a smaller combustion chamber where the fuel/air mix is first ignited. Once ignition is achieved, the larger volume of fuel/air is then burnt in the engine cylinder.

COMBUSTION TURBINES

In the decade following the passage of PURPA and the revitalization of the cogeneration industry in the United States, combustion turbines have gained an increasing share of the market, either as a simple cycle, or when combined with a steam turbine, as a combined cycle. They have been employed in a variety of applications ranging from hospitals, universities and industrial facilities to central power plants. In addition, they have been successfully used both in peaking and in baseloaded applications. A combination of economic and performance characteristics, including availabilities of over 8,600 hours per year, have resulted in the combustion turbine becoming a technology of choice in the 1980s. When used in a cogeneration system, maintenance requirements for other components may reduce the overall system availability below 8,600 hours.

Concept

Turbines are rather simple devices as compared to other types of engines. The simplest *open cycle* combustion turbines operate through a *Brayton Cycle* as shown in Figure 2-10, and consist of three major components (see Figure 2-11): a compressor, a combustor and a gas turbine. The air compressor takes air at atmospheric pressures and compresses it to pressures that are many times atmospheric. In general, smaller turbines of 20 MW or less will boost air pressures up to 250 psig, while larger turbines can produce pressures approaching 600 psig. This process also raises the air temperature by several hundred degrees.

The higher pressure and temperature air is delivered to a combustion chamber where fuel is injected and burned, further increasing the temperature of the mix of air and combustion gases. The temperature of the gas mixture exiting the combustor can be 2,200°F or more. The combustion process takes place at a constant pressure and occurs with large quantities of excess air. The turbine exhaust is usually oxygen rich. With oxygen concentrations approaching 15% or 16%, the turbine exhaust gases can support additional combustion.

Finally, the high temperature, pressurized gas mixture is delivered to an expansion or power turbine, where the expanding

FIGURE 2-10
SIMPLE CYCLE GAS TURBINE
BRAYTON CYCLE

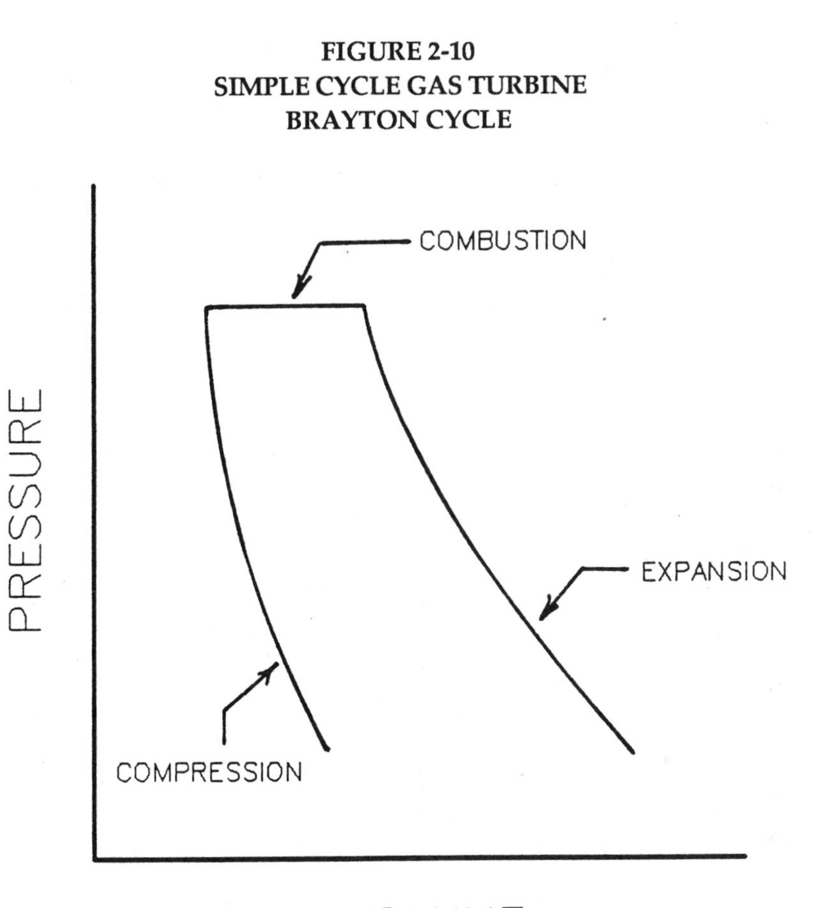

gases perform mechanical work by rotating the turbine shaft. Throughout the expansion process, the gases lose heat and the exhaust from the power turbine is at atmospheric pressure and temperatures that can approach 1,000°F. All mechanical work, including the work required by the air compressor, is delivered by the power turbine. This air compressor can require over one third of the power turbine's output, leaving only a fraction of the turbine's mechanical energy as the simple cycle output.

The turbine's mechanical efficiency is a function of the difference between the turbine inlet temperature and the turbine

FIGURE 2-11
COMBUSTION TURBINE

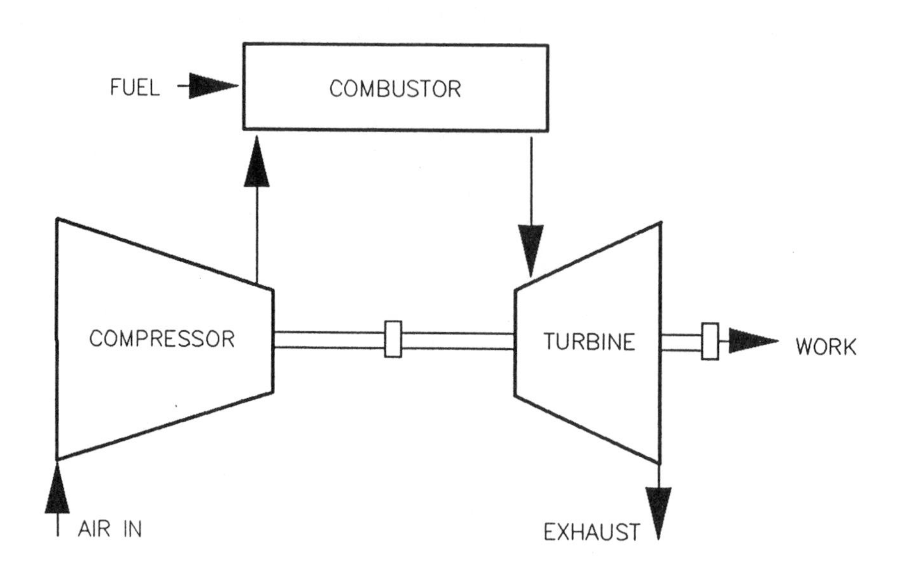

exhaust temperature. Relative increases in the inlet temperature result in improved efficiency, as do decreases in the power turbine's exhaust temperature.

The turbine exhaust, which can be approximately 1,000°F, is then available as a source of heat in a cogeneration system. This exhaust is relatively clean and high in oxygen and can be used directly in drying processes. Alternatively, it can be delivered to a *Heat Recovery Steam Generator* (HRSG) where it is used to produce steam. Turbine heat recovery will be discussed in more detail below.

In general, the temperature of the turbine exhaust will be higher than the temperature of the air as it exits the compressor. In this case, it is possible to increase the cycle efficiency by using some of the heat in the turbine exhaust to preheat the air entering the combustor. This *regenerative cycle* includes an air-to-air heat exchanger as shown in Figure 2-12. The regenerator reduces the amount of fuel required to produce a specified turbine inlet temperature, thus increasing the mechanical efficiency of the cycle. It does, however, reduce the temperature of the turbine exhaust gases and decrease the amount of heat that is recoverable.

FIGURE 2-12
REGENERATIVE TURBINE

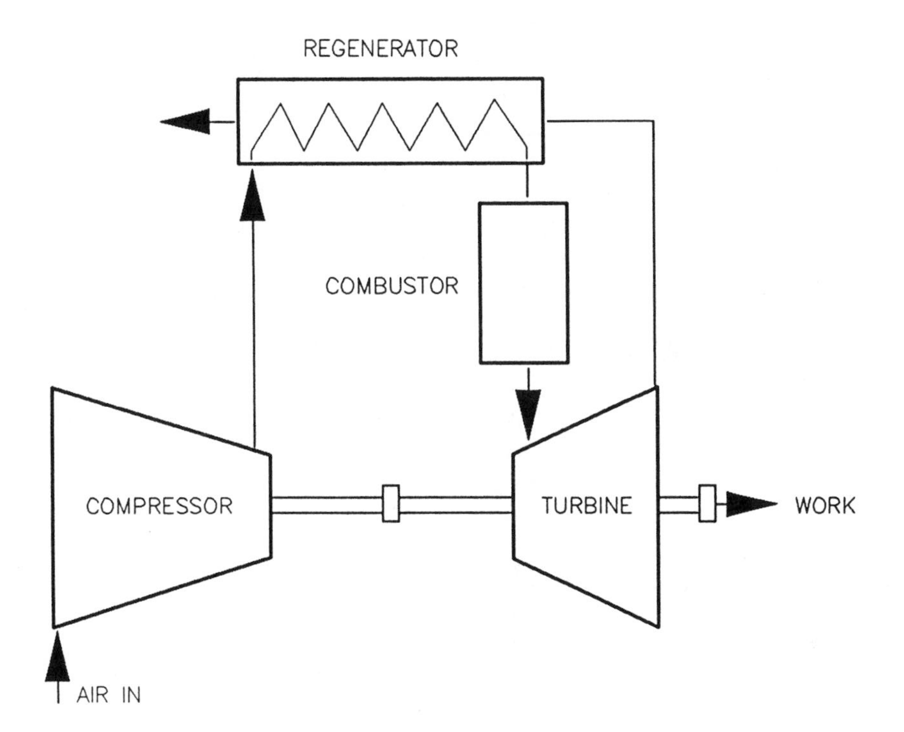

The regenerative cycle uses exhaust heat to preheat air prior to the combustor, thus reducing the total fuel requirement at a constant power output. A second approach is to introduce more fuel into the process in a *reheat cycle.* As shown in Figure 2-13, the combustor gases are only partly expanded in the power turbine. At some intermediate pressure, these gases are exhausted from the turbine and diverted to a second combustor or reheat element, where additional fuel is burned to increase the temperature to the original turbine inlet temperature. The higher temperature gas, which is at a pressure between the compressor exit pressure and atmospheric, is then routed to the second section of the expansion turbine where it

FIGURE 2-13
REHEAT CYCLE

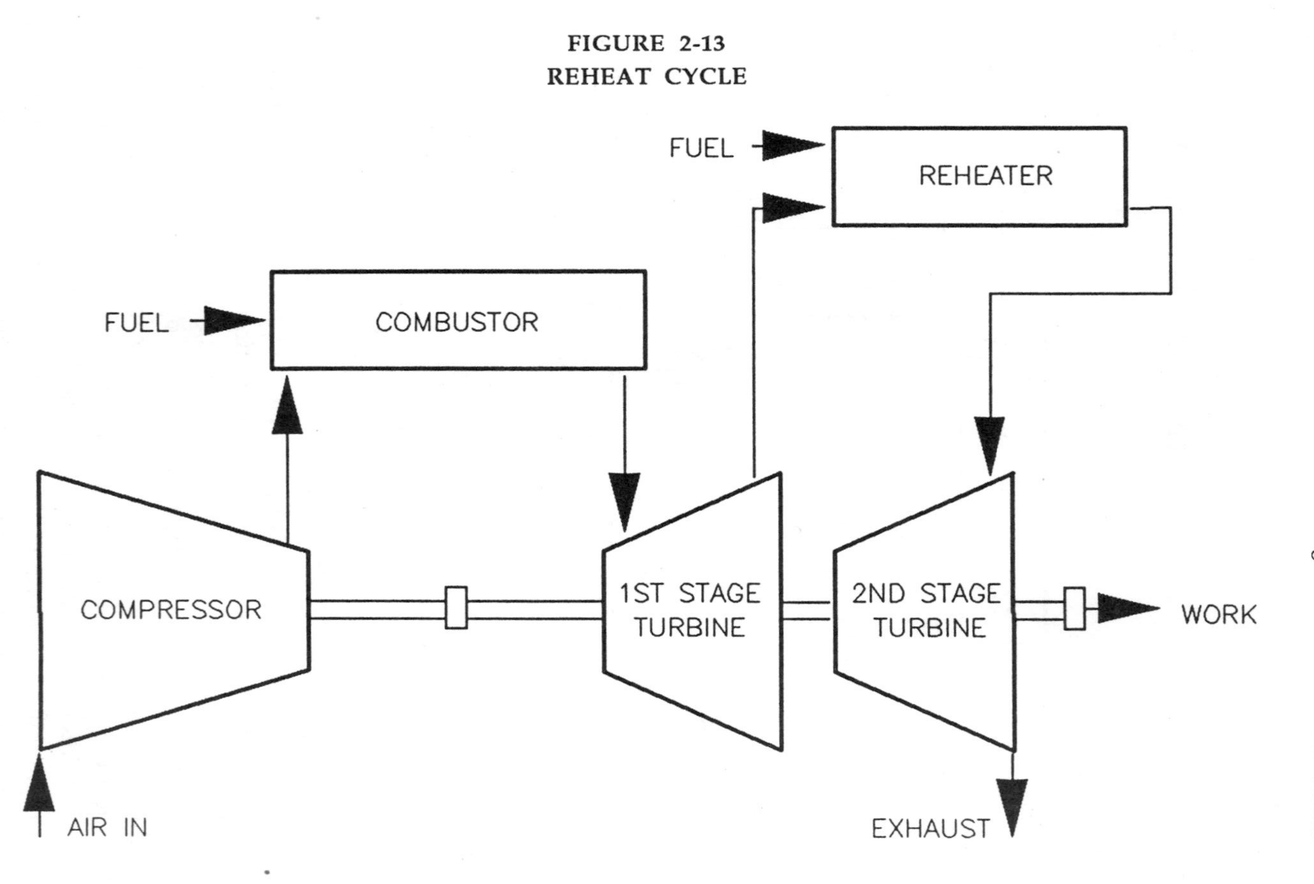

performs additional work. The turbine gases contain adequate oxygen to support the additional fuel firing. This cycle increases the efficiency of the turbine by increasing the overall temperature of the gases entering the turbine.

As with the reheat cycle, where the turbine is divided into two sections to increase the average temperature, it is also possible to split the compressor into two sections in order to lower the average temperature of the air in the compressor. When the compressor is split, an *intercooler* is introduced to cool the air between the two sections of the compressor (see Figure 2-14) which, as a result, decreases the compressor work and lowers the temperature of the air exiting the compressor. When combined with a regenerator, the intercooler allows lower cycle exhaust temperatures, thus increasing both output and efficiency.

FIGURE 2-14
REGENERATIVE/INTERCOOLER CYCLE

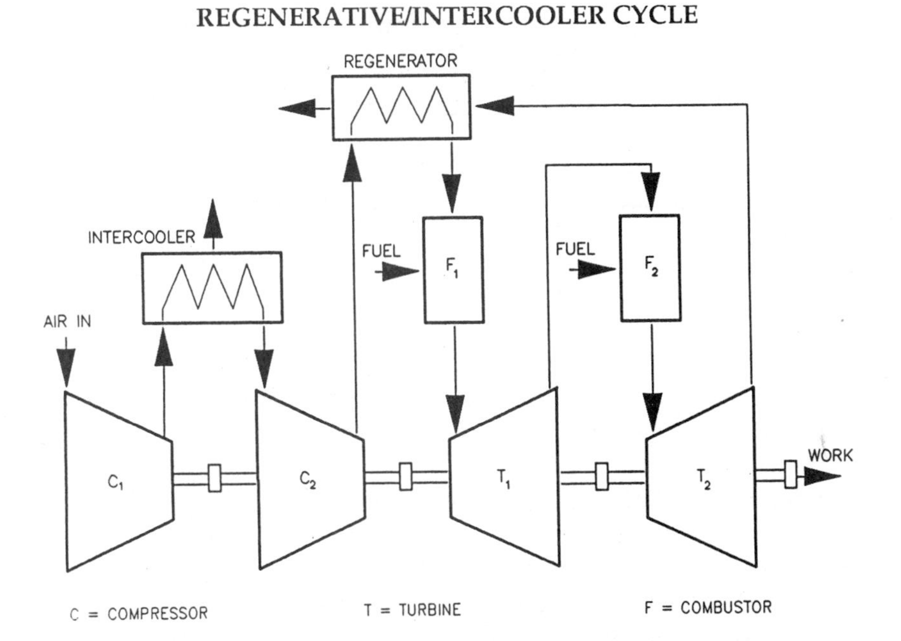

The Brayton cycle is an open cycle wherein the working fluid is exhausted to the atmosphere. Turbines can also be operated in a *closed cycle* as shown in Figure 2-15 where the combustor is replaced by a heat exchanger and the working fluid is not exhausted. This cycle, which is more costly than the open cycle turbine, allows the use of less expensive, heavy or dirty fuels and lower operating costs, both for the fuel and for turbine maintenance.

FIGURE 2-15
CLOSED CYCLE TURBINE

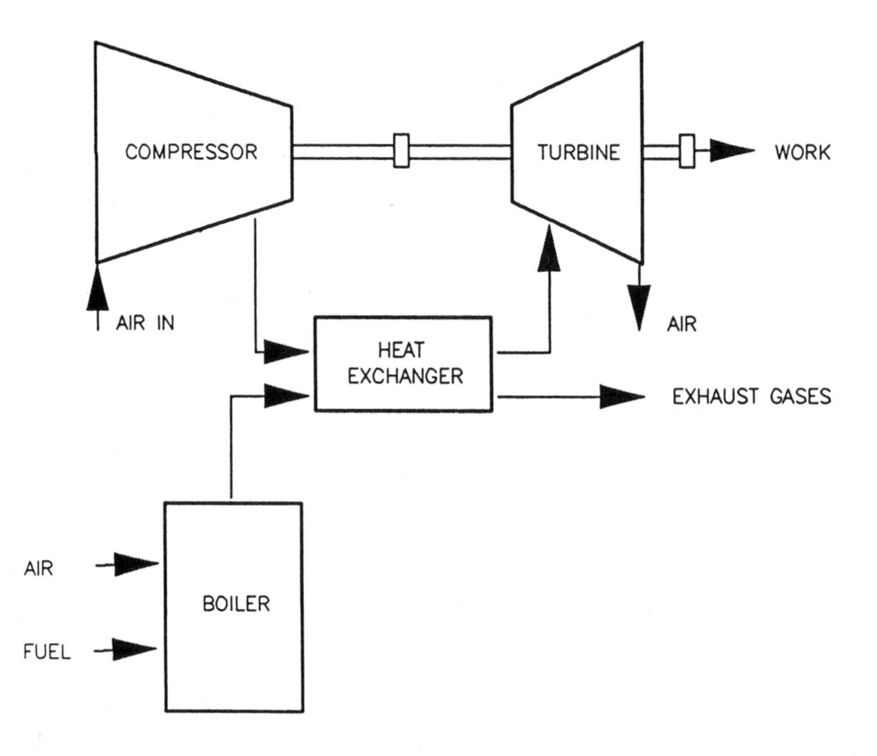

Equipment Availability and Performance Characteristics

Combustion turbines are classified as either aerospace derivative or industrial, with the industrial turbines being larger and heavier than aerospace turbines. While aircraft derivative turbines are less expensive than stationary turbines, they generally require more extensive maintenance programs. With the increased use of aero derivative turbines in cogeneration and central plant, combined cycle systems, market differences between the two types are disappearing. They are commercially available in sizes ranging from a few hundred kilowatts to 150 MW. Turbine efficiency increases with size (see Figure 2-16), and smaller size turbines are inefficient as compared to reciprocating engines.

Combustion turbine prices are also extremely sensitive to size with smaller unit costs averaging $400 to $500 per kilowatt for the basic turbine generator set with controls. Mid-sized turbine (5,000 kW to 25,000 kW) unit costs range from $300 to $400 per kilowatt with larger turbine unit costs ranging from $250 to $300 per kilowatt.

Because all combustion takes place external to the power turbine, combustion turbines are capable of switching fuels instantaneously with no observable change in mechanical output. They are, however, limited to lighter oils with No. 2 fuel oil being the alternative to gaseous fuels. As with the reciprocating engine, the turbine is capable of operating on a number of gaseous fuels, including those with lower calorific values. In the case of low heat content gases, the turbine rating may be significantly reduced. Turbine efficiency decreases slightly when liquid fuels are burned.

Turbine manufacturers normally specify the turbine's capacity and performance under *ISO* conditions consisting of sea level pressure and 59°F, with no losses for inlet or exhaust pressure drops. Additionally, the manufacturer should specify the fuel which is the basis for the turbine rating. It should be noted that some manufacturers claim the specification is within 5% of actual performance and that individual turbines will exhibit variations in both capacity and heat rate. Many factors will influence the turbine's performance, with some of the more significant considerations described below.

The turbine capacity is directly effected by the mass of the combustor exhaust gases and, more specifically, by the ambient air density. As a result, the capacity decreases as the ambient temperature increases. Figure 2-17 illustrates the relationship between *ambient temperature and capacity* for a typical combustion turbine. In those cases where summer turbine capacity is critical or alternative sources of power costly, it may be cost effective to precool the turbine inlet air to increase capacity. Either mechanical, absorption or evaporative chillers may be used, depending on weather and costs.

The density of ambient air is also effected by *altitude* and, in general, a turbine's ISO capacity must be reduced by approximately 3% to 5% per 1,000 feet above sea level.

FIGURE 2-16
TURBINE SIZE VS. HEAT RATE (HHV Btu/kWh)

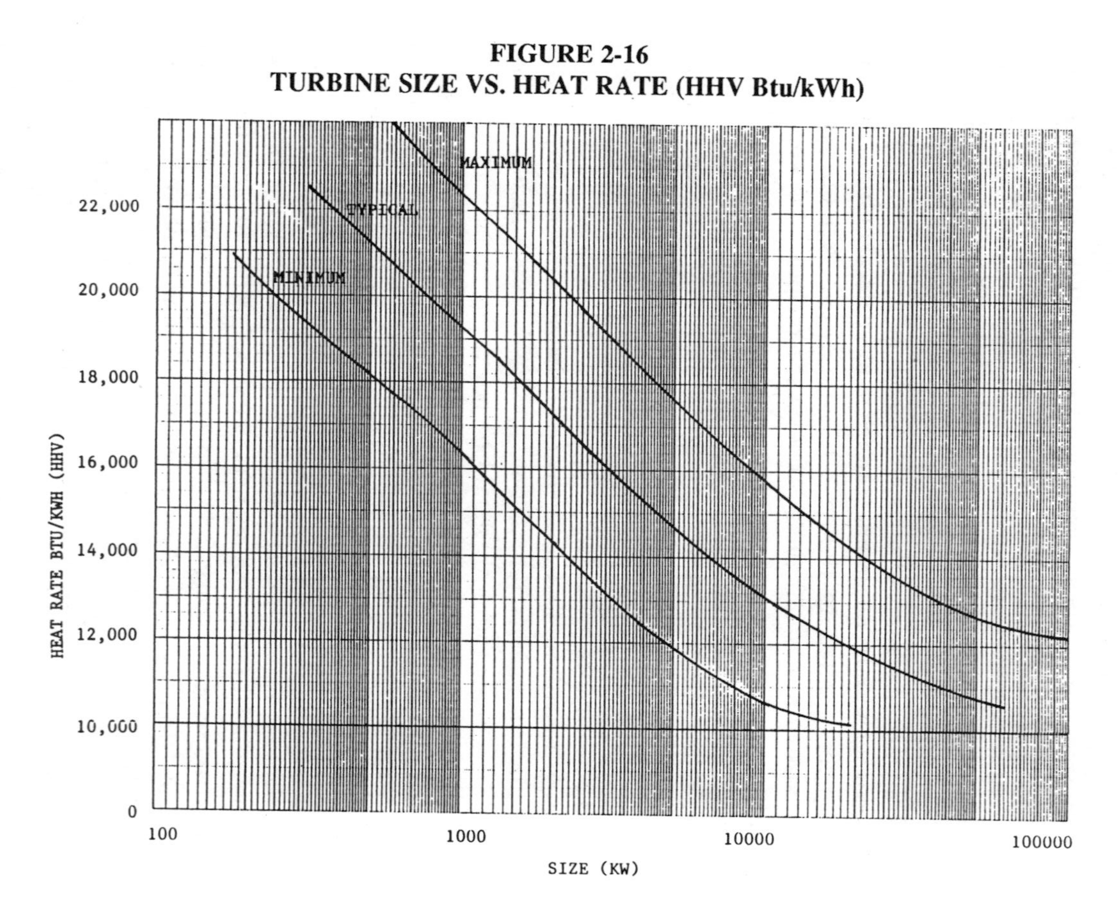

FIGURE 2-17
AMBIENT TEMPERATURE VS CAPACITY

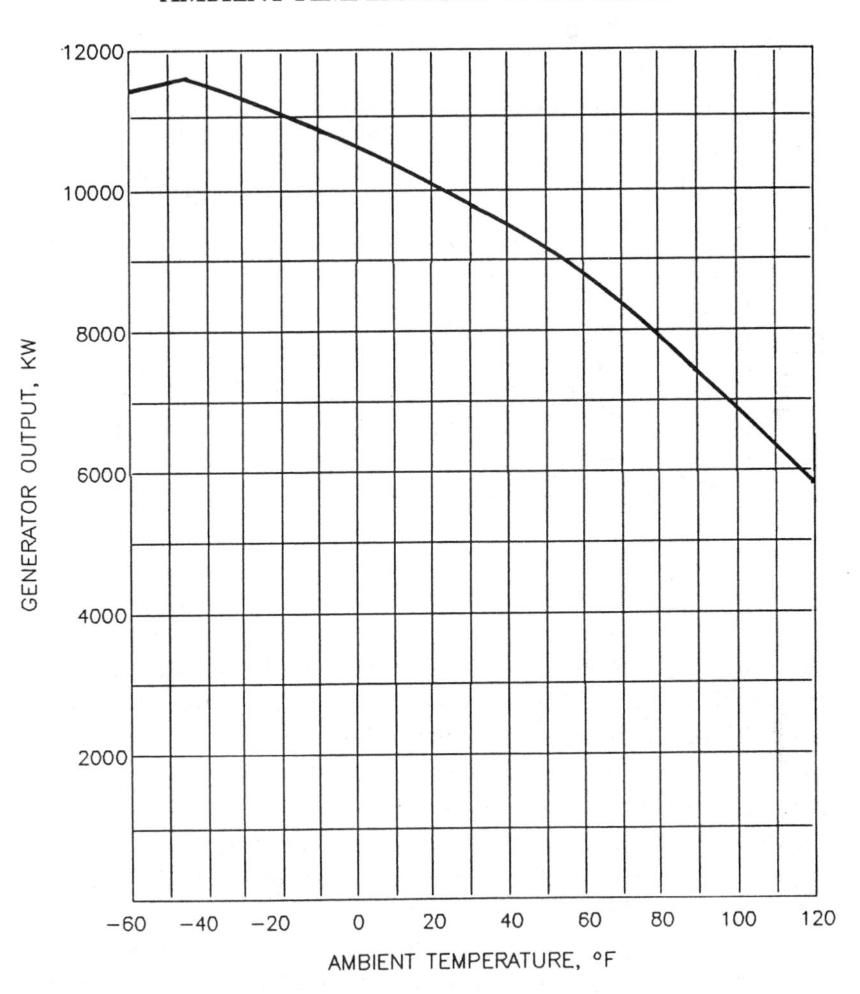

The capacity and heat rate of the turbine will also be reduced by the amount of work required to overcome *inlet and exhaust pressure losses*. Figure 2-18 illustrates the relationship between the pressure losses, expressed in inches of water, and the turbine capacity.

Turbines typically operate at high speeds sometimes approaching 20,000 revolutions per minute. As a result, it is often

FIGURE 2-18
PRESSURE LOSSES VS CAPACITY

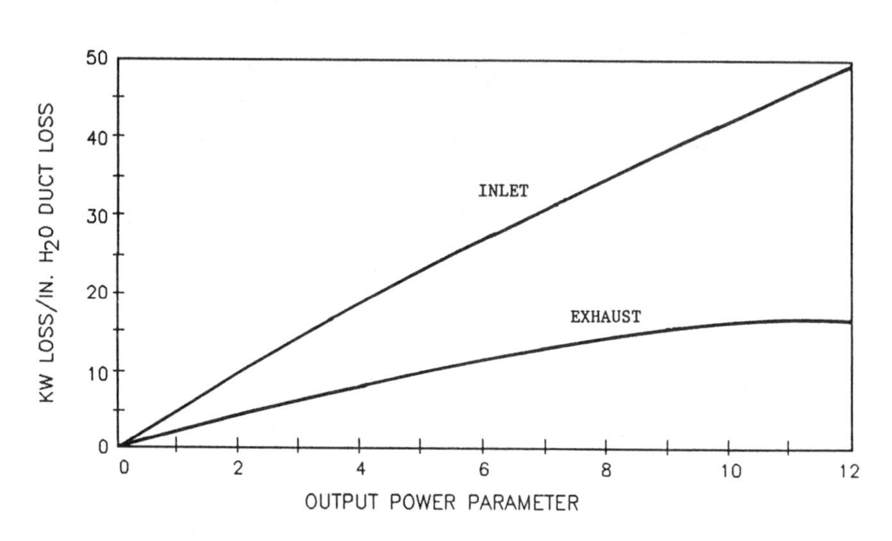

necessary to use a gear box to reduce the shaft rotational speed to that required by the load. The gear box, which can introduce losses of up to 1% or 2%. In addition, if the turbine is used in a cogeneration system it is also necessary to include the generator's losses which can decrease performance by another 2% to 4%. Overall, mechanical losses can be quite significant in deter-mining a turbine's net capacity, and manufacturers vary with regard to including such losses in the published ratings.

As was pointed out above, the pressure of the air entering the combustor may approach several hundred pounds and, when operating on a gaseous fuel, it is necessary to pressurize the gas to slightly more than the turbine's operating pressure. In some cases, high pressure natural gas may be available from the fuel supplier; however, when the available gas pressure is less than the turbine's working pressure, it is necessary to use a supplemental gas compressor. The work required to raise the gas pressure will depend both on the available and the required pressures; however, this *parasitic load* directly reduces the amount of power that is available for load. Under extreme pressure differences, the compressor load may total 5% of the turbine's output. While a gas compressor does reduce the available mechanical power, it does not reduce the amount of recoverable heat.

Finally, the manufacturer's specification usually quotes the turbine's heat rate at the nominal or rated output. This rating must be adjusted for local conditions, including altitude and temperature to determine the site rating. The turbine full load heat rate may require similar adjustments to determine the site performance. However, when a turbine is operated at *part load,* or a capacity other than its full load, the turbine heat rate drops off significantly. Figure 2-19 illustrates the relationship between output and heat rate.

FIGURE 2-19
TYPICAL TURBINE PART LOAD EFFICIENCY

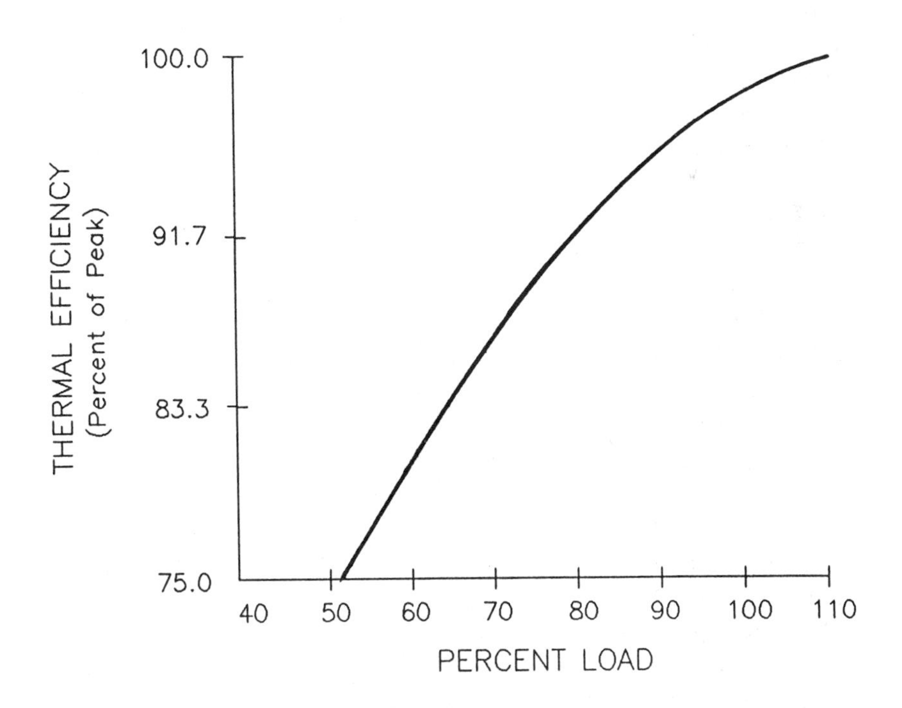

Maintenance Requirements

The turbine's design, which allows rapid disassembly and quick engine change outs, lends itself to a minimum of down time for scheduled maintenance. While all turbomachinery should be visually inspected on a daily basis, contractor-supplied outside maintenance is required on a monthly basis. In general, these monthly procedures can be accomplished without an engine shutdown. Engine shutdowns are required at intervals of 4,000 hours, usually for a borescope inspection of the engine for blade erosion and checkout of fuel handling systems, sensors and controls, burner cleaning and auxiliary systems. Major engine refurbishment is required at intervals of 20,000 hours or more. Typical maintenance costs range from three mills to six mills per kilowatt hour.

Many factors influence the typical maintenance requirements described above including type of fuel, duty cycle and operating environment. Continued firing on No. 2 oil instead of natural gas will reduce the maintenance intervals by 10% to 20%. Peaking or standby operation with frequent start-ups and shutdowns can further reduce the interval between maintenance procedures. Finally, operating in a corrosive or dirty atmosphere, or at high temperatures can reduce the interval between procedures by 40% to 50%.

Heat Recovery Characteristics

While turbines can be mechanically efficient engines, almost all of the energy that is not converted into shaft power is rejected in the turbine's exhaust gases. These gases are relatively clean and can be used directly for process applications or directed to a *Heat Recovery Steam Generator* (HRSG) where they are used to produce steam. Inasmuch as the turbine operates with a large quantity of excess air, these exhaust gases are high in oxygen and can support the combustion of additional fuel to further increase the heat content of the exhaust. The concept is shown in Figure 2-20 which illustrates a turbine, a duct burner where additional fuel is fired, the HRSG and the exhaust stack. The duct burner is usually limited to natural gas or light oils, and operates at an efficiency of 90% to 94%.

The steam from the HRSG can be used in a process thereby reducing or totally eliminating the steam that is required from conventional boilers. If the turbine exhaust gases cannot meet the

FIGURE 2-20
DUCT BURNER

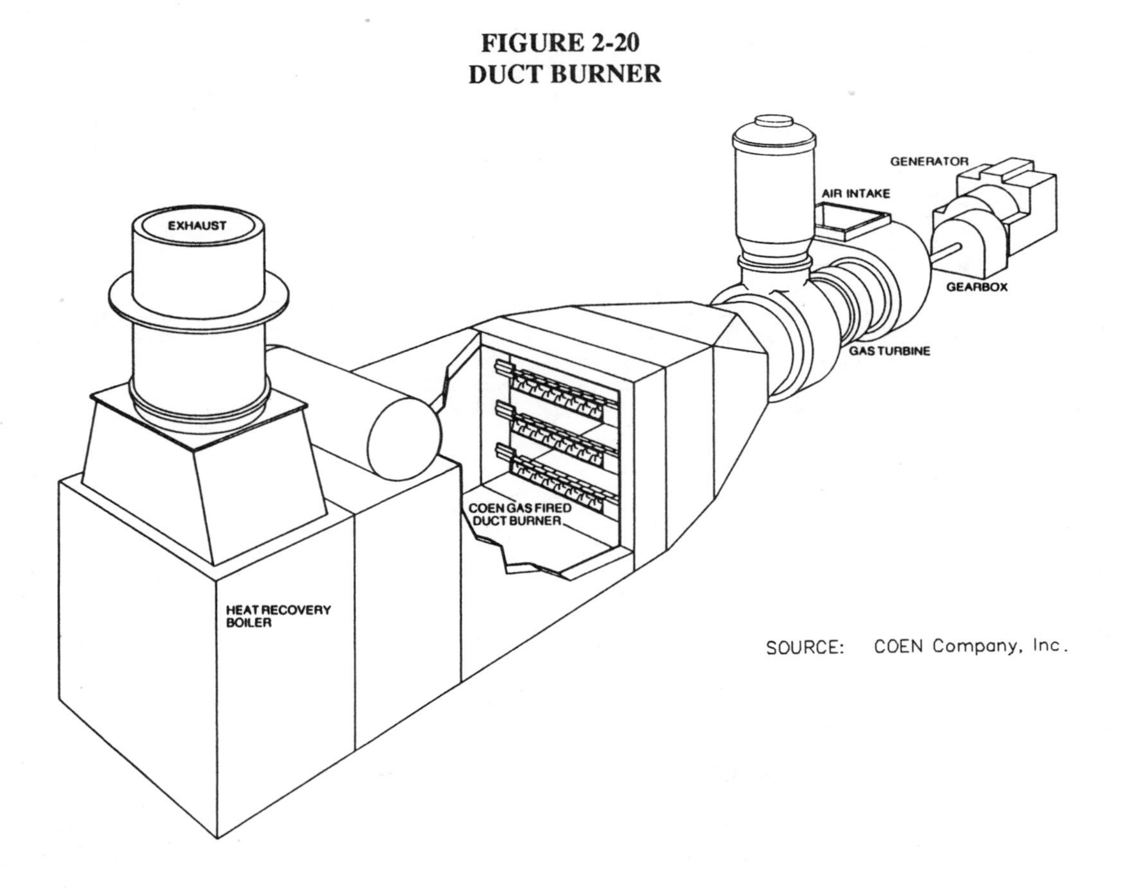

SOURCE: COEN Company, Inc.

facility's thermal requirement, it is necessary to fire supplemental boilers to produce the additional steam that is required. These supplemental boilers can consist of the conventional boilers, a duct burner that increases the HRSG output, or both.

Alternatively, in a *Combined Cycle* the high pressure HRSG steam can be directed to a steam turbine where it is used to produce additional power. When operated in a cogeneration mode, the lower pressure steam that is extracted or exhausted from the steam turbine can be used in process, again reducing conventional boiler requirements. As an alternative, the steam can be condensed and returned to the HRSG. In this latter mode, the system operates as a more efficient electric power plant but, because it produces no useful thermal energy, it is not a cogenerator.

The duct burner provides additional benefits to a cogeneration system. First, it operates at higher efficiencies than does a conventional boiler, thus reducing the amount of fuel that is required. Second, if only minimal amounts of supplemental steam are required, the duct burner can be fired at low levels at efficiencies that are significantly higher than low level operation of a conventional boiler. Finally, the duct burner can provide higher pressure steam required for more efficient operation of the steam turbine.

Environmental Considerations

When operated on natural gas, undesirable turbine emissions are generally limited to NO_X , CO and unburned hydrocarbons. SO_X emissions are only troublesome when higher sulfur content fuel oils are used.

The production of NO_X results from the high firing temperatures in the combustor; CO from incomplete combustion. Within the turbine, NO_X can be reduced by effectively lowering the combustion temperature; however, these lower temperatures can increase CO production. Thus, reduction of turbine emissions can become one of balancing the relative costs of NO_X and CO control.

As one approach to emission control, manufacturers have redesigned the combustors and the burners to modify the firing process. These modifications, which include more complete mixing of the air/fuel mixture and alterations to the combustor to stage the actual combustion, have the potential to reduce NOx emissions by over 80%, reaching levels of 25 to 40 ppm.

A second approach is *water or steam injection* into the combustion zone. This control strategy is based on lowering the flame temperature and the formation of NO_X and has several performance consequences. First, the increased mass flow through the power turbine can increase the turbine's capacity. This added capacity, however, is not without a cost as the water or steam that is injected must be treated to minimize corrosion. While the turbine heat rate is not affected if steam is injected, water injection decreases the turbine efficiency by approximately 5% because of the heat that is needed to vaporize the water. Water injection does reduce the mass that is required as compared to steam injection. While steam injection can reduce NO_X to levels of 25 ppm, it does increase emissions of CO and unburned hydrocarbons.

As pointed out above, limited amounts of steam injection can produce modest increases in turbine capacity. Large amounts of steam injection can significantly increase turbine capacity, and this approach to power augmentation is frequently used when a cogeneration system serves varying steam requirements. When used for power augmentation, steam injection can increase capacity by up to 50% while improving efficiency by 20%.

A third approach to NO_X control is the use of *selective catalytic reduction* (SCR) technology. SCR consists of the injection of ammonia into the turbine exhaust. The two gases then pass through a catalyst where they react to produce elemental nitrogen and water. To be effective, both the rate of ammonia injection and the temperatures at which injection takes place must be controlled. While this approach is downstream of the turbine and does not impose any penalties on turbine operation or increase CO production, it does impose limitations on the design of the HRSG and may limit the use of supplemental duct burners.

SCR can significantly increase the capital cost of the system, and particularly so for systems of 5 MW or less. In addition, SCR can increase operating costs by 20% to 50%.

STEAM TURBINES

Steam turbines, operating through a *Rankine Cycle* (Figure 2-21), have been the backbone of the electric power industry. Since their introduction in the early 1900s, they have dominated the central power plant market, with steady and continuing improvements in the basic technology. With high availabilities and the capacity to fire almost any fuel including process wastes, these systems have been applied in both electric utility and industrial power plants and cogeneration systems. The modern steam power plant can operate at temperatures in excess of 1,000°F with pressures of several thousand pounds.

FIGURE 2-21
RANKINE CYCLE

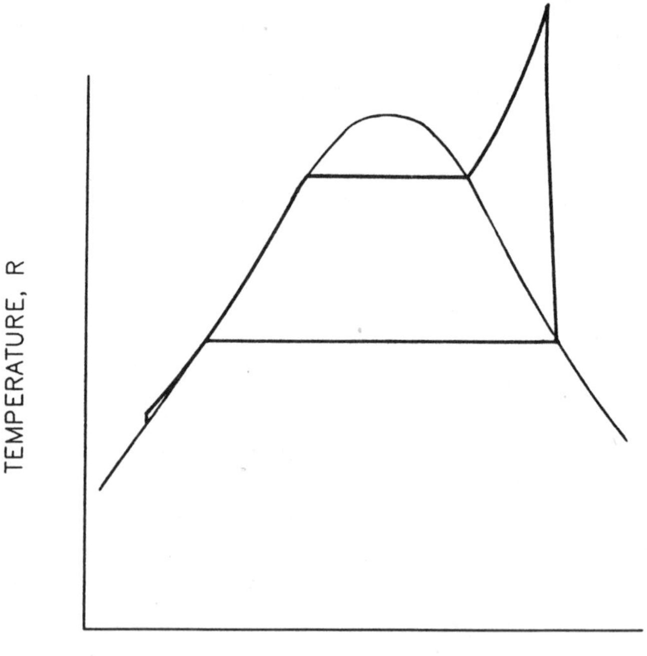

Concept

The steam turbine power plant (Figure 2-22) consists of a fired boiler, which produces high pressure, superheated steam; a steam turbine where the steam is expanded producing mechanical power; and a condenser where the turbine exhaust steam is further cooled producing water. The condenser increases the efficiency of the turbine, although it results in very low temperature heat rejection.

FIGURE 2-22
BOILER STEAM TURBINE SYSTEM

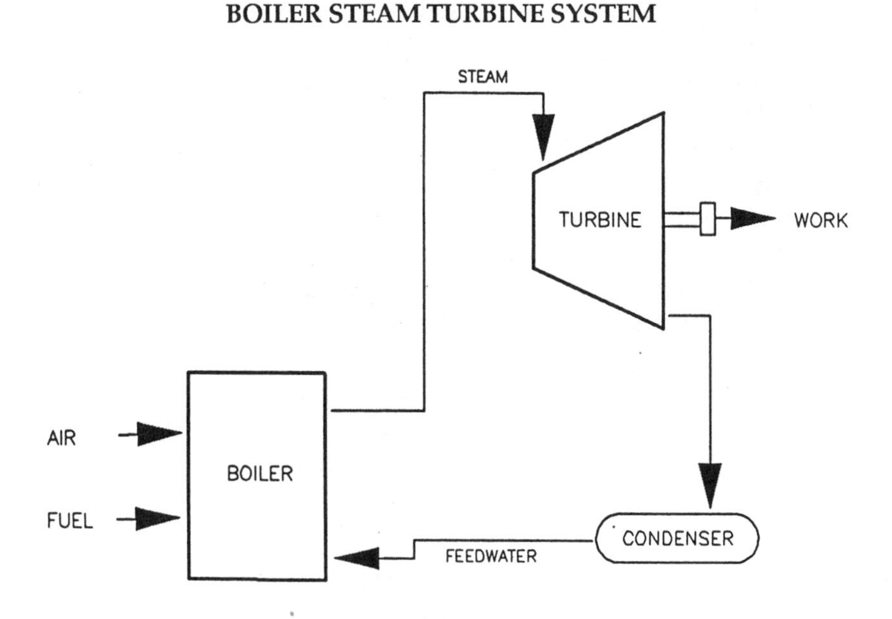

The boiler is the start of the steam cycle, and improvements in boiler design have led to increased efficiency. Because the boiler fuel is kept external to the cycle's working fluid, it is possible to use a variety of fuels ranging from low Btu coal wastes, chemical or process industrial wastes, coal, light and heavy oils and natural gas. Depending on the fuels and boiler design, it may be possible to rapidly switch fuels with no effect on the system's power output.

The fuel's chemical energy is converted to thermal energy in the boiler. In order to minimize energy losses and erosion of the turbine blades, the steam is usually superheated or raised to a temperature above its saturation temperature. Superheating removes any water droplets from the steam. On exiting the boiler, the dry, superheated steam is directed to the steam turbine where the potential energy of the steam is converted to mechanical energy. This latter step is performed as the steam expands and impacts on the turbine blades. The steam turbine's efficiency can be improved by either increasing the temperature and pressure of the incoming steam and/or decreasing the temperature and pressure of the turbine exhaust.

Turbines can be classified based on the manner in which the steam exits the turbine. The simplest cycle consists of a *non-condensing or backpressure turbine* (see Figure 2-23) where the steam is exhausted at atmospheric pressures or higher. Backpressure turbines are capable of operation with very high exhaust steam pressures. However, turbine efficiency is directly related to the difference between steam inlet and exhaust conditions and the *condensing turbine* exhausts steam at less than atmospheric pressures. The condensing turbine's thermal efficiency will be greater than the non-condensing turbine's efficiency for the same steam inlet conditions, and condensing turbines are generally used in electric power plants.

FIGURE 2-23
BACKPRESSURE TURBINE

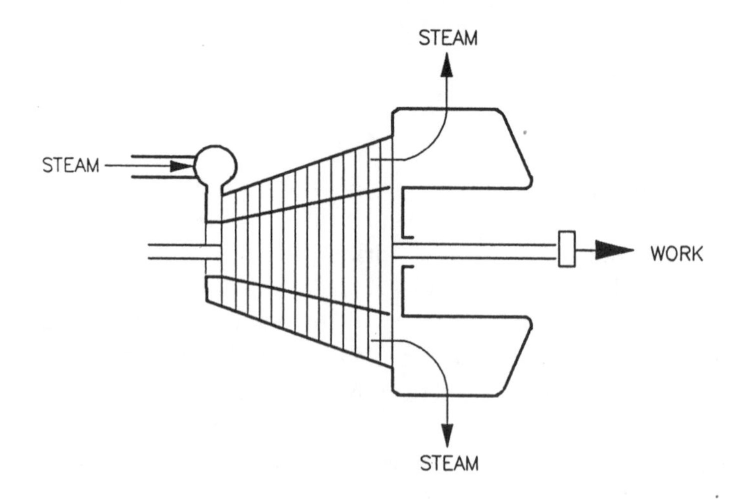

In many industrial applications there is a requirement for both power and higher pressure steam. In these cases, an *extraction turbine* (see Figure 2-24), which has ports where steam is extracted from the turbine prior to full expansion at the turbine exhaust, can be used. This type of turbine expands all the steam to the pressure or pressures required by the process, delivering steam at one or more extraction points. The remainder of the steam is expanded to condenser temperatures and pressures. While this type of turbine is not as mechanically efficient as a condensing turbine, it operates as a cogeneration system producing both high pressure steam for process and electric power.

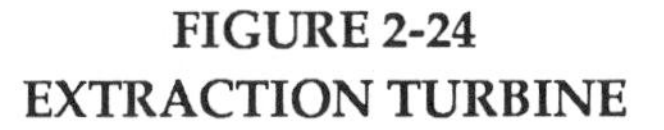
FIGURE 2-24
EXTRACTION TURBINE

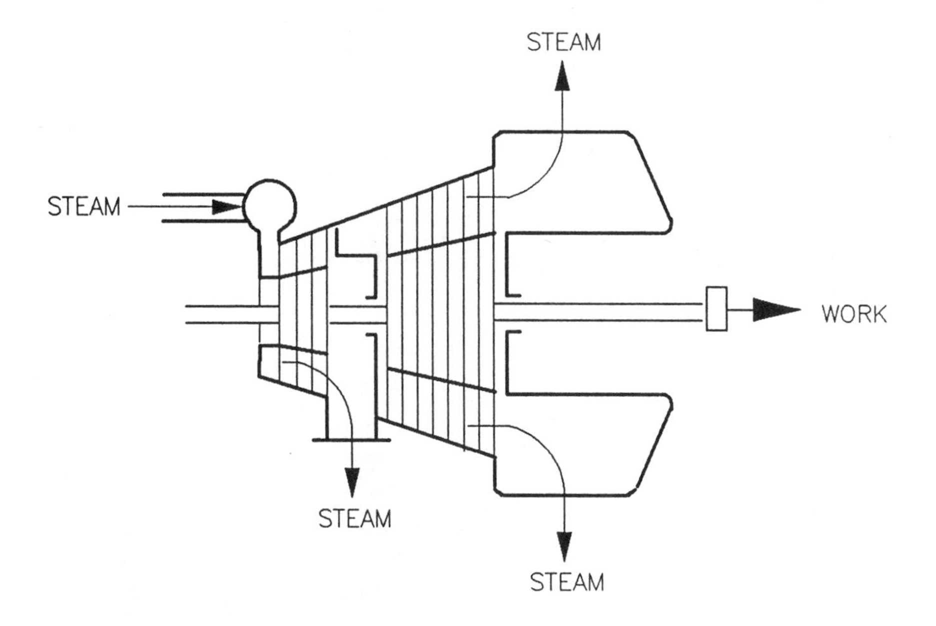

The extraction turbine can be designed with multiple extraction ports and, therefore is quite flexible in satisfying diverse steam requirements. Additionally, the turbine can be readily adapted to supply different quantities of steam per unit of electric output, thus increasing its suitability to industrial cogeneration.

An *induction turbine* (see Figure 2-25) has the capability to accept steam at two or more different pressures. Thus byproduct steam from the process can be used to increase the turbine's power output.

FIGURE 2-25
INDUCTION TURBINE

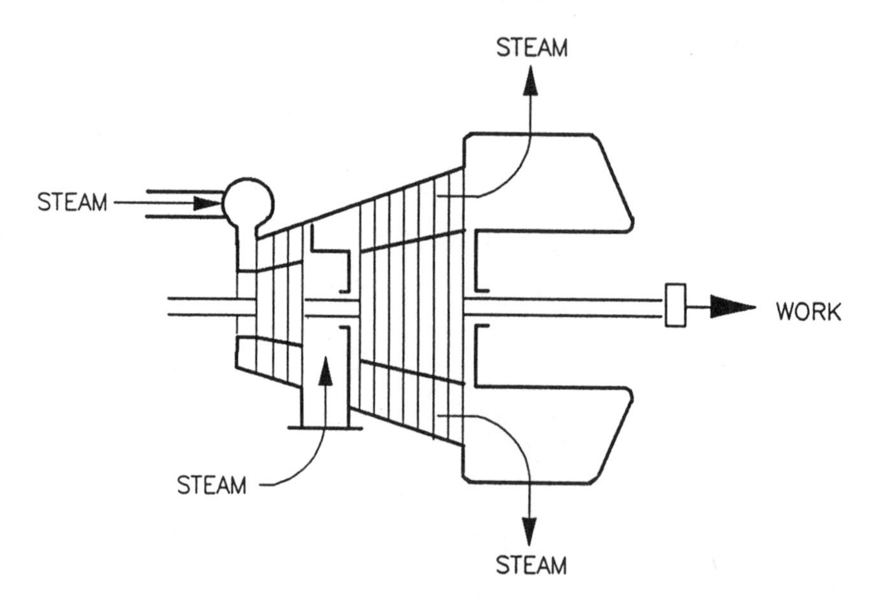

Some steam turbines allow for a *reheat cycle* (see Figure 2-26) wherein the steam is extracted from the turbine and reheated during the expansion process. Reheat cycles, with one or two reheat points, improve the overall thermal efficiency and reduce any moisture that may form as the steam pressure and temperature is lowered in the turbine.

Steam turbines may also include a *regenerative cycle* (see Figure 2-27) where steam is extracted from the turbine and used to preheat the boiler feedwater. The preheating of feedwater increases the overall cycle efficiency, in part because the steam's latent heat of condensation is returned to the process.

FIGURE 2-26
REHEAT CYCLE

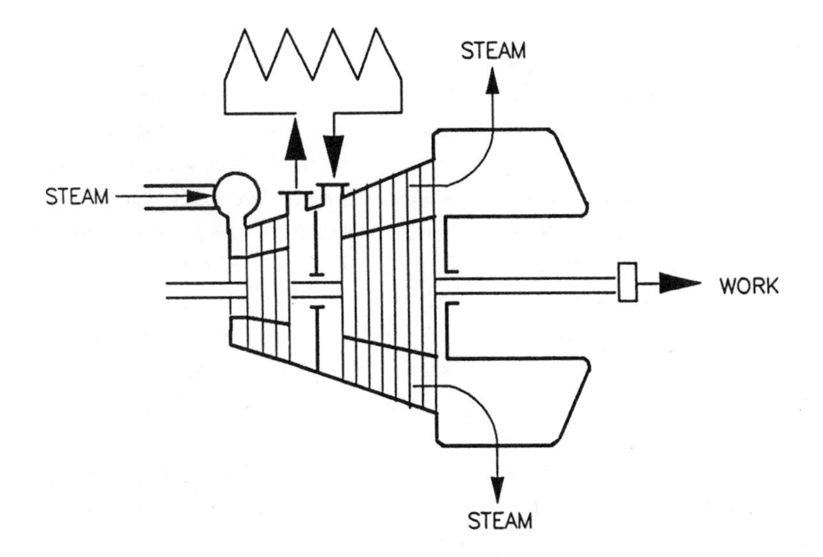

FIGURE 2-27
STEAM TURBINE REGENERATIVE CYCLE

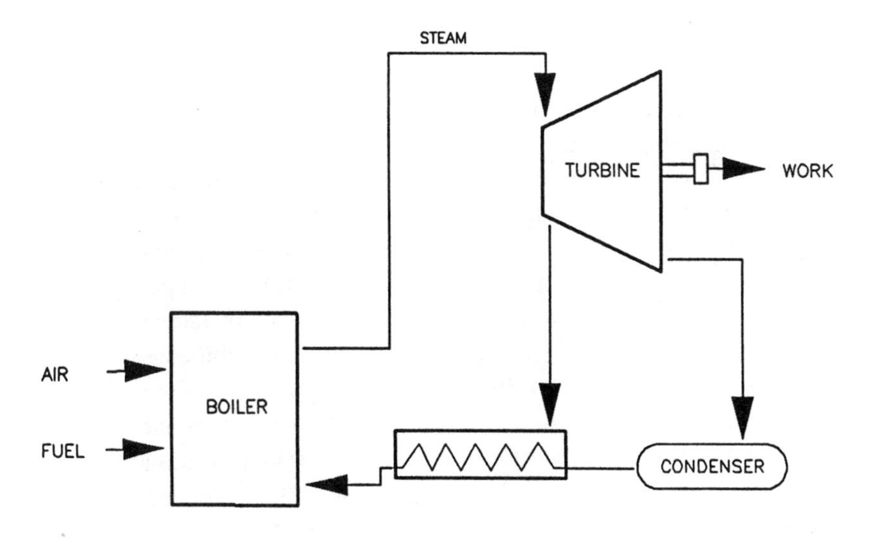

Finally, it should be noted that various combinations of the above configurations are possible including induction/extraction, extraction/noncondensing and extraction/condensing. When compared to the reciprocating engine and the combustion turbine, the steam turbine plant provides the greatest flexibility in matching the cogeneration system's electrical and thermal output to the load. These combinations are possible with reheat and regeneration.

While water is the most commonly used working fluid in a Rankine Cycle power plant, some systems have been developed using organic fluids. These plants usually operate at lower temperatures and pressures, and are found in applications where low grade waste heat is available. When used with the exhaust from a reciprocating engine, an *Organic Rankine Cycle* (ORC) can increase the engine's power output by 10% with no increase in energy input. In some cases, the working medium can be a mixture of two fluids, such as water an ammonia, with a resulting increase in efficiency.

As pointed out in the discussion of combustion turbines, a HRSG can replace the Rankine Cycle's fired boiler forming a Combined Cycle (see Figure 2-28). This system can significantly increase overall thermal efficiency.

Equipment Availability and Performance Characteristics

While steam turbines are available in an almost limitless number of sizes, they are most generally employed in larger applications. Systems as small as several hundred kilowatts are commercially available, although when boilers are fired with solid fuels, materials handling considerations usually result in applications of 10 megawatts or more. The upper range exceeds 1,000 megawatts. The typical cycle efficiency will depend on a number of factors and can range from 15% to over 35%. In general, smaller turbines will be less efficient than the larger units.

While the performance of the steam turbine is not directly affected by ambient temperatures, high temperatures may reduce the heat rejection capacity of the condenser and decrease the steam turbine capacity.

FIGURE 2-28
COMBINED CYCLE

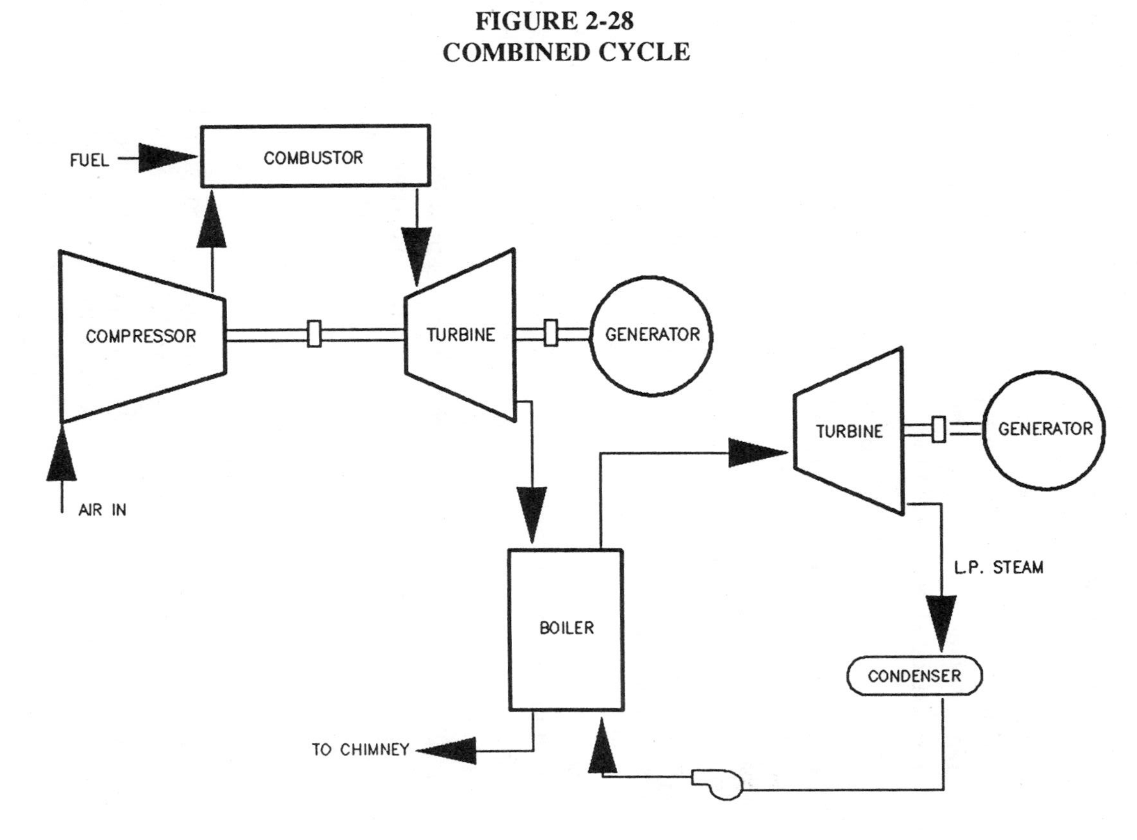

Maintenance Requirements

The steam turbine is capable of availabilities equal to that of a combustion turbine. Unlike the combustion turbine, the use of an external boiler means that the choice of fuel will not effect the turbine's maintenance requirements. Fuel choice will effect the availability of other system components and may impact overall system availability. Unlike the combustion turbine, major overhauls require extended periods of time.

Heat Recovery Characteristics

The Rankine Cycle steam turbine cogeneration system has the greatest flexibility for matching an end user's particular combination of power and thermal requirements. The noncondensing and extraction turbines will provide steam at pressures ranging up to several hundred pounds. In addition, the extraction turbine can be designed to supply different quantities of steam at these higher pressures, thus allowing the plant designer to match the ratio of thermal to electric output to the site requirements. These turbines have also been used in combined cycle systems, allowing maximum power production while still satisfying steam requirements.

Condensing turbines, which exhaust steam at less than atmospheric pressure, provide lower quality thermal energy. Available temperatures rarely exceed 190°F, and the use of this heat is generally limited to hot water applications.

Environmental Considerations

Environmental requirements applicable to steam turbine facilities are primarily targeted on air and water emissions. Air emission concerns are similar to those for other prime movers except that the ability to use solid fuels introduces added concerns regarding particulates and trace materials. Coal and oil fired systems face increasing controls to limit SO_x emissions.

Condensing plants must reject large quantities of low temperature heat, either to the atmosphere through cooling towers or to rivers and lakes. In areas of concern over thermal pollution, dry cooling towers may be required.

EMERGING TECHNOLOGY

The renewed interest in conservation that has accompanied energy cost increases when combined with PURPA's lowering of barriers to cogeneration has resulted in the development of added technologies for the on-site market. These developments span a range of products including prime movers, heat recovery systems, system controls and emission control equipment. Those technologies, which may have a near term impact, are reviewed below.

The prime movers described above were each based on the conversion of chemical energy to heat with the subsequent conversion of heat to mechanical energy and to electrical energy. This process, whether it is based on a reciprocating or rotating engine, introduces mechanical losses into the process. In contrast, *fuel cells* are energy conversion devices which generate electrical energy directly from the fuel, with no intermediate mechanical step. The fuel cell stack produces DC power and, therefore, it is necessary to include an inverter to convert the DC to the commonly used AC. While the inverter introduces some electrical losses, it provides the capability to produce AC power at a broad range of frequencies or voltages as required by the specific end use.

As shown in Figure 2-29, the fuel cell combines hydrogen and oxygen to produce water and direct current electric energy. In practice, these fuel cells include a fuel processor or reformer which strips hydrogen from a fuel such as natural gas, propane, butane or some other hydrocarbon. The fuel's carbon is converted to carbon dioxide. Oxygen is available from the atmosphere, and both are input to the cell stack under high pressure and temperature. In a process that is the reverse of electrolysis, the oxygen and hydrogen combine to produce water and electric power. With water as the primary output, emission control is not anticipated to be a significant problem. Finally, with a minimum of moving parts, the part load efficiency of the fuel cell is approximately equal to the full load value (see Figure 2-30).

Fuel cell technology is relatively old; however, its commercial application for power generation has only been aggressively pursued over the last decade. Today, units ranging from 40 kilowatts to over

FIGURE 2-29
FUEL CELL POWER PLANT

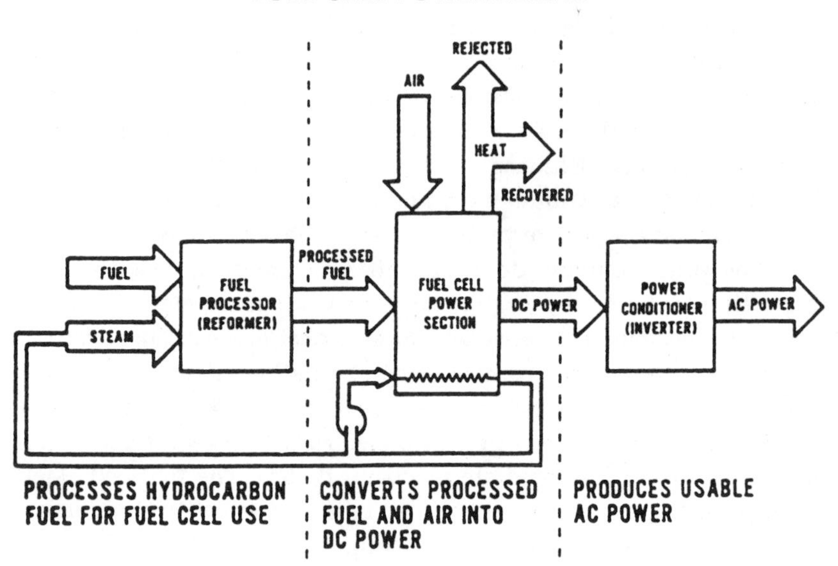

FIGURE 2-30
FUEL CELL PART LOAD PERFORMANCE

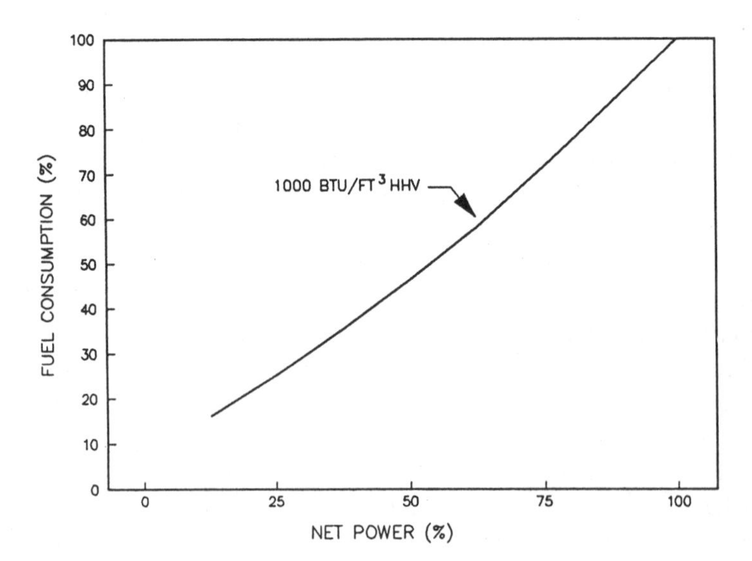

10 megawatts have been field tested. Test sites have included small restaurants, hotels, health clubs, industrial facilities and electric utility substations. The results have been encouraging and indicate that projected efficiency, maintenance and reliability goals are achievable. The most significant challenge is to bring the capital and maintenance costs down to the level of competitive technologies.

As with all prime movers, increased electrical efficiency results in decreased heat recovery potential. Simply, less energy is available. In addition, the quality of the available heat decreases as the quantity of the rejected energy decreases. Early field test fuel cells rejected most of the excess heat at temperatures of 160°F or less, thus limiting their cogeneration potential to those facilities that require low temperature hot water or, alternatively, to facilities where low temperature heat can be used to preheat boiler feedwater.

The *Stirling Cycle* is an externally fired reciprocating engine whose thermodynamic cycle was first identified over 150 years ago. Unlike the internal combustion engine, it is capable of using a solid fuel, or of using gaseous and liquid fuels in a more environmentally acceptable manner. The engine has few moving parts and has the potential to achieve efficiencies of 40%, with good part load performance.

SYSTEM DESIGN CONSIDERATIONS

The design of a cogeneration system will be based on the unique conditions and constraints of each specific application with the development of the ultimate design being an iterative process. Initially many concepts are considered; however, most are discarded and only the most promising are modelled in detail. A complete design analysis considers the time variations of both thermal and electrical requirements, the prime mover performance at rated and part load, the cost and availability of different fuels, and both capital and operating costs as the basis for final selections. However, even at the early stages of design, some generalized rules can be helpful.

Prime Mover Selection

Three types of prime movers are commercially available for use in a cogeneration system: reciprocating engines, combustion turbines and steam turbines. Each has unique characteristics and the choice of a particular type of prime mover for a specific application will depend on a number of technical and economic considerations. The history of the past 10 years indicates that reciprocating engines have most of the small market, while combustion turbines have captured the market above several megawatts. In some cases, the installation may consist of combustion turbines for the cogeneration system and reciprocating, diesel engines for electrical backup to the turbine. Some of the key factors which must be considered in selecting a particular prime mover are reviewed below.

The end user's *thermal requirements* will be a major determinant in the choice of a prime mover. Both the quality of the required thermal energy and the ratio of the thermal and electrical requirements will enter into consideration.

Thermal quality is measured as steam pressure or temperature, and both the combustion and steam turbine can provide high quality steam. The steam turbine can provide higher pressures and temperatures using an extraction port while the combustion turbine may require duct firing to achieve higher pressures or temperatures. The reciprocating engine is least useful, with much of the recoverable heat limited to a maximum of 15 psig.

Steam turbines offer the greatest flexibility with regard to the *ratio of thermal and electrical output*. The ability to extract steam at different pressures allows the system to match almost any thermal and electrical load. Steam turbines have the highest potential for the thermal/electrical ratio. Combustion turbines, in combination with a duct burner, provide a high ratio of thermal to electrical output, although not as high as is possible with the steam turbine. Reciprocating engines provide the lowest ratio.

Electrical *efficiency* is also quite important. Reciprocating engines with efficiencies ranging form 25% to 35% are the most efficient prime mover in sizes of a few megawatts or less. Slower speed, larger reciprocating engines are capable of efficiencies of over 40% and are competitive with medium sized turbines. Combustion turbines of 3,500 kilowatts or more are generally as efficient as reciprocating engines. Steam turbines can be efficient in larger systems of 10 MW or more.

Turbines have extremely high scheduled *availabilities*. Because the major overhaul of a mid-sized combustion turbine can be performed off-site, it may have a higher availability over its economic life than would a steam turbine. While the turbine itself is capable of scheduled availabilities of 8,600 hours, failures with ancillary equipment may reduce overall availability. Reciprocating engines, with their requirement for more frequent and more prolonged maintenance procedures, have the lowest overall availability. Additionally, unscheduled maintenance requirements result in availabilities of between 7,600 and 8,400 hours.

With proper maintenance, including major overhauls, all three prime movers are capable of an *economic life* well in excess of 100,000 hours.

Duty cycle may also be an important consideration. Reciprocating engines have been proven superior for cycling or peak shaving applications. Turbines have been proven superior in baseloaded applications. Turbine cycling tends to increase maintenance costs by decreasing the interval between major overhauls.

Reciprocating engines have proven to be the prime mover of choice in smaller applications, ranging from tens of kilowatts to as high as 1,500 kilowatts. Multiple high speed, reciprocating engines are frequently used in applications ranging up to 3,000 kilowatts, although combustion turbines have also achieved some market penetration in this size. In this size range, the choice between reciprocating engines and turbines is usually based on thermal requirements and equipment availability. Turbines have captured most of the market for applications of several megawatts or more. Combustion turbines, either alone, in multiples, or as part of a combined cycle have been used primarily in larger applications or where a decision to use solid fuels has made combustion turbines unfeasible.

Prime Mover Size

The size of a prime mover will depend on the relative economic return associated with the following operating modes: thermally baseloading, electric baseloading, peak shaving and isolated operation. The baseloaded system can be designed for either of two types of operation. At one extreme for internal-use cogeneration, the prime mover size can be chosen to be so small that the cogeneration system's output is never less than the end user's energy requirements. No power is exported and no heat is rejected. The total prime mover capacity must be less than the minimum thermal and electrical load. At the other extreme, which is typical of projects that sell all power to an electric utility, the capacity of the prime mover is chosen to maximize the amount of power that can be exported. In this case, the recoverable heat may be significantly greater than the site's thermal requirement and the heat may be rejected or used in a combined cycle. In this case, the total cogeneration capacity is based on the economics of a power sale agreement more so than the site's thermal needs.

A system designed to be either *thermally or electrically baseloaded* is sized based on the site's minimum thermal or electric load respectively. The thermally baseloaded system is sized so that the recoverable heat from the prime mover is always, or substantially always, less than or equal to the site's thermal requirement. The byproduct power may be more or less than the site's electrical requirement and either supplemental power may be purchased or excess power may be exported. The thermally baseloaded system produces the maximum energy efficiency since no recoverable heat is lost. The electrically baseloaded system is sized based on the site's minimum electrical requirement and no power is ever exported. The thermal byproduct may be greater or less than the site's thermal requirements and, in some cases, may be rejected.

The peak shaving system's viability is driven by the rate structure with its size based on the difference between the end user's peak and off-peak demands or upon the maximum peak demand. The ultimate size will be based on rate design, and recoverable heat may be used when produced, stored for use later in the day, or rejected. Heat recovery is a less critical concern with peak shaving system economics.

An *isolated* cogeneration system will generally require the greatest installed capacity. At a minimum, the system must have adequate capacity to satisfy the end user's peak requirement with a reserve to accommodate short-term transients. As a rule of thumb, the system must have adequate capacity to start up the largest motor while satisfying all other loads. In addition, if the system is isolated from the electric utility grid, it may be necessary to include additional capacity to allow for both scheduled and unscheduled maintenance.

Number of Prime Movers

The total number of prime movers will depend on the operating mode, the end user's requirements for reliability, equipment availability and overall economics. For those cogeneration systems interconnected to the electric grid, the number of prime movers is a trade off between cost and reliability. When combustion turbines are used, equipment availability may dictate the number of prime movers, particularly in applications of 10 megawatts or less.

For grid isolated systems, a general rule of thumb is to size the engine at one-third of the peak electrical requirement, with at least one engine for on-site backup. In some cases, where high reliability is required, a second backup engine may be used, raising the total number of engines to five, with the total installed capacity equal to 167% of the site electrical load. This approach to sizing can be quite costly, and is seldom economically justified.

REVIEW

The prime mover is the heart of the cogeneration system, serving as the primary source of both mechanical and thermal energy. Most cogeneration systems in place are based on one of three commercially available technologies; reciprocating engines, combustion turbines and steam turbines. All three engine alternatives have proven to be reliable and cost efficient, with the choice from among them based on the unique energy and performance requirements of a specific application.

Over the last decade, reciprocating engines have proved to be very effective in cogeneration systems ranging from a few kilowatts to several megawatts, while combustion turbines have captured most of the larger cogeneration applications. While combined cycle applications were usually only considered for larger applications of several hundred megawatts, their use in applications of ten megawatts or less have also proved to be successful.

The size of a cogeneration system that is interconnected to the electric utility grid will be determined as a trade off between economies of scale, both in capital and operating costs that result from increased size and decreased incremental returns that result from additional power or heat. Larger turbine based cogeneration systems, relying on power sales to an electric utility, have proved viable in plants as large as 450 megawatts. In comparison, smaller reciprocating engine based systems, relying on the reduction of power purchases from an electric utility, have proved viable in applications as small as 45 kilowatts.

Most successful cogeneration applications can be characterized as having first selected a prime mover which matches the mix of electrical and thermal requirements and, secondly, instituted an effective preventive maintenance program. Even the best of system designs can be rendered inoperable by an inadequate operating and maintenance program.

Chapter 3

GENERATORS, INVERTERS & UTILITY INTERTIES

Cogeneration is the sequential production of two forms of energy–power and thermal–from a single source of energy. While the power can be in the form of either electrical energy or mechanical shaft power, most cogeneration systems produce electrical energy. One important benefit of PURPA was the cogenerator's right to interconnect to the electric utility grid and to operate in parallel with that grid. Today most cogeneration systems are interconnected either to sell power to an electric utility, where the power is then resold to the end user or, alternatively, for internal use projects to reduce power purchases from the electric utility. This chapter reviews both the technology by which electric power is produced, and the interconnection of that power source to the electric grid.

GENERATORS

With the exception of the fuel cell, all the prime movers discussed in Chapter 2 produce mechanical energy. This mechanical energy must generally be converted to electrical energy using an electrical generator. In most cases, the engine manufacturer will provide the engine and generator as a single, factory assembled, integrated unit.

Concept

All generators operate on the principle that a conductor moving through a magnetic field will produce an electromotive force or voltage differential between the ends of the conductor. The voltage will be proportional to the strength of the magnetic field and the rotational speed of the conductor. The frequency of the generator's output will be directly proportional both to its rotational speed and to the number of poles. In the United States, where 60 Hertz power is the commercial standard, generators usually operate at a multiple of 60 revolutions per minute (rpm), and typically either 1,200 or 1,800 rpm. Lower speeds are possible; however, the maximum generator speed is 3,600 rpm for a two-pole machine. Where other frequencies such as 25 Hz, 50 Hz or 400 Hz are required, the generator speed must be a multiple of those frequencies. Reciprocating engines operate at speeds that are compatible with generator speeds; turbines, however, operate at speeds that are generally much greater than those of the generator and a gear box is required between the turbine and the generator.

An important characteristic of a generator is that it requires a magnetic field to operate, and generators are classified as to the source of energy creating that magnetic field, either synchronous or induction.

The *synchronous generator* provides its own excitation and usually includes a magnetic core. The *self-excited* synchronous generator includes a second, smaller generator which produces AC power which is, in turn, converted to DC for use in developing the required magnetic field. The *separately excited* synchronous generator uses a permanent magnet to produce the DC voltage required for the exciter rotor windings.

Synchronous generators are used in emergency power applications where the cogenerator wishes to operate in isolation from the utility grid or when the cogenerator wishes to correct the power factor. When operating independently, the cogenerator must provide precise frequency and voltage controls. If the generator is to parallel the electric utility grid, the frequency, voltage and phase angle will be determined by the grid; however, the cogenerator must provide a mechanism for safely synchronizing to the grid.

Induction generators are simply motors that are driven above their synchronous speed. When connected to an electric power source, the device will draw power from the source to produce an internal magnetic field. The interaction of the magnetic fields produces a torque on the motor shaft and will cause the device to rotate and to operate as a motor. The motor will operate at various speeds, up to its synchronous speed, with the power that the motor delivers being dependent on the difference between the motor's rotational speed and its synchronous speed. When mechanical power flows into the device, the speed will exceed the synchronous speed and the device will operate as a generator. When operating as an induction generator, the machine will require a magnetizing current from the external source. An induction device cannot sustain operations as a generator once the source of magnetizing current, usually the utility grid, is lost. Therefore, it cannot be used if isolated or independent operations are required. The generator inductance can interact with nearby capacitance to create an oscillating circuit that will allow the generator to self excite for a period of time. While this type of operation can produce significant voltages, in general it will not operate at the synchronous speed, and frequency or reverse power relays are required to avoid any hazardous operations.

Characteristics

Several performance characteristics are important when evaluating generators including rating, efficiency, voltage, voltage regulation, power factor, harmonics and fault current. Each parameter must be considered in the selection of equipment.

Generator *ratings* are usually specified as *kilovoltamps* (kVA), which is the product of the generator rated voltage and the rated maximum current or total power. As was noted in the discussion of induction generators, these devices require a magnetizing current and some power. The product of this magnetizing current and voltage is defined as *reactive power* (kVAR) which, unlike the mechanical output of an engine, produces no real work. The generator's *real power* is specified in kilowatts (kW), and the ratio of the real power to the total power is defined as the *power factor*. In those instances where the generator rating is specified as its real power output, two factors are required: the total real power as expressed in kilowatts and the power factor expressed as a fraction. Common practice is to specify the generator real power output at an 80% power factor, although other factors are used.

Utilities discourage low power factors either by requiring a minimum power factor or applying a cost penalty to a low load factor customer. Induction generators require reactive power while producing real power and, as a result, the site's power factor will typically worsen when an induction generator is interconnected to the grid. In such cases, it may be necessary to install additional capacitors to correct power factor. Synchronous generators have the ability to vary their power factor and to produce reactive power. They can be used to correct for low site power factor. Utility billing practices should be reviewed in selecting the generator and in determining the extent to which power factor correction is economical.

Both short-term or transient and steady-state conditions must be considered in sizing a generator. Short-term conditions are especially critical when the system does not parallel the utility.

Generators usually operate at a rather high *efficiency,* generally in excess of 90%. Units of 1,000 kVA or more will generally reach efficiencies in excess of 95%, while units below 100 kilowatts or generators operated at low output (see Figure 3-1) will have efficiencies of 85% or less.

The electric grid supplies *voltages* that vary over a broad range (see Table 3-1). Low generation voltages tend to decrease efficiency as internal losses account for a greater fraction of the input energy. Therefore, as a general rule, the highest available voltage should be selected. Higher voltages do result in increased cost, both for the generator and for the associated transformers, relaying and switchgear. Ultimately, the choice of a generator voltage becomes an economic tradeoff between capital and operating costs. Most smaller generators are designed for 480-volt operation, while larger, multi-megawatt generators are designed for 13.2 kV or more.

When interconnected to an electric utility grid, small cogenerators will simply follow the grid voltage, which can generally vary by as much as 5%, and frequency and regulation is not a significant problem. When the cogenerator operates in an isolated or stand-alone mode, the engine and generator controls must provide the capability to maintain the desired voltage and frequency. Voltage regulation may become a more severe problem when site loads vary significantly. In some cases, where the utility may reduce voltage (above the 5% tolerance) during capacity shortages, the cogenerator

FIGURE 3-1
TYPICAL GENERATOR EFFICIENCY VS. LOADING

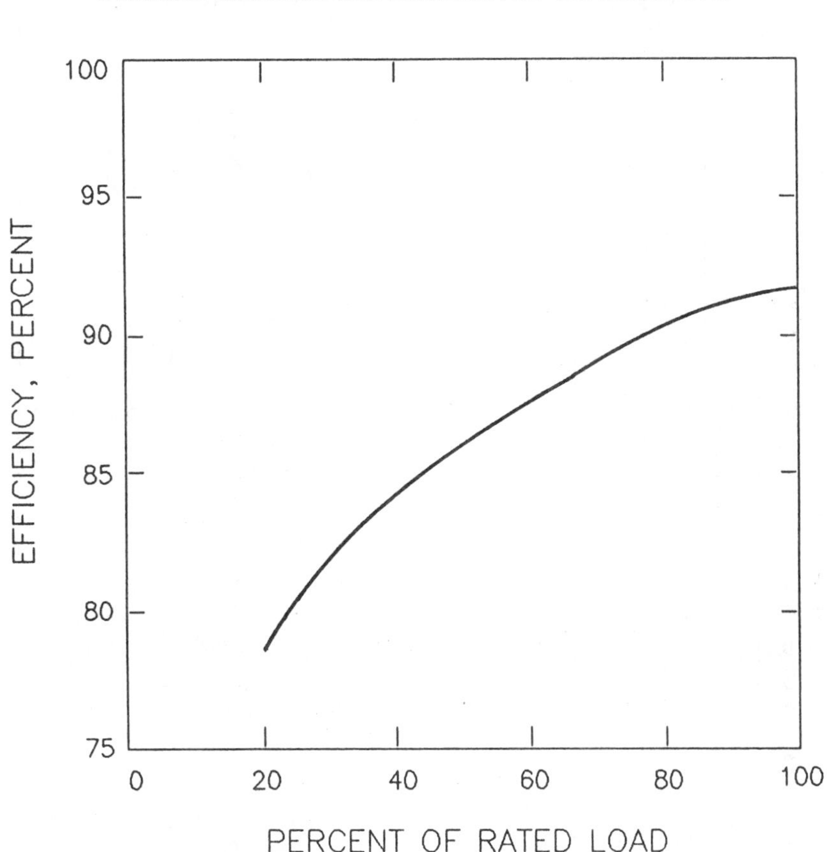

may elect to isolate from the grid and to operate at nominal voltages.

Generator operation during a fault can be a critical concern. When the cogenerator is paralleling the electric utility, the generator can contribute additional fault current and coordinating with the utility is critical. When operating in an isolated mode, the facility's protection system must consider the generator's short circuit or available fault current. Such coordination is required both to protect the generator and to protect the grid or site equipment.

TABLE 3-1
STANDARD SYSTEM VOLTAGES*

NOMINAL SYSTEM VOLTAGE (Note a)			VOLTAGE RANGE A			VOLTAGE RANGE B		
			Minimum		Maximum	Minimum		Maximum
Two-wire	Three-wire	Four-wire	Utilization Voltage (Note b)	Service	Utilization and Service Voltage (Note d)	Utilization Voltage (Note b)	Service	Utilization and Service Voltage (Note d)
Single-Phase Systems								
120			110	114	126	106	110	127
	120/240		110/220	114/228	126/252	106/212	110/220	127/254
Three-Phase Systems								
		208Y/110	191Y/110	197Y/114	218Y/126	184Y/106	191Y/110	220Y/127
						(Note c)	(Note c)	
		240/120	220/110	128/114	252/126	212/106	220/110	254/127
	240		220	228	252	212	220	254
		480Y/277	440Y/254	456Y/263	504Y/291	424Y/245	440Y/254	508Y/293
	480		440	456	504	424	440	508
	600		(Limits Not Extablished)			(Limits Not Established)		
	2400		2160	2340	2520	2080	2280	2540
		4160Y/2400	3740Y/2160	4050Y/2340	4370Y/2520	3600Y/2080	3950Y/2280	4400Y/2540
	4160		3740	4050	4370	3600	3950	4400
	4800		4320	4680	5040	4160	4560	5080
	6900		6210	6730	7240	5940	6560	7260
		8320Y/4800		8110Y/4680	8730Y/5040		7900Y/4560	8800Y/3080
		12000Y/6930		11700Y/6760	12600Y/7270		11400Y/6580	12700Y/7330
		12470Y/7200	(Note e)	12160Y/7020	13090Y/7560	(Note e)	11850Y/6840	13200Y/7620
		13200Y/7620		12870Y/7430	13860Y/8000		12540Y/7240	13970Y/8070
		13800Y/7970		13460Y/7770	14490Y/8370		13110Y/7570	14520Y/8380
	13800		12420	13460	14490	11880	13110	14520
		20780Y/12000		20260Y/11700	21820Y/12600		19740Y/11400	22000Y/12700
		22860Y/13200		22290Y/12870	24000Y/13860		21720Y/12540	24200Y/13970
	23000		(Note e)	22430	24150	(Note e)	21850	24340
		24940Y/14400		24320Y/14040	26190Y/15120		23690Y/13680	26400Y/15240
		34500Y/19920		33640Y/19420	36230Y/20920		32780Y/18930	36510Y/21080
	34500			33640	36230		32780	36510

Higher Voltage
Three-Phase Systems in kV

Nominal System Voltage	Maximum Voltage
46	48.3
69	72.5
115	121
138	145
161	169
230	242

For these systems Range A and Range B limits are not shown because, where they are used as service voltages, the operating voltage level on the user's system is normally adjusted by means of voltage regulation to suit his requirements.

NOTES:

(a) Three-phase, three-wire systems are systems in which only the three-phase conductors are carried out from the source for connection of loads. The source may be derived from any type of three-phase transformer connection, grounded or ungrounded. Three-phase, four-wire systems are systems in which a grounded neutral conductor is also carried out from the source for connection of loads. Four-wire systems are designated by the letter Y (except for the 240/120-volt delta system), a slant line, and the phase-to-neutral voltage. Single-phase services and loads may be supplied from either single-phase or three-phase systems.

(b) Minimum utilization voltages for 120-240 volt circuits not supplying lighting loads are as follows:

Nominal System Voltage	Range A	Range B
120	108	104
208	187	180 (Note c)
240	216	208
480	432	416

(c) Many 220-volt motors were applied on existing 208-volt systems on the assumption that the utilization voltage would not be less than 187 volts. Caution should be exercised in applying the Range B minimum voltages and note (b) to existing 208-volt systems supplying such motors.

(d) For 120-480 volt nominal systems, voltages in this column are maximum service voltages. Maximum utilization voltages would not be expected to exceed 125 volts for the nominal system voltage of 120, nor appropriate multiples thereof for other nominal system voltages through 480 volts.

(e) Utilization equipment does not generally operate directly at these voltages. For equipment supplied through transformers, refer to limits for nominal system voltage of transfer output.

*Information from
American National Standard C92.2-1967

Extra High Voltage
Three-Phase Systems in kW

Preferred Nominal System Voltage	Preferred Maximum Voltage
345	362
500	550
700	765

Source: *Cogeneration and Small Power Production*, American Public Power Association, November, 1980.

A generator's *available fault current* will depend on the type of generator and its design. An induction generator requires an external source of magnetization, and when the external voltage drops during a fault, the reactive power falls and the generator's ability to deliver power to a fault rapidly collapses. A synchronous generator is capable of continuing to deliver power during a fault, and the generator protective relaying must be capable of preventing sustained overloading and overheating of the generator.

Harmonics, which are multiples of the required electrical frequency, can cause several undesirable effects, and a generator's contribution to site harmonics must be considered. Harmonics cause the generator's output waveform to be distorted as shown in Figure 3-2. They can cause excessive heating in motors or they can reinforce each other producing high peak voltages or currents. They can also introduce errors in induction meters and computer equipment.

FIGURE 3-2
VOLTAGE WAVE DISTORTED BY
ODD HARMONIC FREQUENCIES

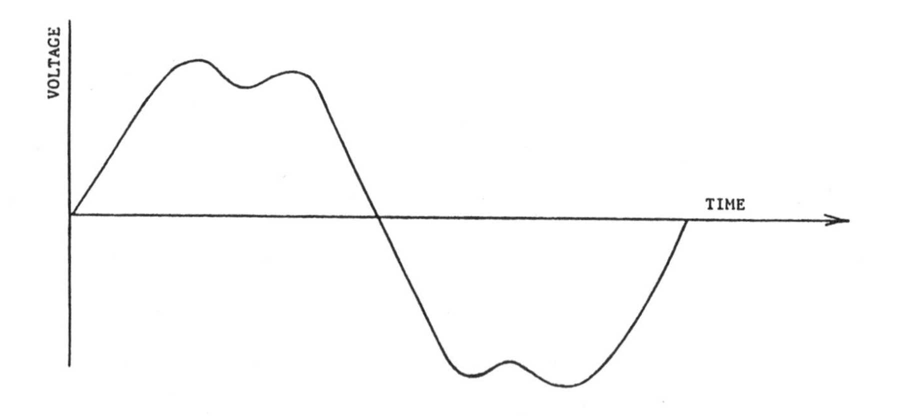

These harmonics are produced by non-linear and rapidly changing end use requirements such as might be typical of welders, X-ray machines and rectifiers, and are classified as either voltage or current. Both the electric utility grid and an on-site generator deliver power with some harmonic content, with synchronous generators producing more harmonics than induction generators.

There is no universal standard for harmonic content; however, most generators can meet a typical requirement for no more than 5% total harmonic distortion (THD), with a maximum of 2% or 3% in any one harmonic.

INVERTERS

Inverters are devices which convert DC power to AC power. They are used when power is produced as DC and required as AC, or sometimes when power is required at a frequency other than the standard 60 Hz. They are frequently applied in computer operations where 400 Hz power is required.

The input to an inverter is direct current power. The inverter operates by chopping the DC power into a series of blocks or square waves of different durations, but each of which being a multiple of the desired output frequency. These discrete square waves are then summed to create a wave shape that approximates the shape of a sinusoidal wave at the desired frequency. The harmonic content of the output and the shape of the wave can be improved by adding both inductance and capacitance to the inverter circuits.

As with generators, inverters may be either self-commutated or line-commutated. The latter type of inverter synchronizes its output to an external power source and is less expensive than the self-commutated inverter. Line-commutated inverters are more likely to interact causing problems with other end uses on the same utility or site distribution grid.

UTILITY INTERCONNECTION

One of the more significant benefits of PURPA was the right of the cogenerator to interconnect to the utility grid; either to export cogenerated power, to import power that was supplemental to the cogenerated power, or both. In some cases, based on the cost of utility supplied power, the cogenerator may choose to isolate from the utility grid. The decision to isolate may be a short-term one, based on rate design and temporary problems with the utility grid, or it may be a long-term option selected due to the high cost of utility supplied power.

Along with the right to interconnect to the electric utility grid comes the responsibility for the cogenerator both to protect the grid and other customers of utility power from any adverse effects of the cogenerator and to protect the cogeneration system and the end user's site equipment from any adverse effects caused by the grid interconnection. While the interconnection design must consider many factors and constraints, it will utilize the same basic equipment that has been used by electric utilities and end users. The components included in the interconnect are relays, circuit breakers, fuses, contactors or switches, transformers, synchronizers and meters. These components allow for the safe interconnection of the cogenerator and the larger utility grid.

The Utility Grid

An electric utility produces power at one or more central power plants which are frequently a considerable distance from the power end users. The system of transmission lines, substations and distribution lines that link the many power plants to the many end users is often referred to as the *electric utility grid*. The utility grid includes multiple power plants, multiple utility companies and frequently multiple states. The cogenerator may interconnect with the grid simply to receive additional power, to deliver power to another end user, to export power to the local utility, or to sell power to some other utility which may even be in a different state.

Those cogenerators who sell part or all of their power to an electric utility may be concerned with the higher voltage portions of that grid. This portion of the grid, which operates at voltages as low as 69 kV and as high as 765 kV, is referred to as the *transmission network* and its function is to link the many central power plants to major power distribution points or substations. The lower voltage portion of the grid, the *distribution system*, links the substations to the end user. Most cogenerators, and particularly those that use power on-site to reduce purchases from the utility, will be concerned with the distribution system. In some areas, a *subtransmission system* links the distribution system and the transmission network.

The distribution system can be categorized as either radial or network. A *radial* system (see Figure 3-3) is defined as a distribution system with a single path for power flow to an end user. This type of system is commonly found in non-urban areas, is less expensive, and is not as reliable as a more complex network system. Interconnection of a cogenerator to a radial system is somewhat straightforward. A *network* system (see Figure 3-4) has several possible paths for power to flow to an end user. Typically, the network is found in urban areas and provides increased reliability, albeit at a higher cost than the radial system. The network system is designed for multiple current flow paths within the distribution grid and, as a result, the distribution substation has special relays and switches which protect the transmission or subtransmission system for reverse power flows. This level of protection can complicate the interconnection of a cogenerator or any active power source to the network.

FIGURE 3-3
RADIAL DISTRIBUTION SYSTEM

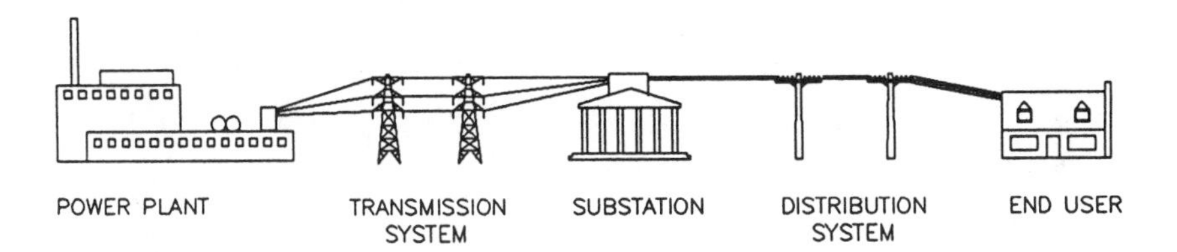

The electric utility's primary concern with the distribution system is the safe, reliable operation of that system in order to deliver power to its customers. The distribution system is designed to satisfy those requirements. Active power sources, such as cogenerators, who are a part of that distribution system, add complexity to the distribution system and these cogenerators must provide the same level of protection while not reducing the reliability of the electric utility grid. Among the electric utility's concerns are:

- The need for the cogenerator to avoid energizing a utility feeder which has been de-energized by the utility, thus

FIGURE 3-4
NETWORK DISTRIBUTION SYSTEM

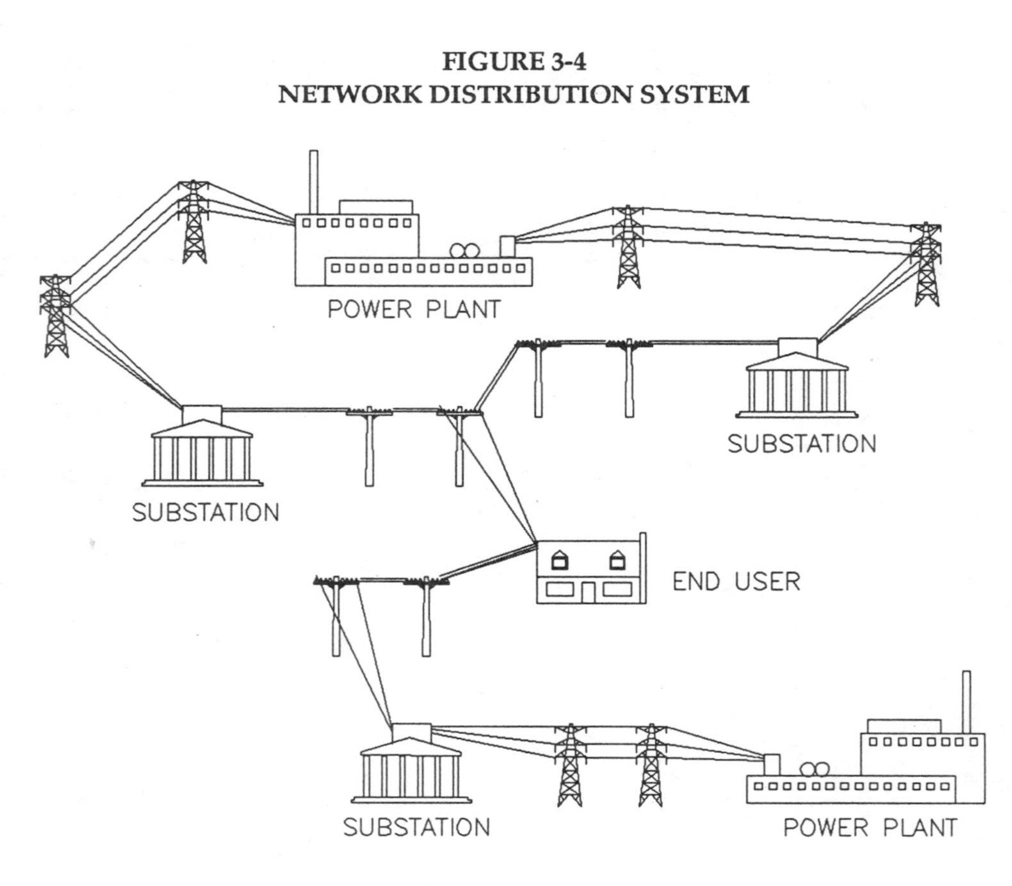

jeopardizing the safety of personnel who may be working on the feeder,

- The need to coordinate the cogenerator's fault contribution with the utility's fault clearing procedures, and

- The need to maintain power quality in the distribution system.

Specific protective devices and alternatives are discussed below.

Utility Interconnect Devices

The intertie introduces an active power source to the distribution system and the cogenerator must not only provide protection for the utility grid and people working on that grid, but also for the cogeneration system. Most utilities require that an active source cannot parallel with the grid unless it has permission to do so, and that all such sources have a *disconnect switch* capable of isolating the grid from the cogenerator under utility control. In general, they are only opened after a circuit has been de-energized.

A second major safety concern of utilities is the effect that the cogenerator can have on the utility's procedures for dealing with faults. At a minimum, introduction of the cogeneration system will increase the available short circuit current in at least some portion of the distribution system. This increase may require that the utility limit the amount of short circuit current available from the utility grid. In some cases, where the distribution system is designed on the assumption that there is a single source of power, it may be necessary to redesign specific portions of that system. Cogenerators with a capacity that is significantly larger than the site load can pose unique problems for the electric utility.

It is helpful to review how utilities manage faults in order to understand the problems that they cause for cogenerators. When a fault occurs on the distribution system, the system acts to identify that a fault condition exists, to determine the location of the fault, and to isolate and remove the portion of the distribution system as quickly as possible. The system then seeks to restore service to as many customers as possible.

Circuit breakers are automatic switches that are designed to open or break a circuit under specified conditions. They are capable of operating at different voltages and can be set to open after some predetermined time. They may be latched close and are triggered by protective relays. *Relays* are devices which sense the electrical condition of a circuit and, based on changes in that condition, cause a breaker to open. Relays detect unsafe operating conditions and typically sense high or low voltage, high current flows, over or under frequency, phase imbalance and the direction of power flow. Electric utilities will vary as to the specificity with which they define the interconnection relay requirements.

Because most faults are temporary in nature, utility grids generally include reclosures which first interrupt a fault current and then restore service after a short outage. In most cases, the fault clears after a short outage and the recloser re-establishes normal service. In cases where the fault does not clear, the breaker will again open, isolating the fault. The proces of interrupting the fault current and then reconnecting to the load can be repeated several times and, in some cases, the length of the outage increases with each successive interruption. Finally, if the fault does not clear, the breaker is locked open.

The use of reclosers causes unique problems for active power sources that are interconnected to the utility distribution system. During the period when the local distribution system is isolated and de-energized, the overall utility grid continues to operate at 60 Hz. As discussed above, an induction generator will lose its source of magnetizing current and both the output voltage and the power frequency will change. When the recloser energizes the system, the generator will be at a different voltage than the grid. Frequency shifts may further aggravate the voltage difference, even for outages of a few seconds or less. While the average voltages of the distribution system and the generator may be approximately equal, the instantaneous voltage difference may be significant, causing high transient currents (see Figure 3-5). Even though synchronous generators may hold their voltage during a short outage, they may also drift out of phase with the same consequences.

FIGURE 3-5
OUT OF PHASE 60 Hz SYSTEM

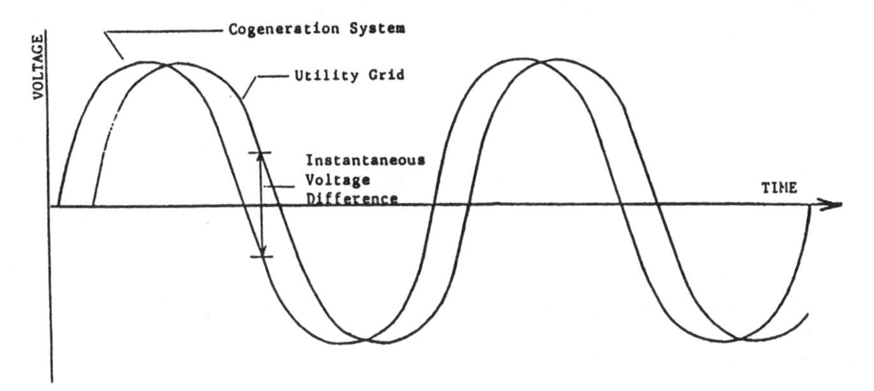

One approach to dealing with this problem is to immediately isolate from the distribution system at the occurrence of the first loss of voltage. The cogenerator can then choose to either shut down or to operate in an isolated mode until the grid is stabilized and the on-site generator can be synchronized to the grid.

Transformers serve several functions in the utility grid and in the interconnection to the grid. First, they provide a mechanism for matching the voltage that is available from the cogenerator to the voltage that is required at the site or by the grid. Second, they can provide a measure of isolation between the cogenerator and the grid and can limit fault currents, harmonics, voltage variations and load imbalances. Finally, small potential transformers (PTs) and current transformers (CTs) are used to sense voltage and current levels both for metering and for protective relays.

Meters are required to measure both the quantity of electrical energy (kilowatt hours) and the rate at which it is delivered (kilowatts). Where power factor is a concern, they can be used to measure reactive energy (kilovoltamp hours) and power (kilovoltamps). Depending on the size of the cogenerator and whether or not power is exported, a simple cogeneration system may require from one to four meters, two each measuring power and energy flow in each direction. Up to four additional meters may be required for reactive power, and a recording meter may be required when time of day rates are in effect.

The design of the electric intertie will depend on many factors and Figures 3-6 through 3-11 illustrate but a few alternatives.

FIGURE 3-6
SINGLE METER

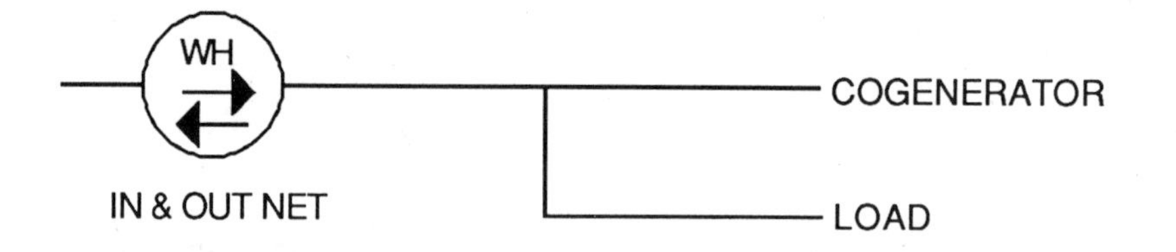

Simplest, assumes rate for billing is the same as the rate for buyback.

Source: *Cogeneration and Small Power Production, Guidelines for Public Power Systems*, prepared by the American Public Power Association, November 1980.

FIGURE 3-7
DUAL METER SYSTEM

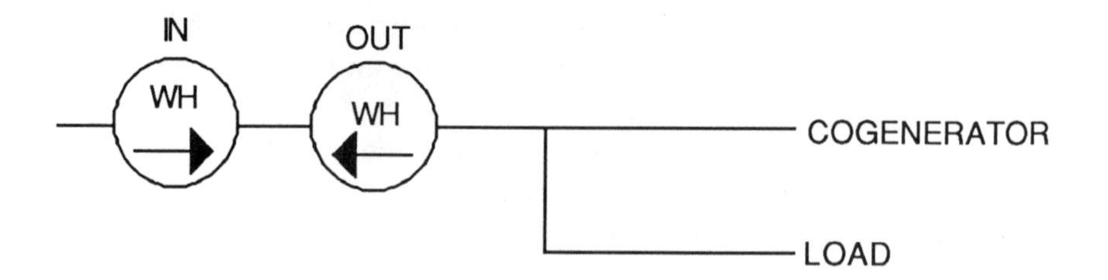

Needed if rate for billing is different than the rate for buyback.

Source: *Cogeneration and Small Power Production, Guidelines for Public Power Systems*, prepared by the American Public Power Association, November 1980.

FIGURE 3-8
MAGNETIC TAPE REACTIVE POWER METERING

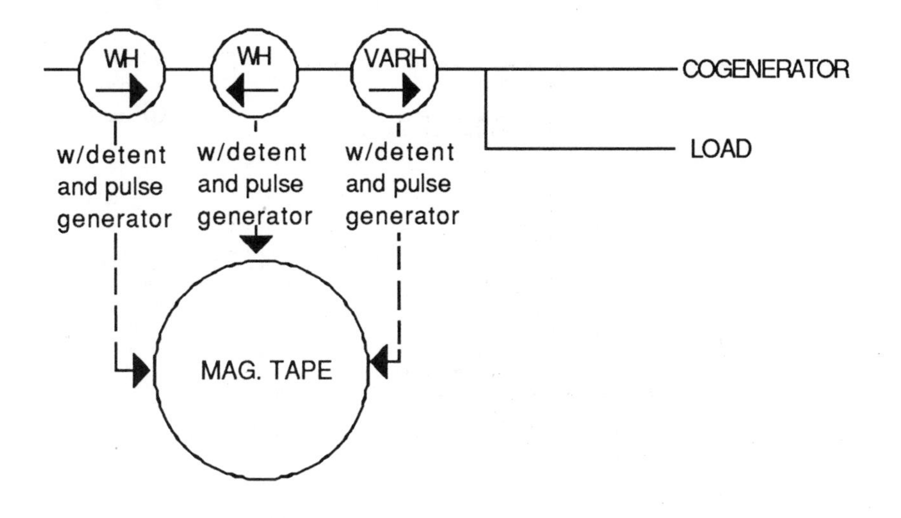

Provides for different rates for billing and buyback with time of day capabilities.

Source: *Cogeneration and Small Power Production, Guidelines for Public Power Systems*, prepared by the American Public Power Association, November 1980.

FIGURE 3-9
REACTIVE POWER METERING

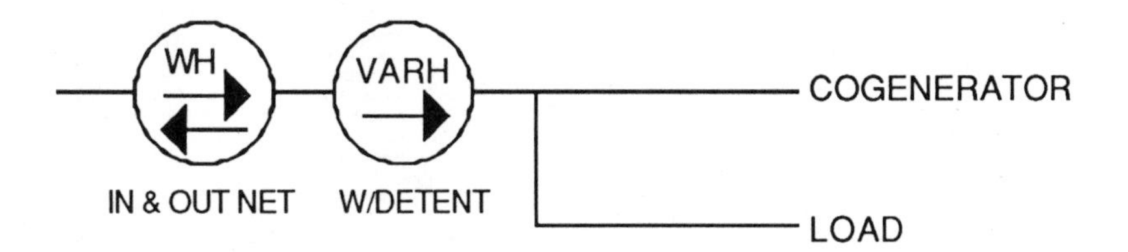

This option is for metering varhours for billing (induction generator).

Source: *Cogeneration and Small Power Production, Guidelines for Public Power Systems*, prepared by the American Public Power Association, November 1980.

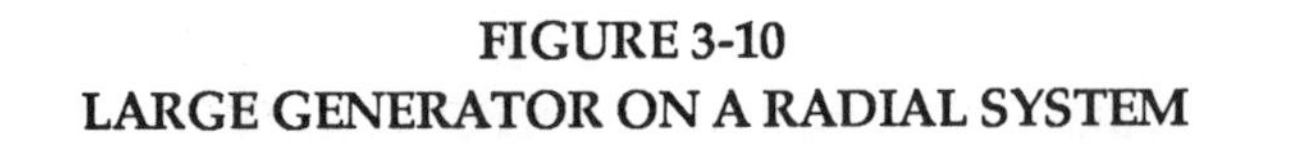

FIGURE 3-10
LARGE GENERATOR ON A RADIAL SYSTEM

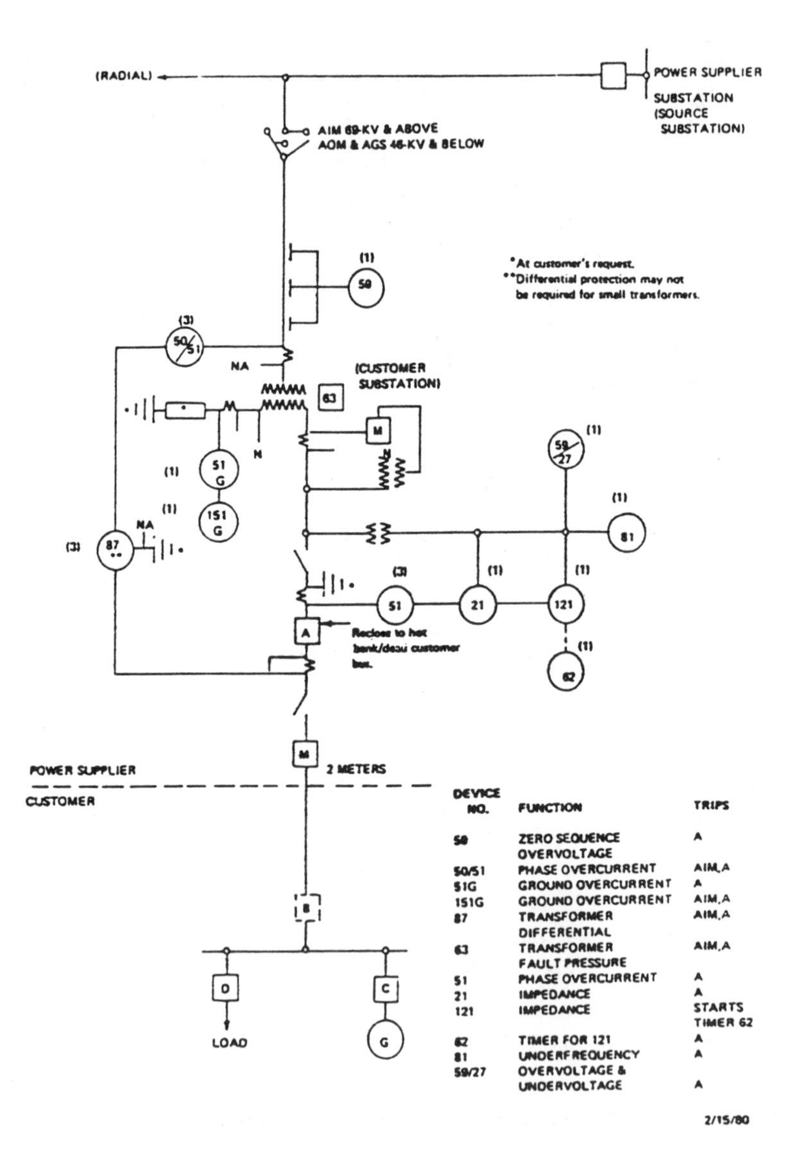

Courtesy of the American Public Power Association, *Cogeneration and Small Power Production*, November 1980.

FIGURE 3-11
LARGE GENERATOR ON A NETWORK SYSTEM

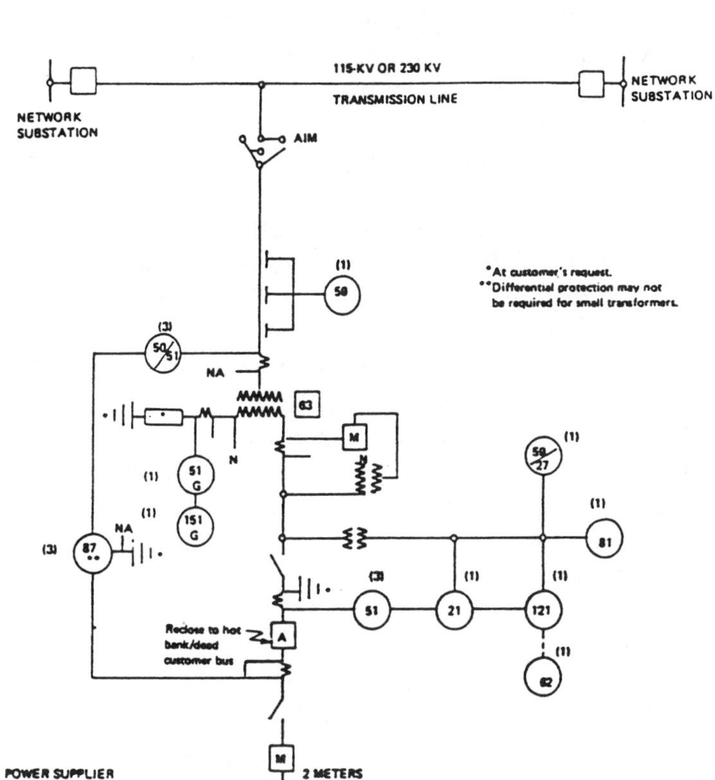

DEVICE NO.	FUNCTION	TRIPS
59	ZERO SEQUENCE OVERVOLTAGE	A
50/51	PHASE OVERCURRENT	AIM,A
51G	GROUND OVERCURRENT	A
151G	GROUND OVERCURRENT	AIM,A
87	TRANSFORMER DIFFERENTIAL	AIM,A
63	TRANSFORMER FAULT PRESSURE	AIM,A
51	PHASE OVERCURRENT	A
81	UNDERFREQUENCY	A
21	IMPEDANCE	A
121	IMPEDANCE	STARTS TIMER 62
62	TIMER FOR 121	A
59/27	OVERVOLTAGE & UNDERVOLTAGE	A

2/15/80

Courtesy of the American Public Power Association, *Cogeneration and Small Power Production*, November 1980.

Interconnection Costs

The cost of the electric interconnection will vary depending on many of the factors reviewed above. In addition, the cost will be a factor of the quality of the equipment, with a distinction being made between "utility grade" and "industrial grade." While there is no standard defining grade, some utilities will only allow the use of relays and other devices which they have tested and have found to be acceptable. These utilities frequently specify the device manufacturer and model and require test results for each relay. While industrial grade relays may offer the same level of protection, the required testing and documentation procedures can significantly increase costs by a factor of two or more.

Interconnection costs for cogenerators of approximately 100 kilowatts or less will vary between $100 per kilowatt and $250 per kilowatt. The cost decreases with increasing size leveling off to approximately $40 to $50 per kilowatt at several hundred kilowatts. High grade protective requirements can double the cost for smaller systems, while having only modest impacts on the cost of interconnecting larger systems.

REVIEW

The ability to interconnect a cogenerator with the electric utility grid to purchase power from that grid, to sell cogenerated power to a utility, or a combination of both, has resulted in significant improvements in cogeneration economic viability. Existing technology available to cogenerators, even in small sizes, is capable of meeting the safety and performance requirements of both the cogenerator and the interconnected utility.

While the components that make up the interconnection are relatively standard, the grid and the requirements for successfully connecting to it for parallel operation will be unique to the utility, and perhaps, even unique to different portions of that distribution grid. It is important for those cogeneration systems which will interconnect to the utility to identify the requirements as early as possible, both to define design requirements and to adequately estimate cost.

Chapter 4

PACKAGED COGENERATION SYSTEMS

Cogeneration systems have proven to be economically successful in electric utility owned central generating plants, industrial plants and in larger building complexes such as military bases, universities and hospitals. Designed for the requirements and economics of the specific site, each was unique with the prime mover, heat recovery system, heat rejection system and controls individually specified and the system "built up" at the site. These systems proved successful, even though these larger customers generally paid less for energy than did other retail customers (see Figure 4-1). However, prior to 1980, attempts to apply cogeneration concepts in smaller installations with higher cost energy, had generally proven unsuccessful, primarily because of the high initial fixed costs associated with any size system.

PCS CONCEPT DEVELOPMENT

With the passage of PURPA and the seeming removal of barriers to internal use cogeneration, many equipment packagers and system developers re-examined the smaller market. As shown in Figure 4-2, which is based on data for one utility, while the size of the typical installation may be small, the large number of potential applications could result in as much capacity in small units as in the larger built-up systems. The immediate response was to identify

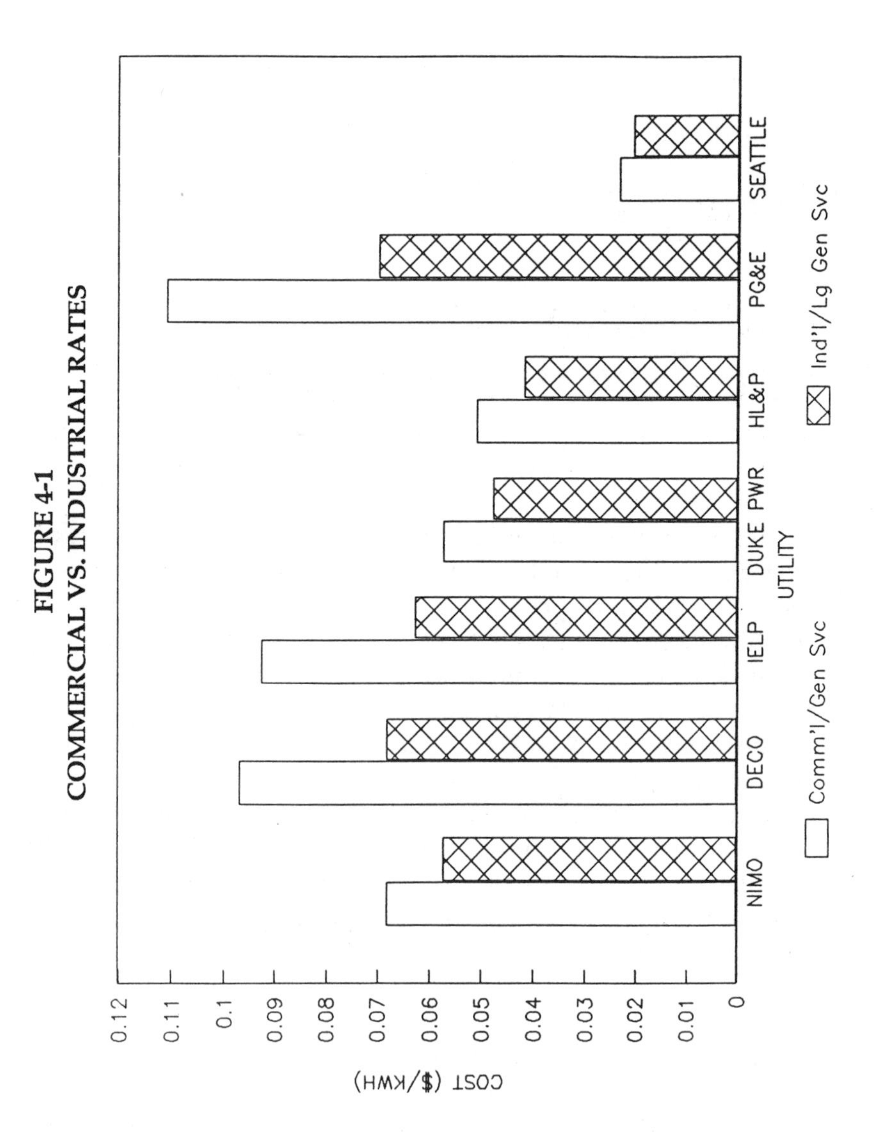

FIGURE 4-1
COMMERCIAL VS. INDUSTRIAL RATES

FIGURE 4-2
COGENERATOR SIZE DISTRIBUTION

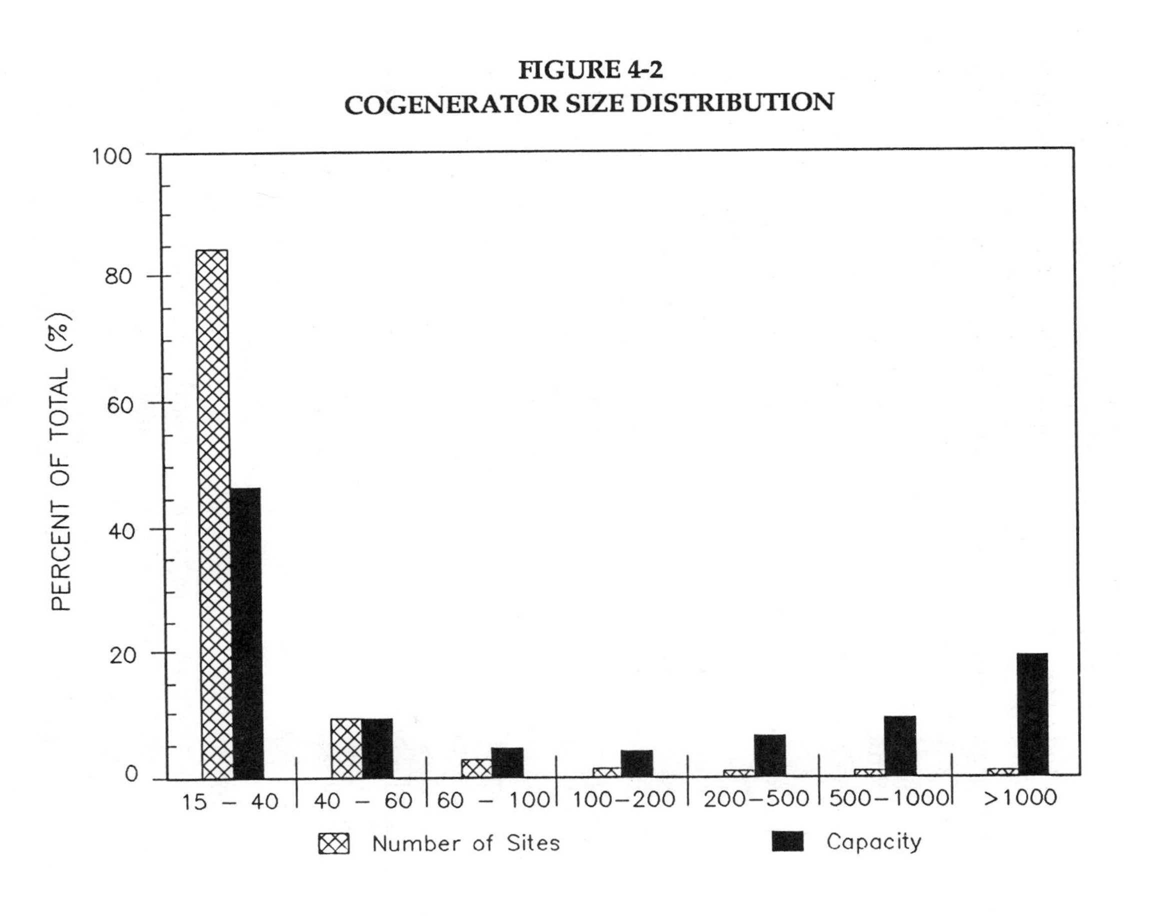

mechanisms for reducing the high fixed costs associated with cogeneration system development including the cost of system financing, design and construction, and for equipment vendors, system developers, and marketing. Additionally, once the system was in place, operating and maintenance costs had to be reduced to rates consistent with those for larger cogeneration applications without imposing specialized staffing requirements on the end user.

While the problems were challenging, the cogeneration industry's perception was that the market and, therefore, the rewards were significant. As a result, early in the 1980s, the concept of a packaged or factory-assembled cogeneration system emerged.

The packaged cogeneration system (PCS) consists of an integrated unit (see Figure 4-3) that is assembled at a factory and shipped to the site as a unit. Design costs are spread over multiple units rather than single units. The application of assembly line techniques reduces unit fabrication costs and, in addition, a minimum amount of site work is required further lowering costs. Since the introduction of the PCS concept in the early 1980s, the small cogeneration industry has continued to evolve and to offer an increasing number of products. Today the inventory of PCS modules ranges in size from 6 kilowatts to over one megawatt.

The PCS or module concept has several advantages which led to its initial market penetration. First and foremost is the capital cost. With unit costs of $1,000 per kilowatt or less, the system cost is competitive with unit costs for larger cogeneration systems. A second advantage is the factory assembly and checkout which minimizes the need for more expensive field trouble shooting and repairs. Most PCS vendors have the capability to test the entire module (including heat rejection/recovery system, controls, etc.) prior to shipment, thus correcting problems at the factory where parts and expertise are more readily available. Third, the standardized design and components combined with minimal site work results in short development periods. A PCS can be operational with a month or two of the decision to proceed with the installation, leading to immediate benefits from the decision to install a PCS. In addition, the rather short development period reduces the uncertainty associated with the total cost of the PCS. A fourth advantage derives from the fact that most PCS vendors provide installation services

FIGURE 4-3
FACTORY-ASSEMBLED COGENERATION MODULE

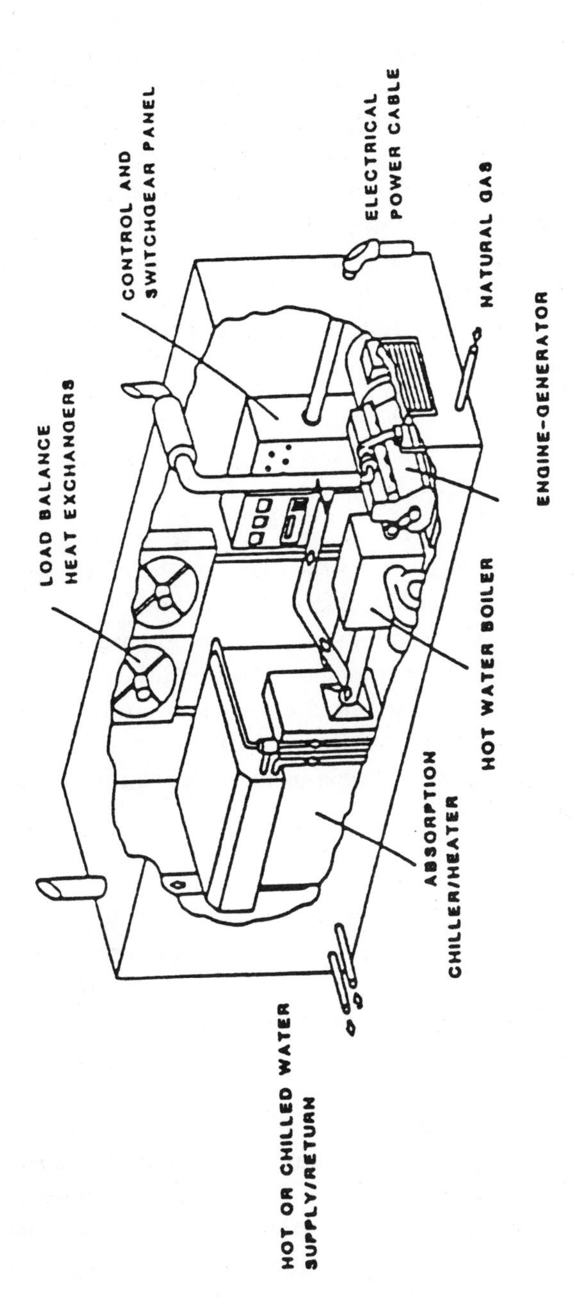

and, therefore, the purchaser has a single source which is responsible for all performance and cost aspects of the system. If the module is purchased under a turnkey agreement with the PCS vendor and then does not perform as projected, it is not necessary to determine where responsibility lies: the engineer, the equipment vendor or the installing contractor. A final advantage derives from the standardized concept which allows the vendor to obtain interconnect approval for the PCS module's interconnection components, thereby reducing the time and cost required for utility approvals.

PCS COMPONENTS

While the PCS concept is based on the notion of a standardized design, there is considerable variation, both from vendor to vendor and within the product line of a single vendor. While all PCS modules contain an engine generator with heat recovery as the basis of the cogeneration system, there are substantial differences in these primary components as well as in the availability and type of ancillary components. Some of the most significant differences and alternatives are discussed below.

Prime Movers

Reciprocating engines are the predominate prime mover for PCS modules, with two exceptions. One major turbine packager offers a multi-megawatt turbine in a standardized unit, while another vendor offers a small steam turbine package. Depending on steam conditions, the steam turbine capacity will vary from a few hundred to six or seven hundred kilowatts.

Additionally, most of the PCS packages are based on automotive derivative engines and particularly so for units of 75 kilowatts or less. Industrial and off-road engines have achieved some market penetration starting at 75 to 100 kilowatts, and have captured almost all applications of several hundred kilowatts or more. In general, PCS engines operate at 1,200 or 1,800 rpm, are fueled with natural gas, and are naturally aspirated.

Most smaller PCS modules (usually less than 150 kilowatts) recover the engine heat as hot water. Ebullient cooling and steam heat recovery are usually not found in modules of less than a few hundred kilowatts.

Generators

Smaller PCS modules use induction generators and, therefore, are not capable of operating independently, although some vendors are offering this capability through the addition of static exciters. Synchronous generators are used in PCS modules of several hundred kilowatts or more; above 500 kilowatts they are almost exclusive.

Heat Rejection

Smaller PCS modules are generally offered without an internal heat balancing radiator and, therefore, cannot operate at rated capacity if the site thermal load is not available. Heat rejection radiators are generally offered in PCS modules of 100 kilowatts or more.

Controls

Controls represent one area of major technological innovation within the PCS industry, with increased reliance on microprocessor based systems. In general, these systems are considered proprietary by the vendor, with little information readily available. Commercial control system vendors are offering microprocessor control packages thus increasing the industry capability to provide adequate control system maintenance.

The microprocessor based control system provides several functions including:

- engine safety,
- monitoring of generator output and sometimes serving as part of the utility interconnect,
- providing data required to schedule maintenance procedures,
- monitoring engine performance to help predict unscheduled outages, and
- scheduling the module operations to maximize operating cost savings.

Many of these systems include a telecommunications capability thus allowing the service organization to monitor performance and minimize operating and maintenance costs.

Chillers

Some PCS modules are offered with an absorption or desiccant chiller integrated into the basic package. The absorption chiller provides additional thermal load during non-heating periods and helps to increase the annual hours of operation and the savings that result from the use of recovered heat. Only single effect absorption chillers have been used in PCS modules.

PCS COSTS

PCS unit costs are comparable to those for smaller, field built up cogeneration systems. Figure 4-4 presents vendor FOB costs for a sample of modules. However, when installation and "soft" costs are considered, the PCS module cost may be only a fraction of the ultimate system cost.

Installation Costs

Installation costs are dependent on a number of factors including space availability, the type of fuel, and the interconnection with the site's thermal systems and the electric utility intertie. The installation of small induction generator PCS modules incurs minimal intertie costs. When the recovered heat is used in a process or space heating loop, these systems will have modest installation costs of $15,000 to $20,000. In fact, this type system defines the cost threshold for PCS applications. There are little opportunities for economies of scale which will produce costs of less than $15,000 and sites that require multiple small units will incur installation costs that are multiples of the single unit installation cost.

A more complex synchronous generator PCS installation operated in parallel with the electric grid, and with recovered heat used for multiple applications including potable water, will result in installation costs that can exceed $100,000 for a system with a capacity of 200 kilowatts or 300 kilowatts.

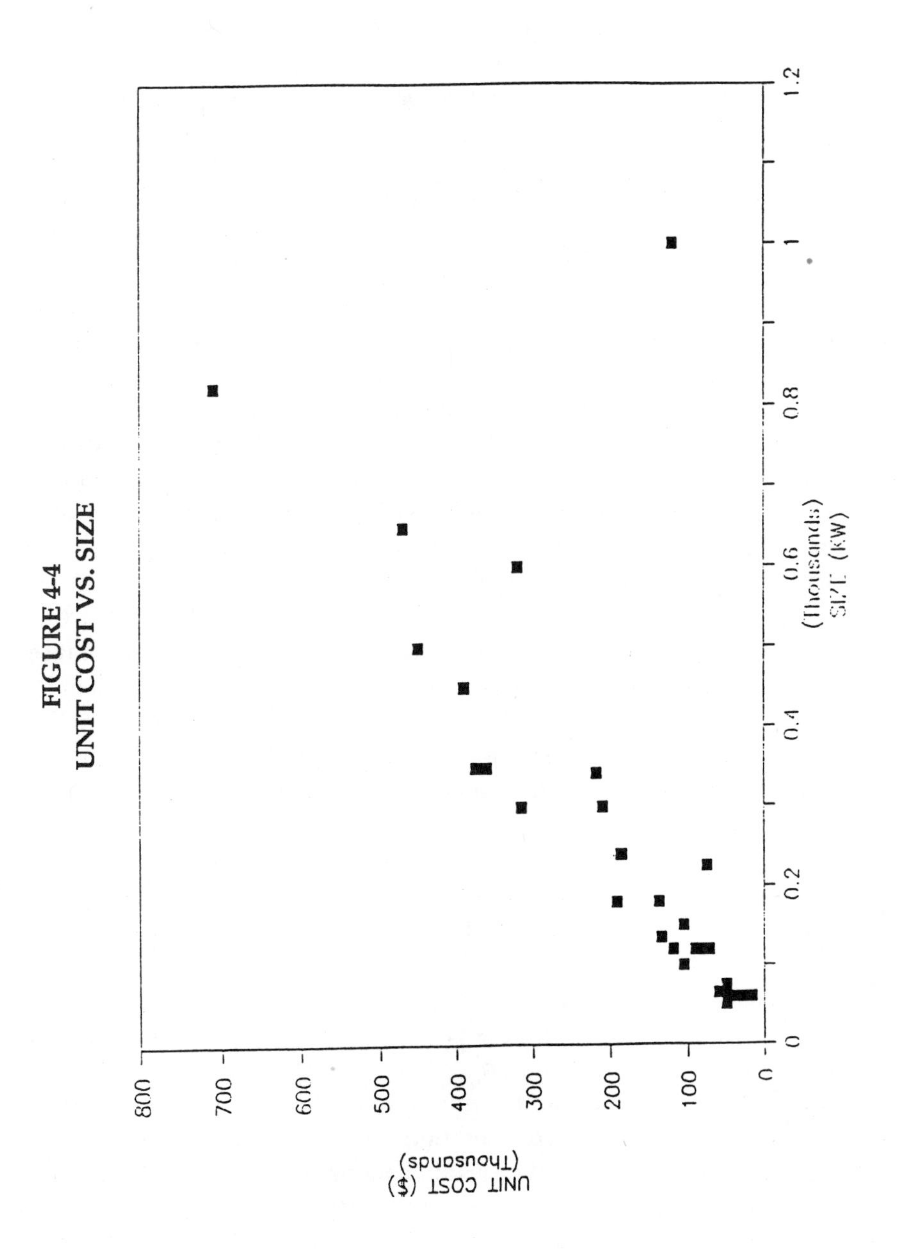

FIGURE 4-4
UNIT COST VS. SIZE

Figure 4-5 provides installation unit costs as a function of application size. As can be seen, these costs vary over a broad range for smaller sizes, only narrowing for larger sizes. In general, a mid-range value can be used for preliminary estimates for typical or simpler applications. As factors such as the use of recovered heat for two or more purposes, an absorption chiller, or the use of synchronous generators, each of which adds to the installation's complexity, are identified higher than average cost estimates should be used.

Electrical interconnection costs, which can be 50% or more of the installation charges, will be dependent on the specific utility interconnection requirements and the choice of a generator. The synchronous generator cost premium for smaller cogeneration systems is approximately 10% to 13% of the electrical costs. *Mechanical interconnection costs* are primarily dependent on the use of the recovered heat. When recovered heat is used for potable water, heat exchanger costs can be considerably more than the case when the water is used for non-potable purposes.

Fuel system costs will generally be modest for retrofit applications where the fuel is already available on site. At those sites where no fuel capability exists, the cost of a new service can be several thousand dollars. While fuel system costs are essentially independent of the PCS size for natural gas fired engines, oil burning fuel system costs will be dependent on size and the amount of on-site storage that is required.

For an internal PCS installation, *stack costs* will depend on whether or not an existing stack can be used. If a new stack is required this cost can be quite significant.

Soft Costs

"Soft" costs are those fees paid for professional services such as engineering, zoning, permitting, financing, legal review, construction monitoring, etc. They can be quite low for those applications where the end user is also the cogeneration system developer, taking on all implementation responsibilities and using either equity or general financing. At the other extreme, and particularly where these soft costs become part of the project budget, which is then financed by a third party, they can be quite significant adding 33% or more to the cost of the equipment and its installation.

FIGURE 4-5
INSTALLATION COST ($/kW) VS. SIZE

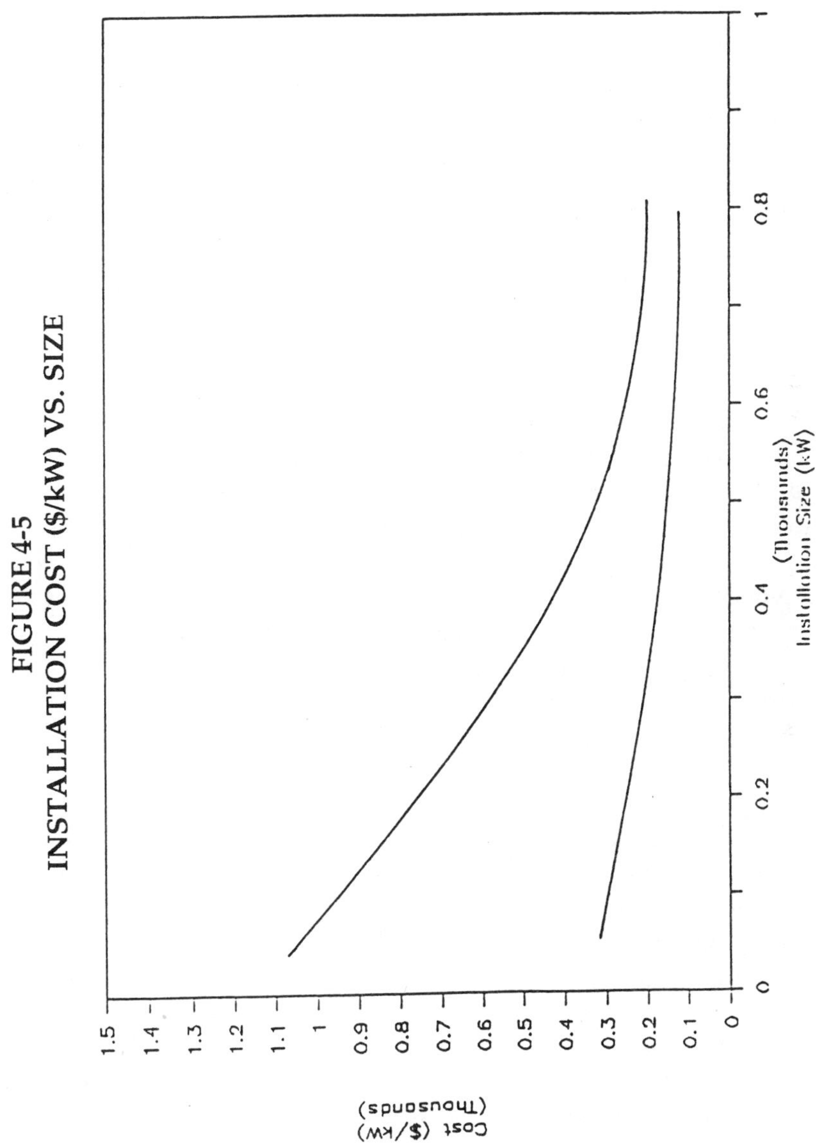

Figure 4-6 can be used to estimate soft costs for typical installations. As with installation costs, a range of values are shown. Specific cost elements are discussed below.

Engineering fees for small PCS systems may range from 12% to 15% of the cost of the equipment plus installation, with a minimum of approximately $10,000. In many cases, the project cannot afford the cost of an independent engineer and the PCS vendor provides these services at varying fees.

Legal fees are somewhat independent of size and can be quite significant for project financed or shared savings installations. They can range from a minimum of a few thousand dollars to approximately 10% of the project budget for mid- and large-sized PCS installations.

For small and mid-sized PCS installations where only zoning and construction permits are required, *permitting costs* will be a few hundred dollars. Should air quality modelling be required, the costs for an emission permit can increase to over $15,000.

Management and administrative costs, whether incurred as a professional service or internalized by the project developer, can be another 5% to 10% of the total installation cost.

PCS TRENDS

The PCS concept has achieved initial market penetration and is recognized as an energy cost reduction alternative for commercial and smaller industrial facilities. Today, some electric utilities are actively attempting to determine the role of PCS modules within the utility's system. In general, the industry is seeking to identify techniques and applications which increase the perceived value of their particular PCS module.

The most straightforward approach is through improvement of the cogeneration module itself, with a second being an expansion of the PCS to include additional energy-related functions. Major engine manufacturers are promoting higher-speed engines for cogeneration applications as one mechanism for bringing down the cost per kilowatt while maintaining engine efficiency. Other efforts target heat recovery system costs, emission controls and module controls.

FIGURE 4-6
SOFT COST ($/kW) VS. SIZE

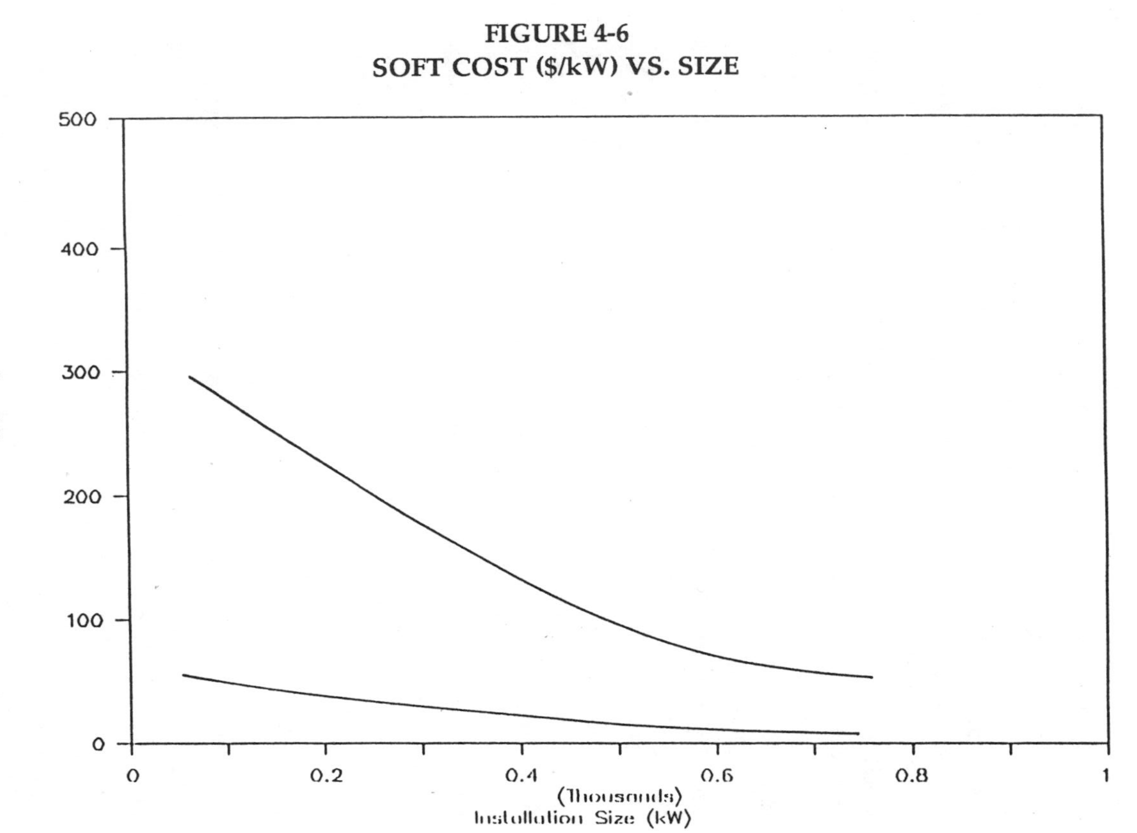

As discussed above, module controls are a major target for developmental efforts as microprocessor technology is applied to engine and system optimization. Engine manufacturers now incorporate computer monitoring and controls into fuel supply systems as a means to adjust the fuel system to varying fuel quality. Another application for microprocessor technology is in actual system operation where operating costs are compared to power purchase costs and real time decisions are made as to the most cost efficient operation of the system. This same microprocessor can also select between an absorption or an electric chiller based on the amount of recoverable heat available and the real time operating cost differentials. Perhaps most importantly, the microprocessor can monitor critical engine parameters and act as an early warning system, identifying potential problems before they cause an unscheduled outage.

The most innovative aspect of PCS marketing consists of efforts to include other energy functions within a cogeneration context. Specifically, absorption chillers and desiccant dehumidifiers are being offered as part of the PCS system in part to expand the market directly by increasing summer use of thermal energy and indirectly by increasing the savings through electric peak shaving that the project produces.

REVIEW

The concept of a small standardized cogeneration system with modest unit costs was developed during the early 1980s. These systems, which are fabricated and tested at a central factory, have reduced site installation costs, and have been viable in many small commercial facilities such as hotels, nursing homes, health clubs, restaurants and schools.

Many of these PCS modules have incorporated innovative technology, and particularly with regard to operating controls. The use of microprocessor based control and monitoring systems has resulted in high availability and low maintenance costs, thus improving overall economics.

The "inventory" of available PCS modules is quite diverse, with various engine, generator, heat recovery and control system options. Costs also vary significantly from module to module, and the primary task faced by the purchaser of a PCS system is to identify that package which is most cost effective for the particular application.

Chapter 5

COOLING AND REFRIGERATION

Cogeneration systems are often developed at facilities characterized by high electric and thermal load factors and frequently by central boiler plant operations. Many of these industrial plants, university, military and health care complexes also satisfy the site's cooling requirements from a single central chilled water plant. Because of the proximity of this second type of thermal process–cooling–this function can also become an application for the thermal output of a cogeneration system. Cooling can also be a viable extension of a cogeneration system in single buildings.

Mechanically driven vapor-compression cycles are used for most chillers, with absorption chillers and, more recently, desiccant coolers being heat driven. Mechanical chillers can be driven by electric motors, engines, and steam turbines. In some climates, especially those characterized by low humidity, evaporative coolers are used. Because the cogeneration system can produce mechanical or electrical power as well as heat, its performance and economic viability can be directly linked to chiller selection and operation.

MECHANICAL CHILLERS

Mechanical chillers operate through a vapor-compression cycle that can be driven by either an engine or an electric motor. The basic *vapor compression cycle*, as illustrated by Figure 5-1, consists of the following steps.

1. Use of a mechanically driven *compressor* to compress the working fluid producing a high pressure vapor.

2. Use of a *condenser* to remove heat from the working fluid and in the process convert the vapor to a high pressure liquid. The condenser removes both the sensible heat that result from the work done by the compressor and the heat that is removed from the conditioned space or process from the cooling loop. This heat is transferred to the environment using either a radiator or a cooling tower, and the radiator or cooling tower must be sized to reject both the cooling load and the compressor work.

3. Use of an *expansion valve* to reduce the pressure of the liquid.

4. Use of an *evaporator* to vaporize the low pressure liquid by extracting heat from the space or process that is to be cooled. The conditioned space provides the heat of vaporization required by the working fluid, thus lowering the temperature of the conditioned space or medium.

Mechanical chillers can be categorized according to the type of compressor as either a reciprocating, a centrifugal or a screw compressor. In addition, they are categorized as either hermetically sealed (driven by an electric motor) or open (driven by an electric motor or an engine) according to the type of drive. Engine-driven, open-drive compressors can be equipped with heat recovery, just as engine-driven generators.

Reciprocating compressors (Figure 5-2) are derived from reciprocating engines and have reasonable part-load efficiency over a range of 40% of rated load to 100%. They are available in a range of sizes up to 200 tons. When operated at part load they can reduce their output by reducing engine speed or by unloading cylinders and holding speed at some specified minimum level.

Centrifugal compressors (Figure 5-3) operate at speeds of several thousand revolutions per minute and are available in sizes up to several thousand tons. They also modulate output by varying

FIGURE 5-1
MECHANICAL VAPOR-COMPRESSION CYCLE

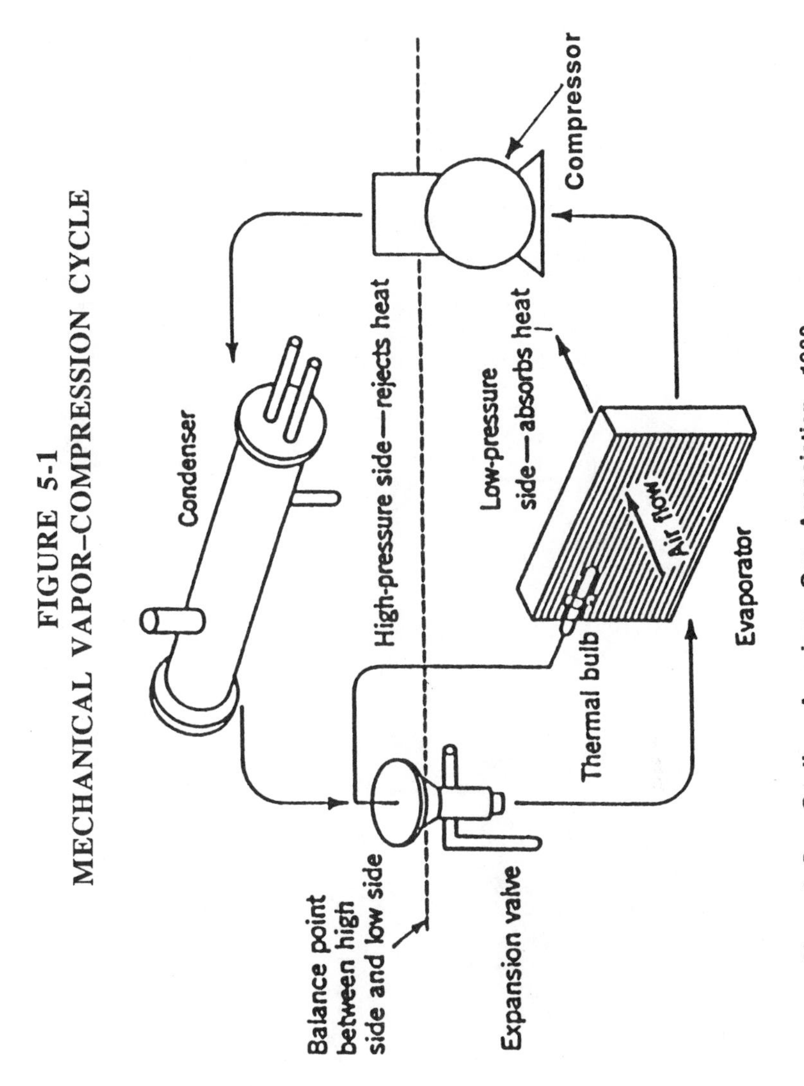

Source: *Natural Gas Cooling*, American Gas Association, 1982.

FIGURE 5-2
RECIPROCATING COMPRESSOR

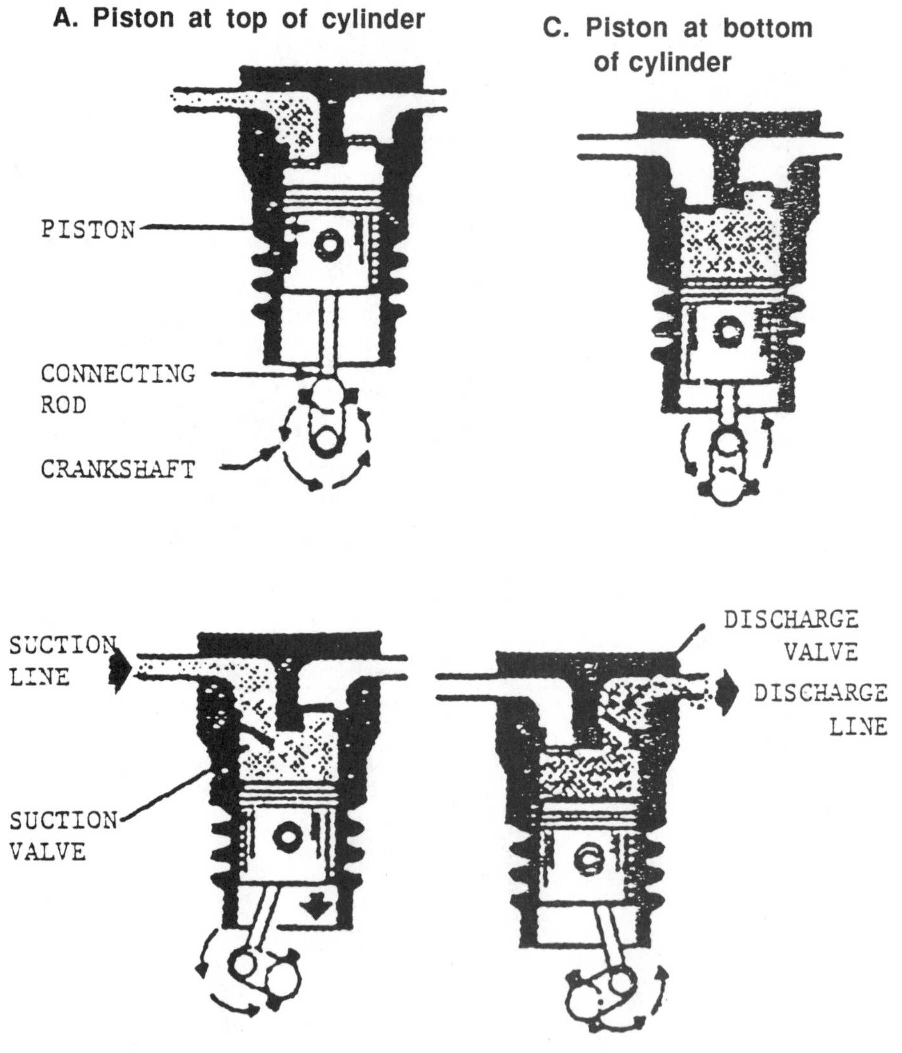

Source: *Natural Gas Cooling*, American Gas Association, 1982.

FIGURE 5-3
CENTRIFUGAL COMPRESSOR

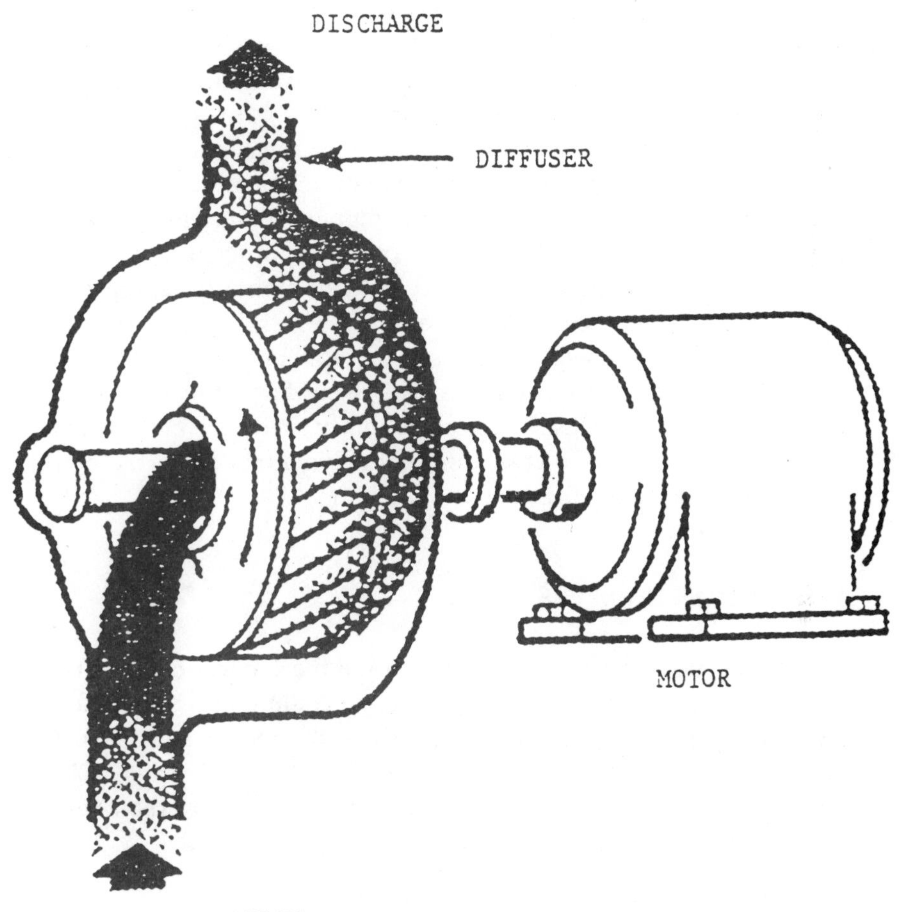

Source: *Natural Gas Cooling*, American Gas Association, 1982.

engine or motor speed. While most centrifugal chillers are motor driven, some are mechanically powered using steam turbines.

Screw compressors (Figure 5-4) operate by compressing the working fluid between two meshed screws and are available in sizes of up to 300 tons.

The working fluid for most vapor compression chillers is usually a chlorofluorocarbon (CFC). The most frequently used CFCs react with oxygen in the lower atmosphere creating ozone and have been identified as major contributors to this form of pollution. Starting in the 1990s, the production of these refrigerants including R-11, R-12, R-13 and halon will be severely restricted and their use in chillers will be lessened. Chemical companies and chiller manufacturers are working to develop and certify equipment that will be capable of operating with more environmentally acceptable refrigerants. During the interim period when the older CFCs are being phased out, new chillers must be capable of operating with either of two different refrigerants. The likely impact of the change in refrigerants is decreased efficiency and capacity for chillers that must operate on either of two different refrigerants or on the new, non-ozone producing refrigerants.

ABSORPTION CHILLERS

Absorption chillers, in contrast to the mechanical chillers, use heat as the source of input energy for the cooling cycle. The basic cycle consists of four steps that produce thermodynamic results equivalent to the four steps of the mechanically driven chiller. The cycle operates with two fluids, a refrigerant and an absorbent, and includes the following:

1. Use of a *generator* where heat is added to the cycle and is used in a distillation process to boil off refrigerant from a mixture of absorbent and refrigerant (Figure 5-5). The refrigerant vapor is at a high pressure, as was the case for the working fluid leaving the mechanical chiller's compressor. The fluid remaining in the generator consists of a strong solution of absorbent.

FIGURE 5-4
SCREW COMPRESSOR

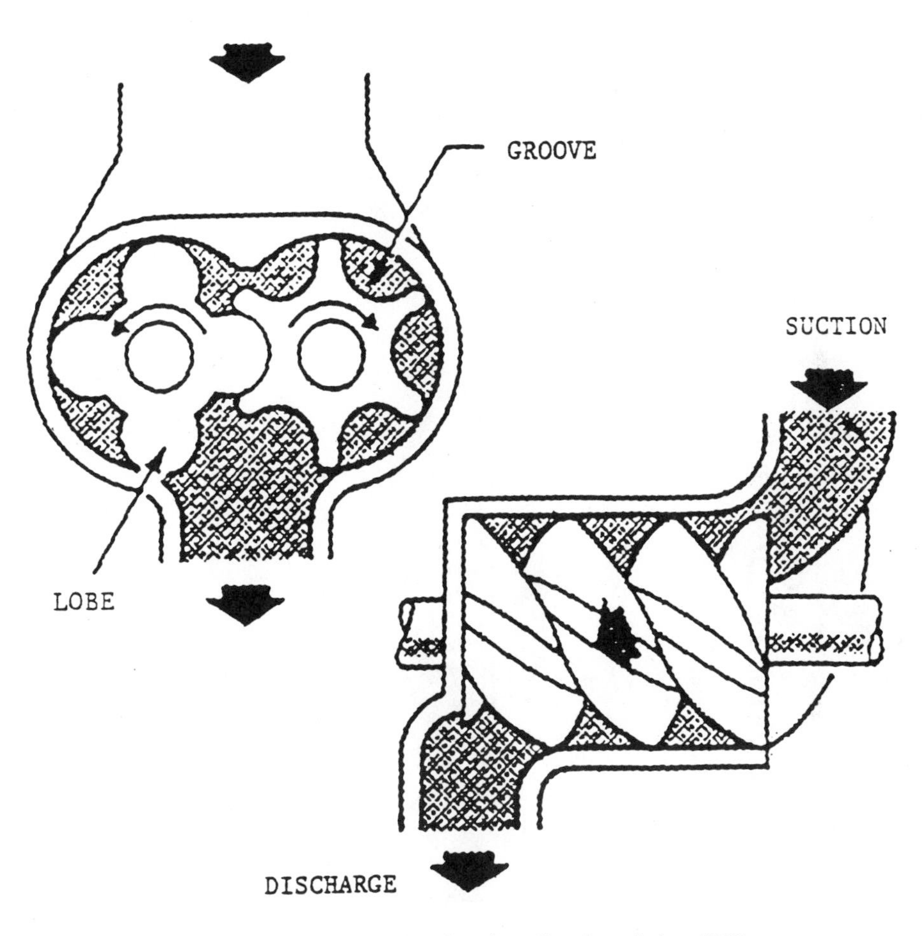

Source: *Natural Gas Cooling*, American Gas Association, 1982.

FIGURE 5-5
ABSORPTION CHILLER GENERATOR

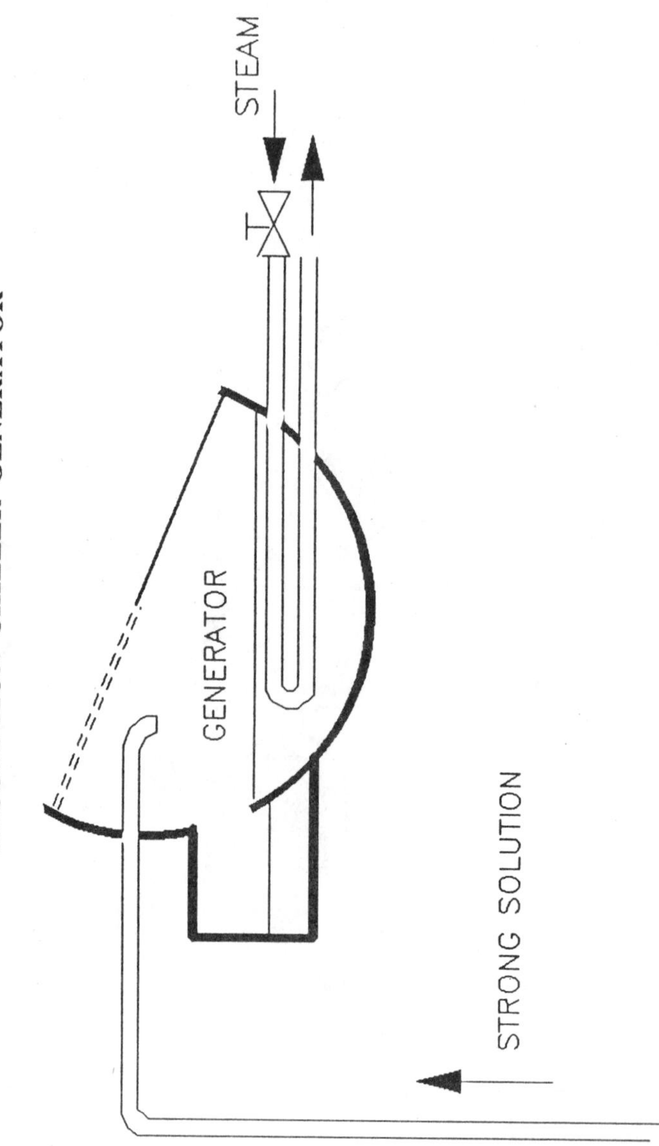

2. Use of a *condenser* to remove heat from the high pressure, vaporized refrigerant which is cooled and liquified (Figure 5-6). The condenser removes both the heat of vaporization introduced in the generator and the heat which the refrigerant has removed from the conditioned space or process. This heat must be rejected using a radiator or a cooling tower, which must be appropriately sized.

3. Use of an *evaporator* where the liquified, low temperature refrigerant is sprayed over tubes containing the fluid to be cooled (Figure 5-7). This process takes place at very low pressures and the heat of vaporization that is required to vaporize the refrigerant cools the fluid in the tubes.

4. Use of an *absorber* to recombine the refrigerant vapor with the absorbent (Figure 5-8). The refrigerant and the absorbent must be chosen from those chemicals which have a high affinity for each other, and thus easily enter into solution. The refrigerant vapor is readily absorbed liberating some heat. More importantly, because the refrigerant is readily drawn into solution, its vapor pressure is kept low, thus producing low pressure in the evaporator.

Absorption chillers require the use of two chemicals that have a high affinity for each other, thus allowing the refrigerant to readily enter into a strong solution of the refrigerant and absorbent. Commercial absorption chillers can be categorized as being either a mixture of *lithium bromide and water* (with water as the refrigerant) or a combination of *water and ammonia* (with ammonia as the refrigerant). Because the lithium bromide system uses water as the refrigerant, the system is limited to temperatures above 32°F, the freezing point for water. If lower temperatures are required, the water ammonia system must be used. The ammonia water system is generally restricted to larger, low temperature applications.

Absorption chillers can also be classified as either single-stage or two-stage. The *single-stage* system is most common and was illustrated by Figure 5-8. It is capable of a seasonal Coefficient of Per-

FIGURE 5-6
ABSORPTION CHILLER CONDENSER

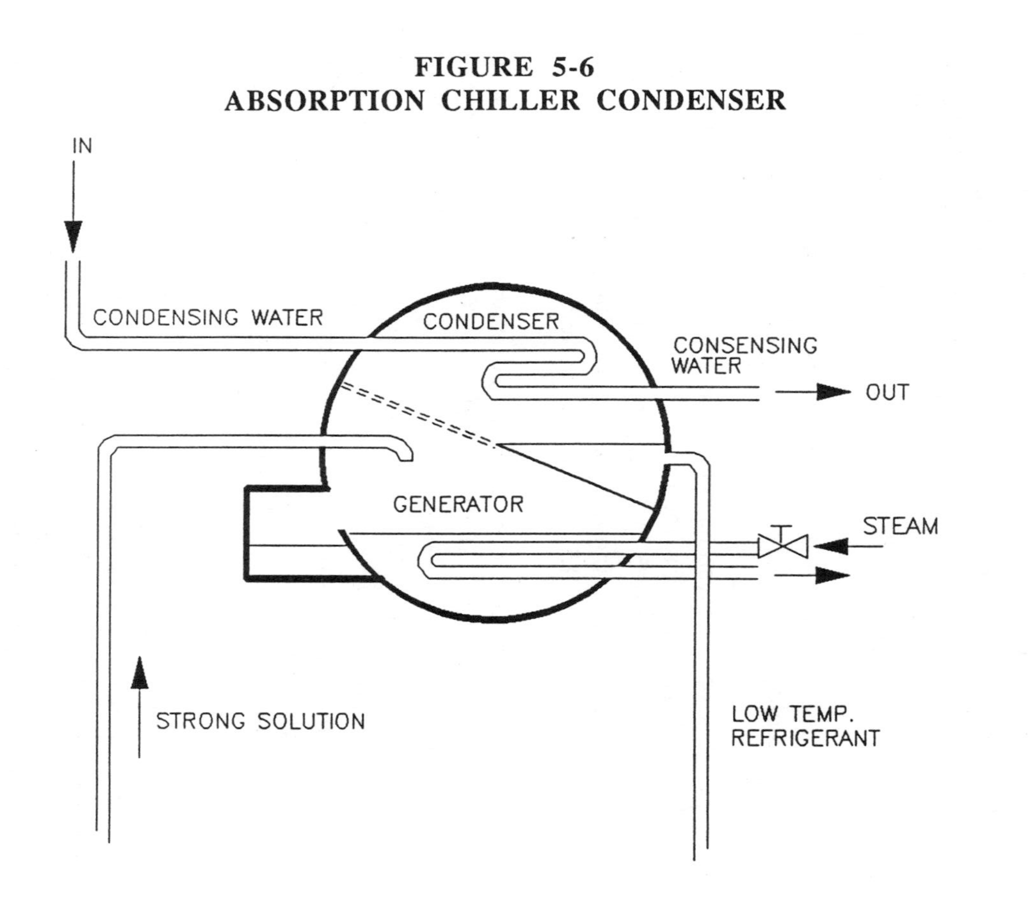

FIGURE 5-7
ABSORPTION CHILLER EVAPORATOR

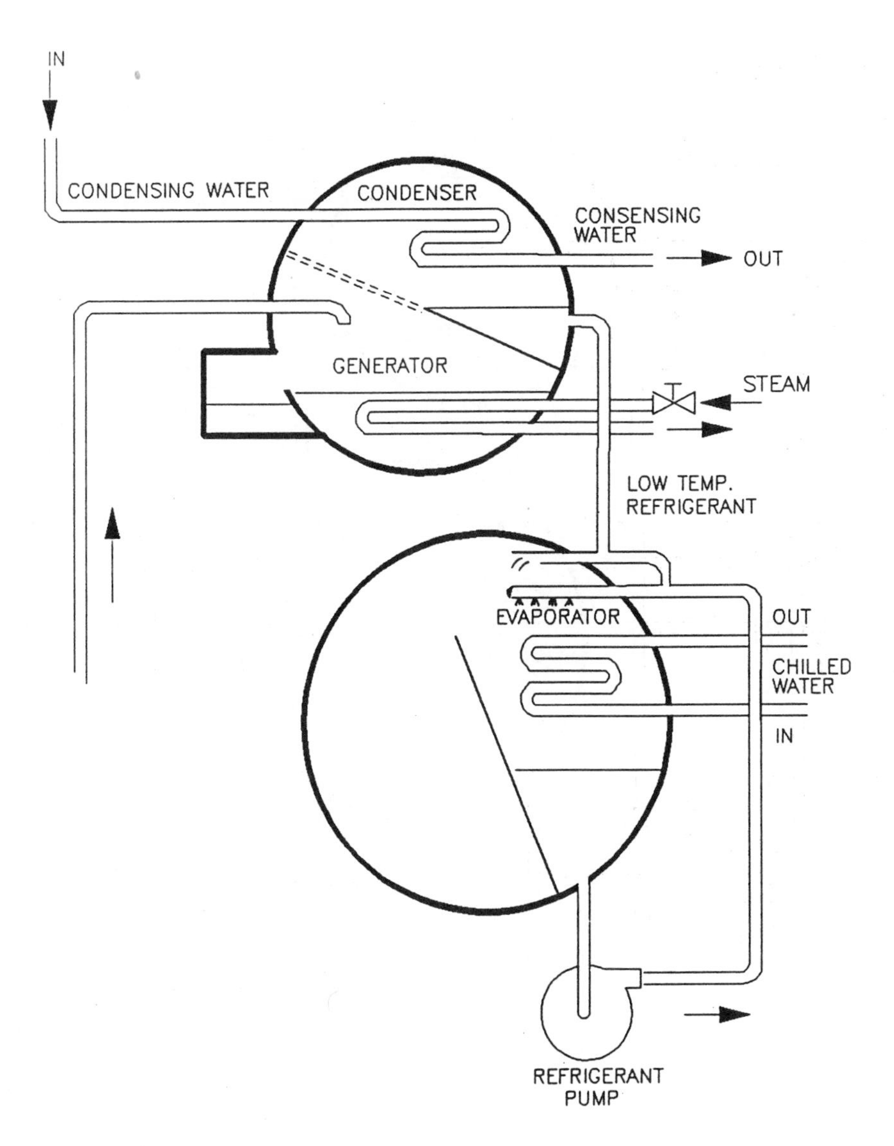

FIGURE 5-8
ABSORPTION CHILLER ABSORBER

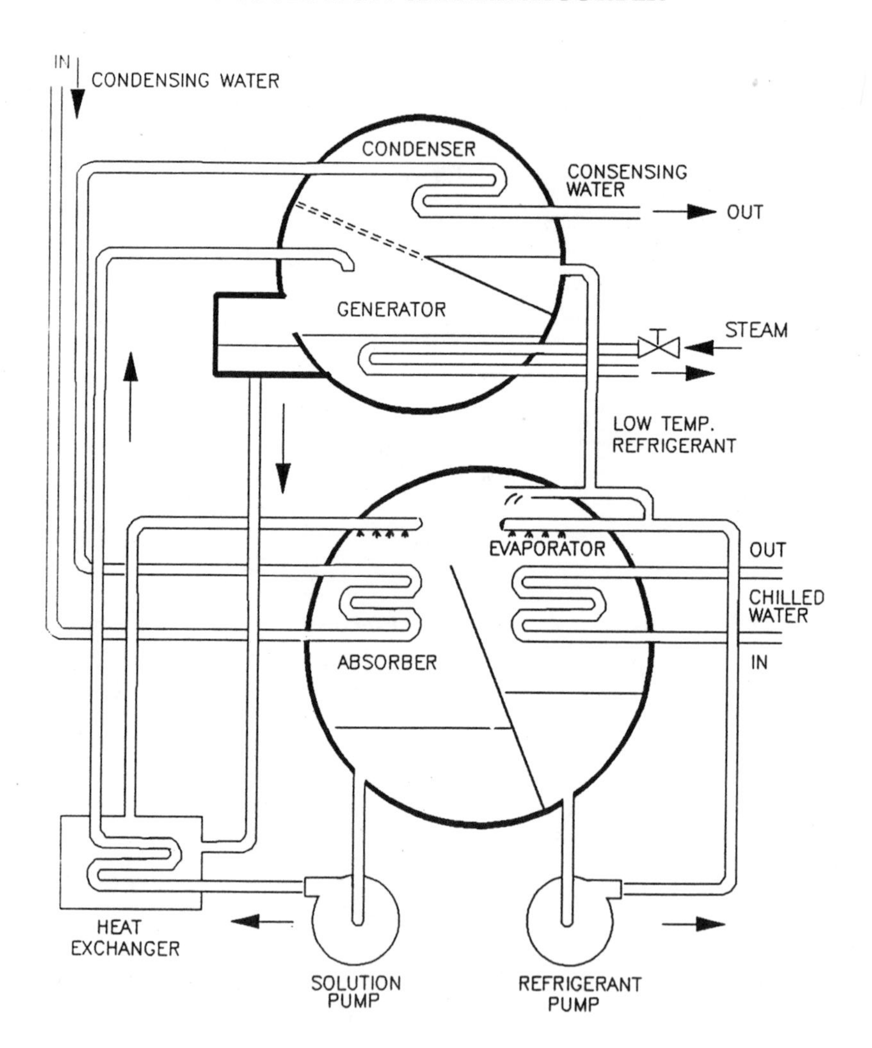

formance (COP) of 0.60 to 0.65 and can be operated with either steam (usually at 15 psig) or hot water as an input medium. When used in conjunction with a cogeneration system, this system can be operated with heat recovered from the engine jacket and/or the exhaust. If only lower temperatures or pressures are available, the capacity of the chiller must be reduced. Finally, absorption chillers have excellent part-load performance characteristics (see Figures 5-9).

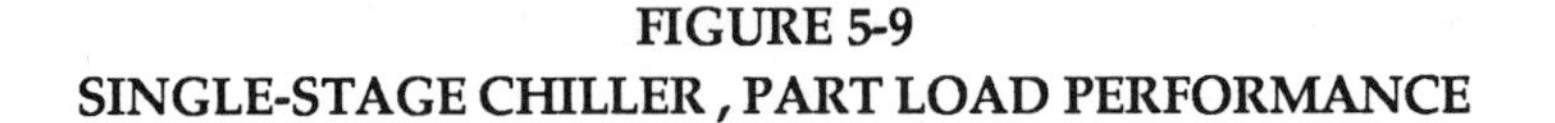

FIGURE 5-9
SINGLE-STAGE CHILLER , PART LOAD PERFORMANCE

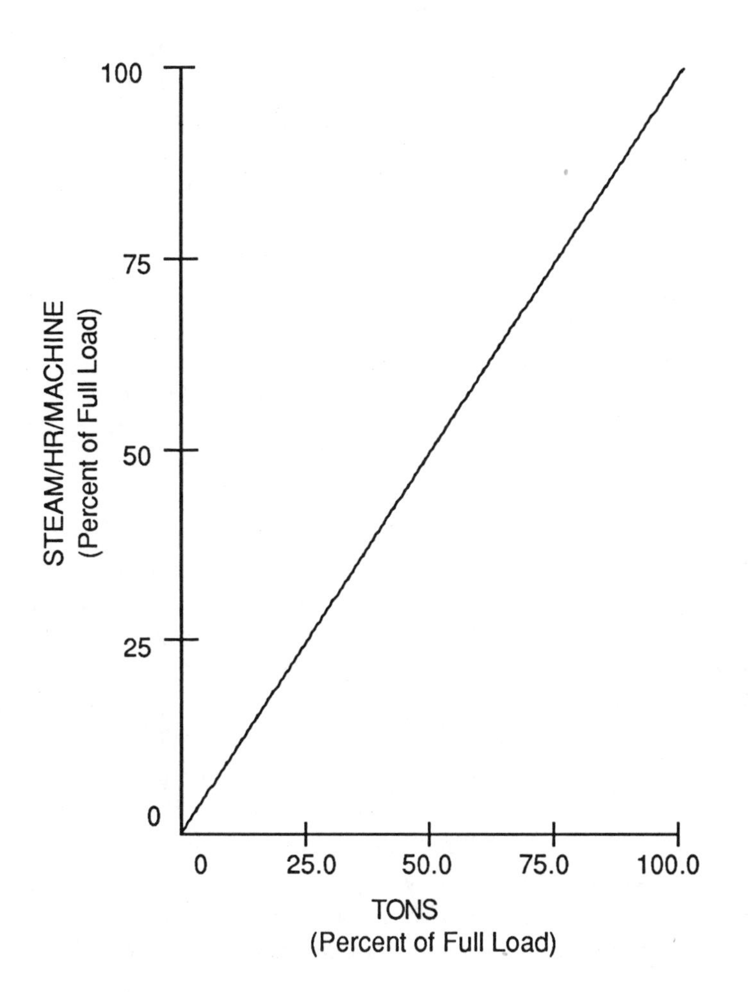

Chiller efficiency can be significantly improved by adding a second generator to the process. The *two-stage* or *double effect* chiller consists of two generators, with the first being referred to as the high temperature generator (see Figure 5-10). This system requires steam at approximately 115 psig although lower pressures can be used. However, as was the case with the single effect chiller, the use of lower pressure steam does require derating the chiller (see Figure 5-11). The high temperature generator produces high temperature water vapor, which serves as the primary heat source for the low temperature generator, where additional water vapor is boiled off. If effect, some of the heat that would normally be rejected in the condenser is first used to produce additional refrigerant vapor. The two-stage chiller is capable of achieving a COP approaching 1.1 and the Hitachi steam-fired chiller requires less than 10 pounds of steam per ton hour of cooling. Because of the requirement for high steam pressures, the two-stage chiller is most compatible with combustion turbine based cogeneration systems.

The double effect chiller can also be categorized by fuel or heat input. The *direct fired* chiller is fired with natural gas, and is typically configured as a chiller/heater capable of producing both hot and chilled water. The total output, including both hot and chilled water, is limited by the fuel input of the chiller; however, the chiller/heater can be operated with the output at 100% heating, 100% cooling or at any combination. The two-stage absorber can also be operated directly off the *exhaust gases* from the engine. When configured in this mode, the turbine exhaust is routed directly to the chiller/heater where it can be used to produce hot water, chilled water or both. Finally, the chiller can be *steam fired* and typically requires steam pressures in excess of 100 psig. The steam fired chiller/heater can be used in a cogeneration system, and typically with a combustion turbine. However, unlike the exhaust gas fired chiller, the cogeneration system must include a HRSG. One advantage of the steam fired machine is that it can be operated with steam produced in conventional boilers during those periods when the turbine may not be in operation.

When considered as part of an internal use cogeneration system, absorption chillers provide two important functions. First, they provide an economic use for recoverable heat during months when

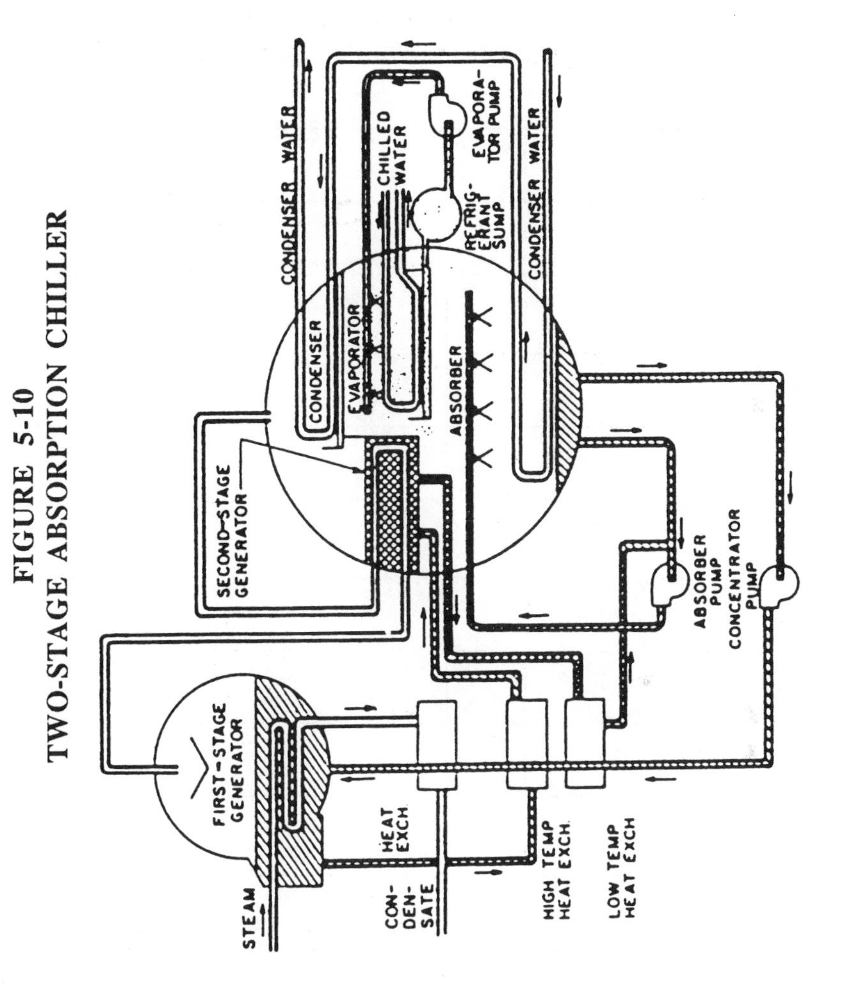

FIGURE 5-10
TWO-STAGE ABSORPTION CHILLER

FIGURE 5-11
TWO-STAGE CHILLER,
PART-LOAD PERFORMANCE

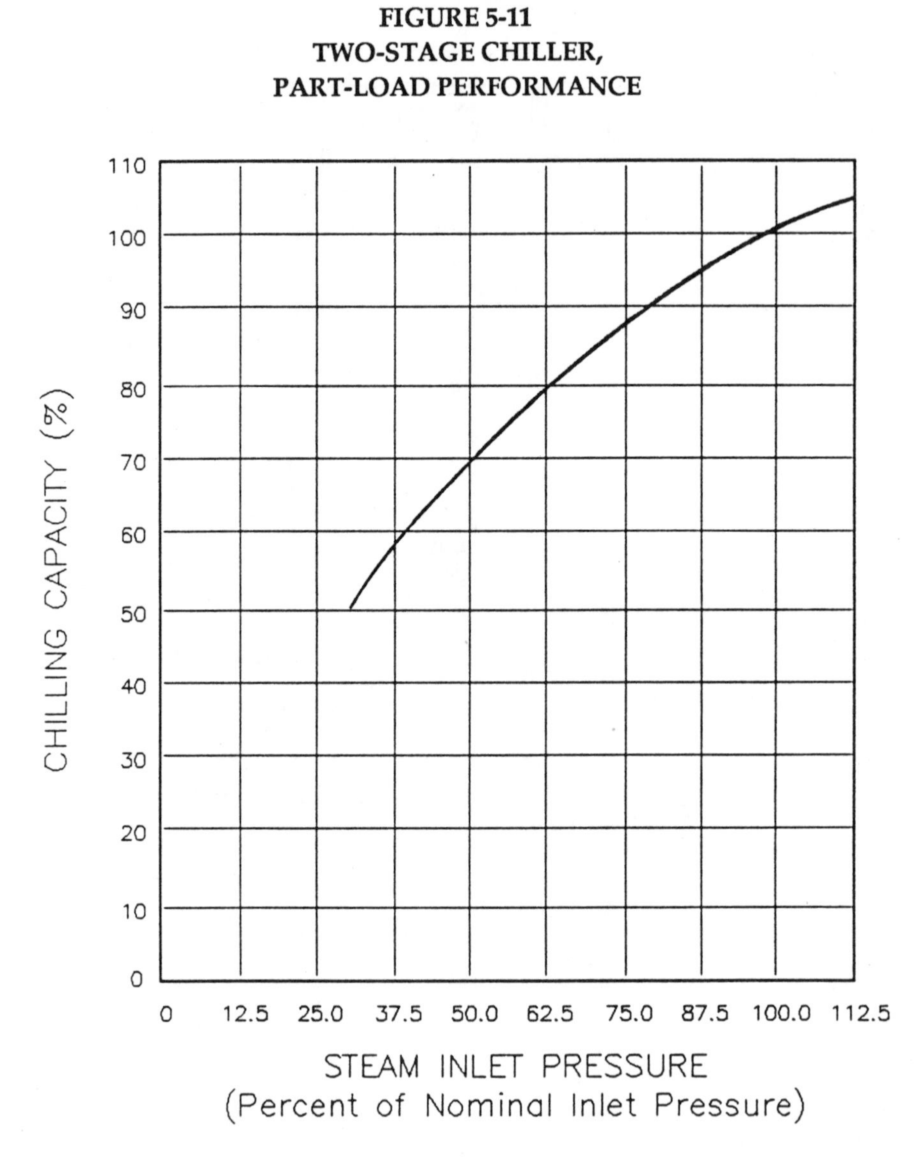

space and water heating requirements may be minimal. Second, because they require negligible amounts of power, they reduce the seasonal differences in power requirements by reducing summer cooling power requirements. This leveling of the site power requirements can lead to better load factors, both for the cogeneration system and for any supplemental power requirements.

Absorption chillers can be either baseloaded, providing all required chiller capacity, or they can be used as peak shaving units in combination with electric chillers to minimize overall costs. In a mixed system, the electric chiller may be baseloaded, thus achieving high electric load factors, while the absorption chiller is used for peaking, thus avoiding high demand charges for limited use of demand. Alternatively, if the electric utility has a time of day rate, the absorption chiller may be used during on-peak periods, thus minimizing more expensive on-peak energy and demand charges, while the electric chiller would be used during off-peak periods.

Various operating modes are possible, and an economic analysis is required to determine which plant mix will provide the lowest overall costs.

In examining the economics of electric driven versus heat driven chillers as part of a cogeneration system, it is important to first examine the economics of mechanical versus absorption chillers, independent of the source of heat. The economics of cogeneration are dependent on economic applications of recoverable heat. Thus, it is the economics of purchased power versus steam chillers that will establish the value of recovered heat. In performing the analysis, it is possible to use the cost of purchased power and the performance of electric and absorption chillers to determine the price that can be paid for natural gas or steam. If natural gas or steam is available at prices less than the threshold, then an absorption chiller can produce operating cost savings, and it is then necessary to determine if the incremental investment is justified. More importantly, in analyzing a cogeneration system this exercise establishes the maximum value for recoverable heat used in an absorption chiller. The value of that heat will be limited to the lower of the actual cost for gas or steam or the price that can be paid for natural gas or steam based on the electric versus absorption chiller analysis. As an extreme case, if an absorption chiller cannot be competitive with an electric chiller, unless steam is available at no cost the economic value of recovered steam used in an absorption chiller is zero.

DESICCANT COOLERS

Desiccant coolers operate by removing water vapor from the air and, therefore, reducing the latent (dehumidification) cooling load. They do not reduce the air temperature or act on the sensible cooling load.

The basic concept consists of passing air through a bed which contains a material with a high affinity for water. Water vapor is removed from the air, being attracted to the desiccant. When the desiccant has reached saturation, it is then heated, driving the moisture off and recharging the desiccant.

Desiccants can be categorized according to the process by which they attract moisture. *Adsorbents* hold water on their physical structure and do not chemically or physically combine with the water which they attract. The water vapor is captured and held on the adsorbent's surface. In contrast, *absorbents* act by physically or chemically interacting with the water vapor. They either attract water vapor and ultimately form a solution in the liquid water, or chemically combine with water vapor to form a distinct chemical.

This type of cooler is most effective in facilities such as supermarkets, where the latent cooling load may exceed the sensible cooling load or where moisture in the air leads to ice build-up in freezers. They are also used in manufacturing or health care facilities where dry air is critical. Desiccants can be re-charged directly using a natural gas burner or they can be re-charged with heat recovered from a cogeneration system's prime mover. Because they require low-grade heat for recharging the desiccant, they are often used in combination with reciprocating engines.

As was the case with absorption chillers, it is necessary to examine the economics of desiccant cooling, both with and without cogeneration. The quantity and cost of fuel required for the desiccant will establish the size of the cogeneration system and the value for recovered heat respectively.

REVIEW

Heat driven cooling, either mechanical, absorption or desiccant based, can be a viable augmentation to a cogeneration system. These systems are particularly useful in those applications where recovered heat is used for either space or water heating and where the heating load is highly seasonal. In those applications, the use of either a single or double effect absorption chiller can level the annual thermal load, thus increasing the overall energy efficiency and cost effectiveness of the cogeneration system. In contrast, electric chillers can exaggerate the seasonality of the site's electrical requirements and may erode operating cost savings by causing an annual imbalance in electrical and thermal requirements.

In considering heat driven cooling as an addition to cogeneration, it is first necessary to evaluate the economics of different chiller options, independent of any decision regarding cogeneration. Where absorption or desiccant cooling is competitive with electric chillers, either in a baseloaded or peak shaving mode, then they can be economically viable additions to a cogeneration system. Where electric chillers are economically attractive, the addition of either an absorption or desiccant chiller to a cogeneration system is not likely to improve the cogeneration system's economics.

Chapter 6

THERMAL ENERGY DISTRIBUTION

Cogeneration systems produce two products, and if the system is to be economically viable, it is usually necessary for each of these products to have some economic or market value. In this chapter, the technology for the distribution of one product, thermal energy, is reviewed.

In most cases, a cogeneration system can be sited at the existing facility's power plant and the distribution of the thermal energy is not a significant problem. It can more properly be considered a mechanical interconnection design issue. In other cases, either because the cogeneration system cannot be conveniently or economically sited close to the thermal load, or because multiple thermal loads exist, or because the "thermal load" consists of a district heating loop, the distribution of thermal energy itself becomes an issue.

A cogeneration plant can be successfully integrated into a central steam or district heating/cooling system. Many electric utilities, both in the United States and Europe, operate district steam loops with the steam produced by a cogeneration plant that may be several hundred megawatts or larger. In the analysis of cogeneration as a component of these systems, it is necessary to separately examine the economics of the district heating loop and to determine whether it is economically viable. If the central heating/cooling concept is viable, it is then necessary to determine whether cogeneration can be a viable component of that system.

District heating (DH) systems consist of the following major components (see Figure 6-1): the *power plant* where the thermal energy is first produced, the *transmission system* which links the power plant to the area served by the district heating system, and the *distribution system* which links the transmission system to the individual thermal energy end users. The terminology and functions are identical to those of the electric power grid, with the major exception being that DH networks are not as highly interconnected as the electric grid. However, as with the electric utility grid, the district heating system may be either *radial* or *loop*. While the modern, well maintained DH distribution system can limit losses to 5% to 10%, older district heating loops may experience losses totalling 25% to 30% of sendout.

FIGURE 6-1
DH SYSTEM

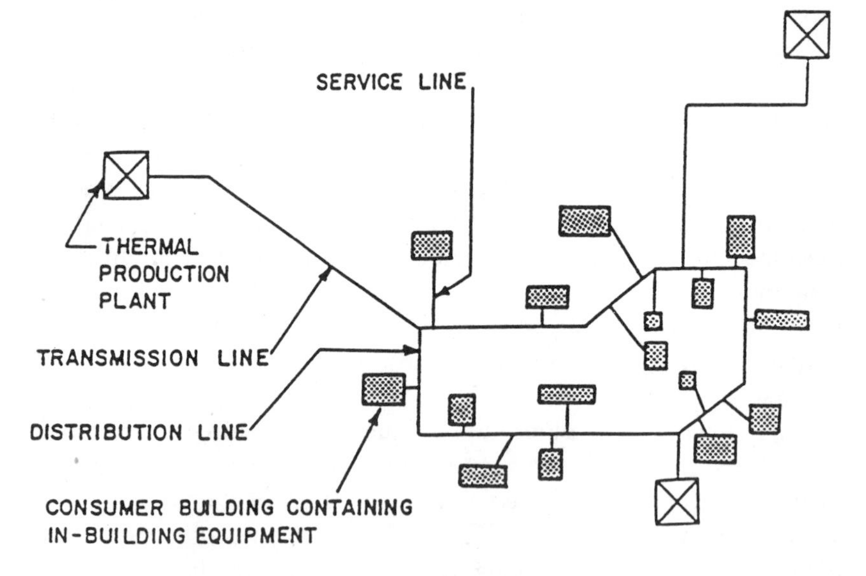

Courtesy of IDHA's District Heating Handbook, 4th Edition.

The thermal system may also be somewhat unique in that, in some systems, the equipment required to interconnect the thermal loop with the building heating/cooling systems may be part of the district heating loop. Central heating systems, both with and without cogeneration, are referred to by a number of terms including District Heating and Cooling (DHC) and Combined Heat and Power (CHP).

SYSTEM MEDIUM

In general, thermal energy can be transported in one of three forms, steam, hot water or chilled water, and the specifics of each type of system are discussed below.

Steam Distribution Systems

The use of steam as the medium for a district heating (DH) system is quite common, and it is believed that the first DH system in the United States used steam. This trend was reinforced as electric utilities became the primary proponents of central heating systems, using cogenerated steam produced from power plants sited in or near the major load centers. Over time, with the retirement of these older, smaller plants, and the siting of plants at significant distances from the load centers, the use of cogeneration in a DH has declined. Many DH loops now operate with heat-only boilers, fueled with either conventional fossil fuels, usually oil and gas, or with a renewable resource such as agricultural or municipal solid wastes.

The steam pipes in the district heating loop can be quite large and costly and, in order to minimize the system's costs, steam systems do not generally return the steam condensate to the power plant, thus eliminating the need for two sets of pipes, one for steam and one for condensate. While this approach can minimize the system's capital cost, it does impose several operating cost penalties including:

- Loss of thermal energy in the condensate,
- Loss of water and the cost of make-up water,
- Added costs, including the costs of chemicals for water treatment,
- Added use and cost of water required to reduce the condensate temperature before it can be discharged to the sewer system, and

- Charges for condensate discharged to the sewage system.

Even if condensate is not returned to the steam plant, it must be removed from the steam system, as it can cause a hammering effect in the steam lines. Therefore, condensate removal points, including drip legs and steam traps, must be provided at all low points in the system.

Steam systems are somewhat older and because of the high corrosive power of the high pressure steam, system repairs can be a major expense. They also impose both design and operational constraints, including:

- High thermal losses that result from high steam pressures and temperatures,
- Loss of steam and condensate through steam traps and drains,
- Complex and lengthy start-up and shutdown procedures, and
- More complex operating systems resulting from the low thermal inertia of steam and the need to balance steam production with demand.

The temperature changes that occur during start-up and shutdown can be significant, particularly if higher temperature steam is used. In order to avoid rupture of the steam lines, the system must be constructed to allow for expansion and contraction. Typical practice includes the use of expansion loops which are designed to allow for changes. These loops may allow for longitudinal, lateral or rotational expansion, or a combination of all three. While expansion loops allow for expansion and contraction to relieve stress, the steam system must also include anchors which minimize unwanted steam line size changes.

Because of the higher losses associated with steam distribution, these systems are usually limited to more dense urban areas or to building complexes such as universities, military bases and medical complexes. In general, a steam system will be 5 miles or less. Because of the need for steam in many industrial processes, they are frequently used in facilities such as chemical plants, petroleum refineries, pulp and paper mills and other industrial processes.

Steam distribution systems can be quite costly in areas where changes in terrain require lift stations to change the steam loop elevation.

Hot Water Distribution Systems

Hot water distribution systems have become increasingly popular, offering several advantages over the older steam systems. The most significant disadvantage can be the capital cost, as a hot water system generally requires two pipes of an equal average size; one for supply and one for return. As with steam, a hot water system may include a cogeneration plant. Hot water systems may be classified as *high temperature* (temperatures above 300°F or 400°F), *medium temperature* or *low temperature* (temperatures below 160°F to 180°F) according to the temperature of the sendout. While temperature ranges are indicated, there are no precise, generally accepted temperature limits.

The following are some of the advantages of hot water systems:

- Because they operate at lower temperatures, hot water systems generally are more efficient than steam distribution systems.
- System temperature can be adjusted to load, thus further minimizing energy losses.
- They are designed for the return of all water, thus avoiding the energy and water losses and costs associated with steam condensate.
- The hot water system includes a significant mass of water, thus providing high thermal inertia and the ability to use the hot water as a storage medium.
- Because both supply and return pipes are normally full, start-up and shut-down are much simpler than for the steam system.
- System leakage is generally less than for the steam system.

With their lower heat losses, hot water systems can be significantly larger than steam systems and, in some cases, may extend for 10 to 20 miles or more. As with steam systems, the hot water distribution system must be designed to allow for pipe expansion and contraction that results from changes in the water temperature.

While hot water may be preferable for new systems, it is costly and sometimes not possible to convert an existing steam system to a hot water system. Where such conversions are possible, the end user's building systems may be based on the direct use of district steam, and the cost of the conversion of the building systems or the installation of water to steam exchangers must be considered.

Chilled Water Distribution Systems

Although they are frequently found in major building complexes such as universities and military bases, chilled water distribution systems (operating at temperatures of 35°F to 45°F) are not as common as are steam and hot water systems. A chilled water system can reduce cooling costs through the use of highly efficient, larger chillers in place of smaller, decentralized units. In addition, the ability to perform central maintenance and to increase control of the system may provide additional savings. When absorption chillers are used in the central chilled water plant, these systems can provide an economical use for cogenerated steam produced during summer months when space and water heating requirements are minimum.

Chilled water system pipe sizes and, therefore, system costs are greater than for a hot water system. These larger sizes are the result of the lower differences between supply and return temperatures for a central chilled water plant as compared to a central heating plant. One benefit of the lower temperature changes is the decrease in the amount of temperature-related pipe expansion and contraction and the reduced need for expansion loops. Chilled water systems generally return all water to the central chiller plant.

SYSTEM DESIGN CONSIDERATIONS

The design of the heating system, including the choice of the medium, will depend on various technical and economic factors. Some of the more significant budgetary considerations are reviewed below.

As discussed above, the selection of the heating medium, either *steam* or *hot water* will have a pivotal impact both, on system design and costs and on operating procedures and costs. Some of the design considerations are discussed above; however, the final decision may be based on key constraints such as the existing distribution system medium and the end user's current thermal requirements and the cost of changes to the end user facilities should the distribution medium be changed. If a steam system is selected, it is then necessary to determine whether or not the condensate will be returned to the central power plant.

A second important decision is whether the DHC system will be *above* or *below ground.* Above ground systems are generally less expensive and, being both visible and accessible, they are somewhat easier to maintain. However, because the ambient air temperatures will vary significantly more than ground temperatures the potential for heat loss increases. Above ground systems are usually limited to applications such as military bases where the interference with public right of ways is at a minimum and where the energy end user has a high level of control of the site and individuals on that site. Below ground systems are generally more costly and, depending on the method of burial, they can be more costly to maintain. While heat losses are lower, corrosion can be a significant threat and identifying the location of steam or hot water leaks and their repair can be quite expensive.

There are several available alternatives if a below ground system is selected. The first consists of a utility tunnel such as that shown in Figure 6-2. While the tunnel is generally the most costly alternative, it can be used for other utility services such as chilled water, electric power, telephone systems and for materials and personnel transport. The primary benefit of a tunnel is the ease of access to the distribution system, facilitating routine inspections and minimizing maintenance costs. Elevating the steam pipes as shown in Figure 6-2 also reduces the potential for corrosion. Finally, because tunnel temperatures are more stable than ambient temperatures, heat losses can be minimized.

A second approach which has many of the advantages of a tunnel is the waterproof concrete trench (see Figure 6-3). The trench is less expensive than the tunnel and the pipe can be mounted on supports to minimize corrosion. While access for inspections and repairs is not as convenient as with the tunnel, the cover can be removed if required.

FIGURE 6-2
UTILITY TUNNEL

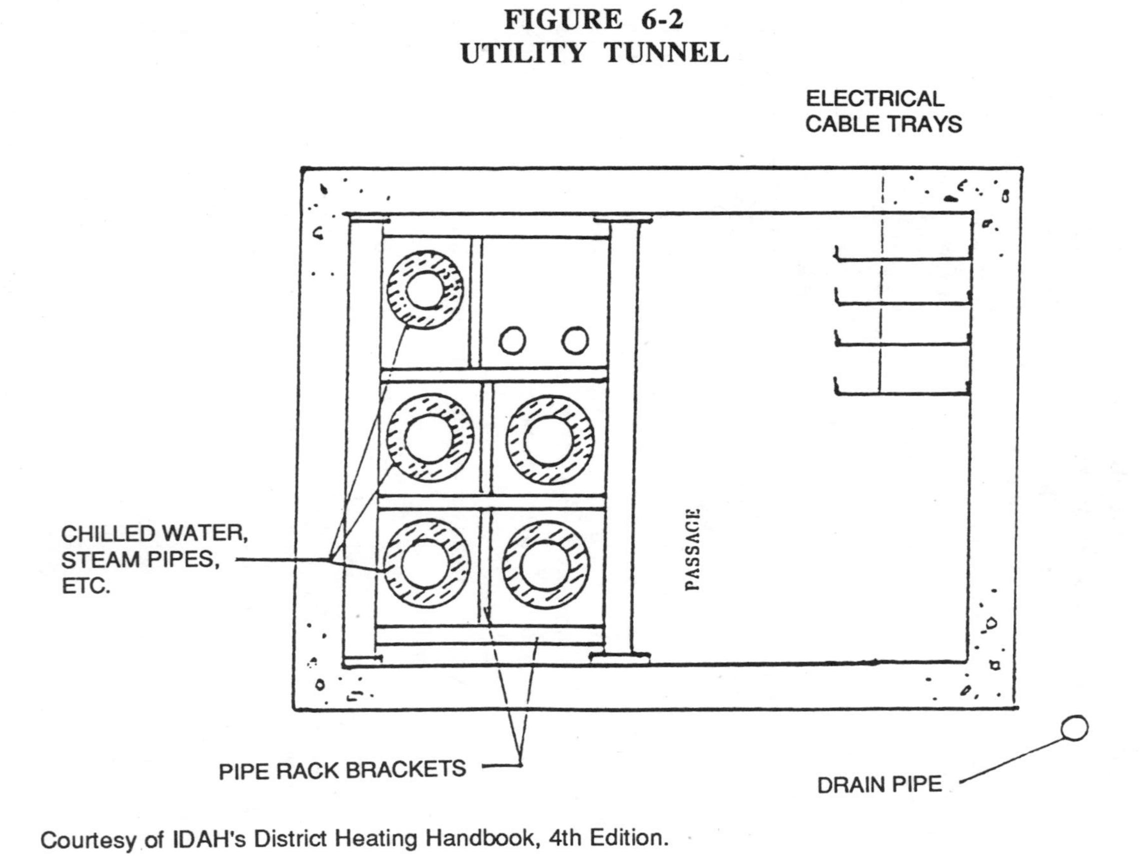

Courtesy of IDAH's District Heating Handbook, 4th Edition.

FIGURE 6-3
CONCRETE TRENCH

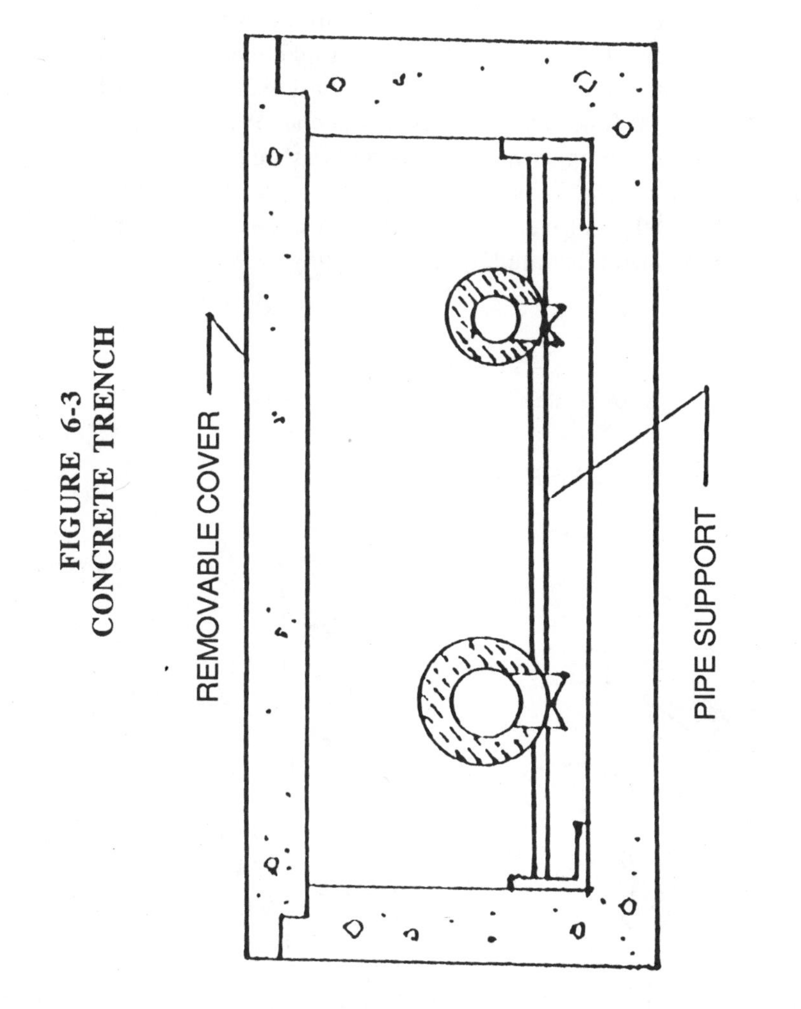

Courtesy of IDAH's District Heating Handbook, 4th Edition.

Another approach is the use of box type concrete conduits as shown in Figure 6-4. The system can be made water tight with internal pipe supports that are used to minimize exposure to water and corrosion. The air space provides for insulation and drying of any accumulated water; however, it increases the cost of the system. This system includes an air space and the system may be pressurized after installation is complete. Loss of air pressure can then be used as an indicator of system failure. Concrete conduits can be either field poured or factory prefabricated, thus minimizing system cost.

FIGURE 6-4
BOX TYPE CONCRETE CONDUIT

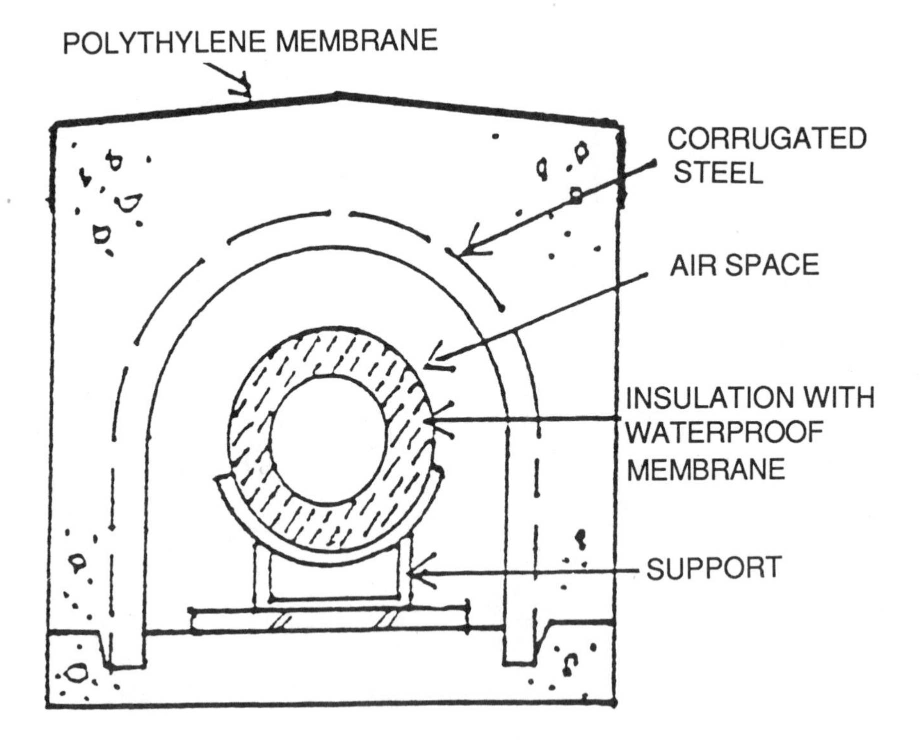

Courtesy of IDHA's District Heating Handbook, 4th Edition.

The last alternative consists of the direct burial of pipes enclosed in metal casings such as shown in Figure 6-5. These systems are least costly; however, because of the direct burial corrosion can be a significant problem and they can incur high maintenance costs. Identifying the location of a leak in the system can be quite difficult despite the increased availability of leak detection systems.

FIGURE 6-5
METAL CASING CONDUIT

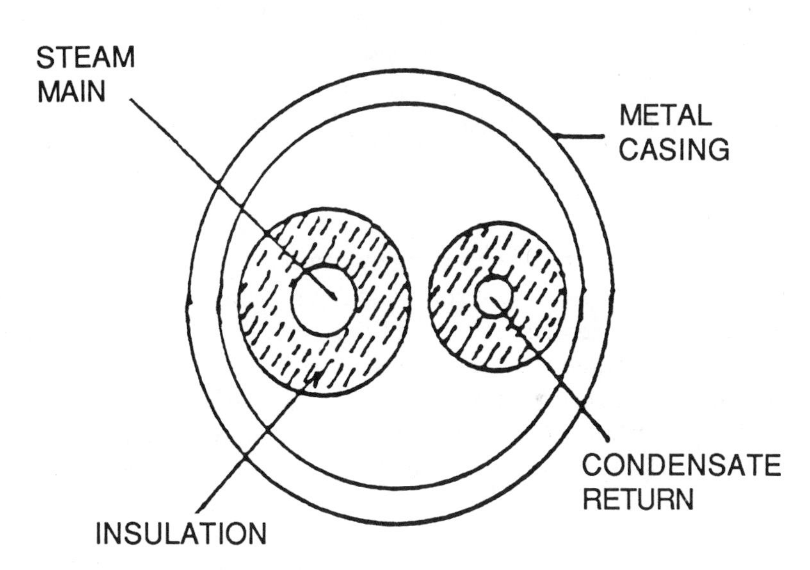

Courtesy of IDHA's District Heating Handbook, 4th Edition.

Corrosion, which is the deterioration of the iron pipe because of the difference in electric potential between the pipe and the soil or another metal, can be a significant problem in any system. The problem can be worsened by standing water, and particularly by salts in the water. There are three general approaches to dealing with corrosion including:

- Elevating the pipe to lessen any contact with standing water,
- Use of protective coatings on the pipe, thus insulating them from water, or

- Cathodic protection, which reduces the impact of differences in electrical potential and also protects the entire pipe.

Most district heating and cooling systems have manholes throughout the loop. These manholes can be used for expansion loops, valves, lifts and traps. Water accumulation in manholes can accelerate the breakdown of the pipe insulation and lead to increased corrosion. In addition, if both heating and cooling pipes are routed through the same manhole, the accumulated water will conduct heat from the heating pipes to the cooling pipes and reduce the system's thermal efficiency. All manholes should be designed to drain off water or to allow for pumping of accumulating water.

REVIEW

District Heating and Cooling systems can provide a significant thermal load for large, central plant cogeneration systems, as were typical of urban centers throughout the United States during the first half of the twentieth century. These district heating systems were generally based on steam distribution, with little or no return of the condensate. While these older steam systems have been costly to maintain and operate, many electric utilities continue to own and operate these systems frequently replacing the older, centrally located cogeneration plants with refuse fired boilers. Today, the trend is to utilize hot water as the medium in a district heating loop. These systems have proved easier to operate, less costly and more efficient.

The evaluation of cogeneration at a facility served by a district heating loop requires a separate analysis of the incremental economics of district heating as compared to on-site boilers. In these cases conversion from central steam to on-site boilers may provide most of the operating cost savings as would an on-site boiler and cogeneration plant at a small fraction of the cost of the installation of the complete cogeneration system. In these instances, the most cost effective solution may be a simple on-site boiler plant with no cogeneration.

Chapter 7

COGENERATION ECONOMIC CONCEPTS

The energy end user typically purchases all required power from the local electric utility and fuel from either a regulated gas distribution company, a broker, or a fuel oil supplier. Some larger end users have developed alternatives to these conventional systems, including on-site power systems, heat recovery from compressors, incinerators and other processes. This Handbook focuses on one process–cogeneration–and this chapter details procedures for determining whether cogeneration is viable at a specific site.

A conventional electric utility central power plant will operate at an efficiency ranging from as low as 25% to as high as 40%. The typical industrial or small commercial boiler operates at a maximum annual efficiency ranging from 65% to 85%. Depending on the utility's power plant mix and the end user's ratio of electrical to thermal energy, the overall end user's fuel efficiency will typically range between 30% and 65%. In contrast, a cogeneration system is capable of operating at an overall energy efficiency approaching 75%. Because of this high efficiency, the cogeneration system requires a lesser amount of fuel, or source energy, to satisfy the same end use energy requirements and, therefore, cogeneration can significantly reduce an end user's utility costs.

The objective of a feasibility analysis is to quantify both the operating cost savings and the required investment to determine if the cogeneration system provides an adequate return on investment.

Historically, most successful cogeneration applications have been sited at industrial facilities, universities and hospitals served by central boiler plants with fairly constant and coincident requirements for both thermal energy and power. In addition, most of these facilities were staffed by full-time operating engineers thus reducing the incremental costs of cogeneration.

In order to better understand cogeneration, it is first necessary to review basic cogeneration economics.

A cogeneration system uses a single form of energy in a sequential process to provide two energy products. As pointed out above, cogeneration systems can operate at relatively high overall efficiencies, producing significant reductions in the energy required to satisfy a site's heat and power requirements. To be economically viable, it is necessary to turn these higher energy efficiencies into operating cost savings.

The viability of a cogeneration system will depend on the following factors:

- the value of the cogenerated power and heat relative to the cost of conventional power and heat,
- the cost of fuel to the cogeneration system, and
- the ability of the energy end user to use *all* the recoverable heat that is available from the power plant.

Each factor is discussed below.

VALUE OF INTERNALLY USED COGENERATED POWER

The value of cogenerated power will depend on the disposition of that power and the design of the electric utility rates. For those applications where the power is used on-site displacing power purchased from the utility, the design considerations include the relative demand and energy costs, time-of-day considerations, demand ratchets and partial service rates. For projects selling cogenerated power to the utility the rate considerations include capacity credits, energy credits and seasonal variations in these credits. The determination of the value of cogenerated power under different conditions is shown in the following examples.

Demand-Energy Rate Design

First, consider an end user purchasing power on a simple industrial or large user rate consisting of both demand ($15 per kilowatt per month) and energy ($0.050 per kilowatt hour) charges. Neither demand ratchets, seasonal or time of day variations in the rate are considered. Figure 7-1 is a representation of a typical rate, with the average cost of power, including all service charges and adjustments, plotted as a function of the load factor.

Assume that a mid-sized industrial customer has a peak load of 10 megawatts and 5,000,000 kilowatt hours for a load factor of 500 hours use of demand. The cost of that power, as determined from Figure 7-2 (see Line Cost 1), would be approximately $.080 per kilowatt hour for a total of $400,000 per month. Based on 500 hours use of demand, the $15 per kilowatt demand charge is equivalent to $.030 per kilowatt hour.

If that end user were to cogenerate all the power that it would require, and only that amount of power, the value of cogenerated power would be the average cost of $.080 per kilowatt hour. If the plant were to produce less than the total site load, the value of the cogenerated power would be the value of the decremental block of power that was not purchased.

A baseloaded cogeneration system can operate at a high electric load factor, with turbine load factors often exceeding 650 hours per month. For illustrative purposes, assume that a 5-megawatt cogeneration plant were installed and operated for 684 hours (95% availability), producing 3,420,000 kilowatt hours of cogenerated electricity. In this example, a cogeneration system sized at only 50% of the peak requirement would supply over 68% of the site's monthly electrical energy requirements. The remainder would be purchased from the local electric utility. At a supplemental demand of 5 megawatts, the resulting load factor for the 1,580,000 kilowatt hours is 316 hours use of demand. The cost of purchased power as determined from Figure 7-2 (Line Cost 2) for this load factor is almost $.096 per kilowatt hour for a monthly total of $154,000 per month for a reduction of $246,000.

FIGURE 7-1
SIMPLE DEMAND ENERGY RATE

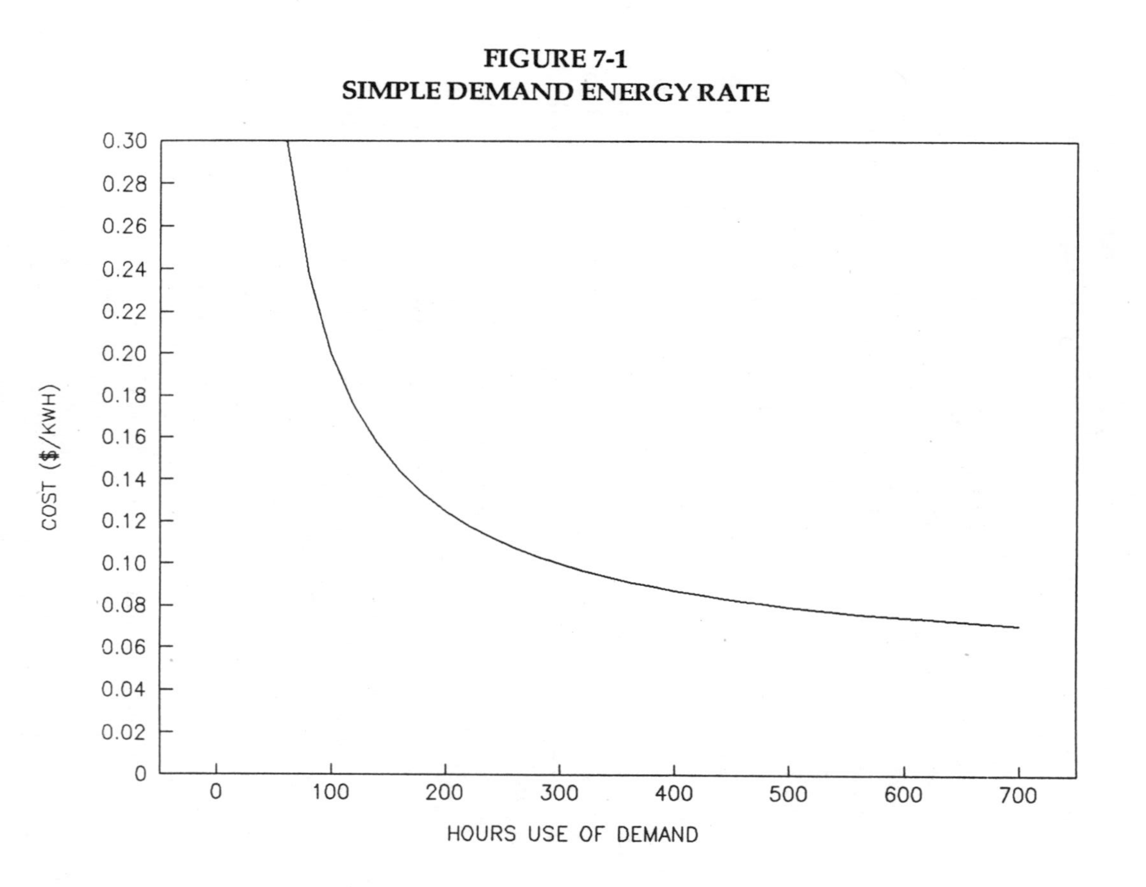

FIGURE 7-2
DEMAND-ENERGY RATE EXAMPLE

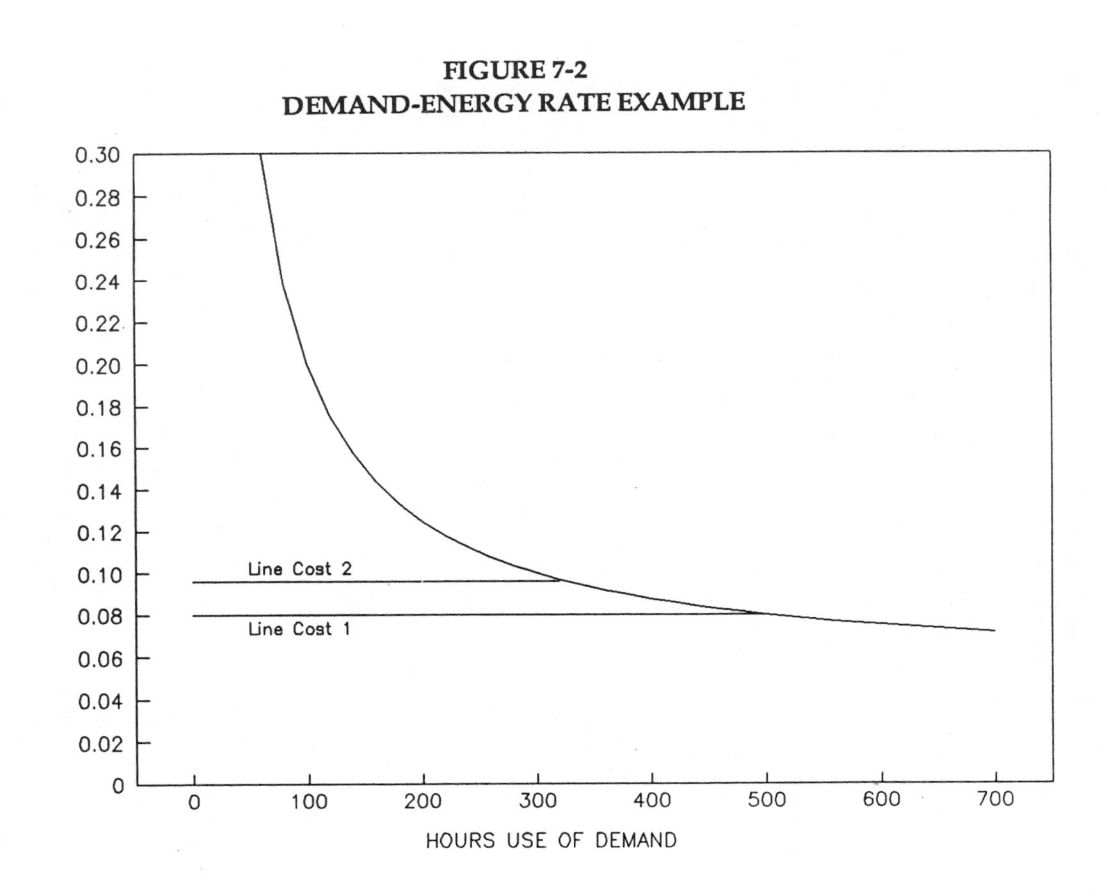

The cogenerated electricity is worth $246,000 or just less than $.072 per kilowatt hour and not the average cost of $.080 per kilowatt hour.

Time of Day Rate Design

The above example illustrates the interaction between load factor and supplemental costs for simple rate designs. Time of Day rates can further complicate the analysis. A baseloaded cogeneration system may very well displace more lower priced, off-peak kilowatt hours than more costly, on-peak kilowatt hours, further decreasing the value of cogenerated power. Alternatively, a peak-shaving cogeneration system can be operated to displace more on-peak kilowatt hours, thus increasing the value of cogenerated power.

Figure 7-3 shows the cost of power for a typical time-of-day option. Both on- and off-peak costs are shown with an average rate based on the peak demand and 30% of the energy consumption occurring during the 40 on-peak hours per week (8 hours per day). The on-peak demand charge is $11 per kilowatt, with an on-peak energy charge of $0.10 per kilowatt hour and an off-peak energy charge of $.040 per kilowatt hour. The total monthly cost of purchased power would be $400,000.

Using the above example of 5,000,000 kilowatt hours and 10,000 kilowatts, the average cost of power is again $.080 per kilowatt hour. Assuming an availability of 684 hours, with no on-peak shutdowns, during the 20 on-peak days, a baseloaded 5-megawatt cogeneration system could produce 800,000 on-peak kilowatt hours (8 on-peak hours per day times 20 on-peak days per month times 5,000 kilowatts), which would result in a need to purchase 700,000 kilowatt hours of supplemental on-peak power (5,000,000 kilowatt hours times 30% on-peak usage minus 800,000 kilowatt hours of cogenerated power). The cogeneration system would produce another 2,620,000 kilowatt hours during off-peak hours, requiring a minimum purchase of 880,000 kilowatt hours of supplemental off-peak power. The fraction of on-peak power increases from 30% to 44%. The load factor is again 316 hours, and the total cost of purchased power decreases to $160,200 for an average cost of over $.101 per kilowatt hour. The cost of purchased power is reduced by $239,800 and the average value of the 3,420,000 kilowatts of cogenerated power is $.070

FIGURE 7-3
TIME-OF-DAY RATE

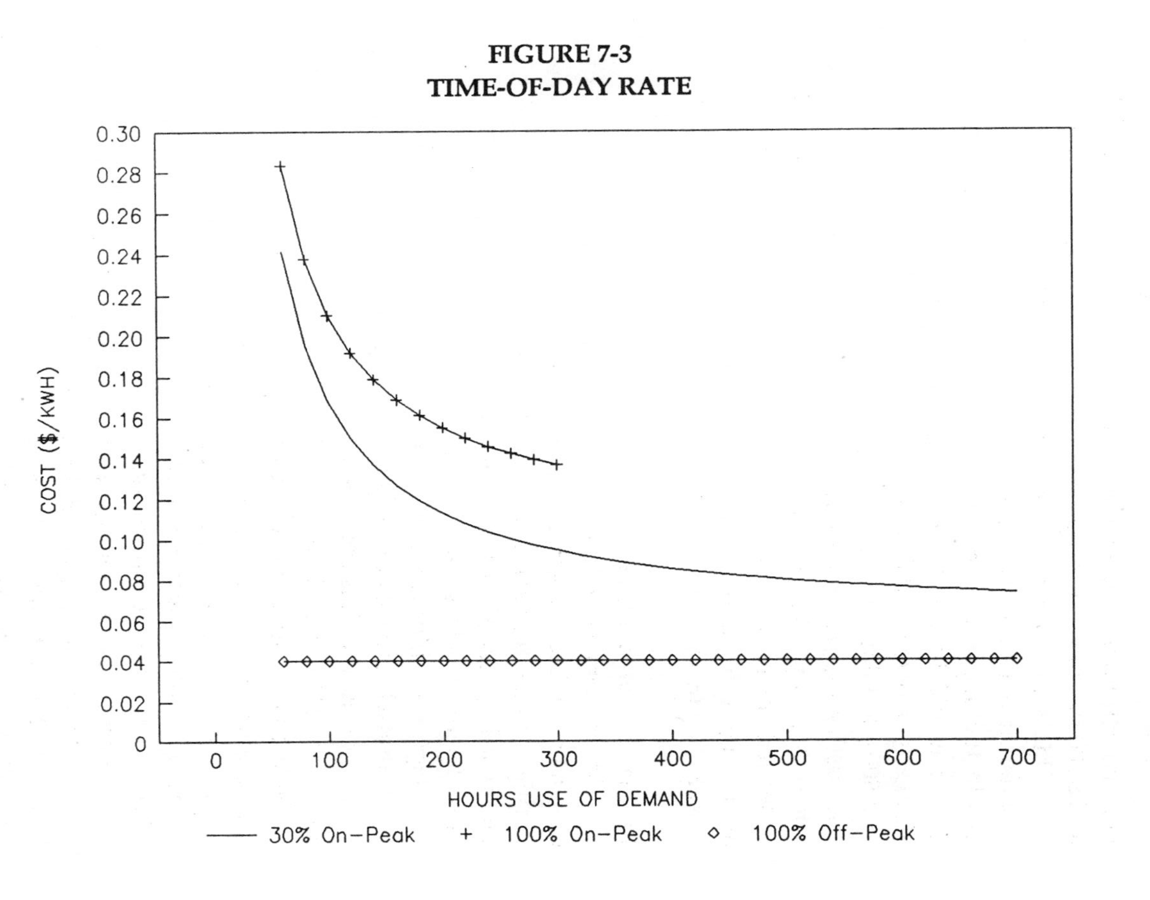

per kilowatt hour as compared to $.072 for the non-time-of-day case and $.080 for conventional power.

This same system can be operated in a peak shaving mode, wherein it only produces power during the 8 on-peak hours each day.In this mode the amount of supplemental on-peak purchased power would be 700,000 kilowatt hours, with a demand of 5,000 kilowatts. However, the off-peak energy would remain unchanged at 3,500,000 kilowatt hours for a total of 4,200,000 kilowatt hours. The cost of supplemental power would be $265,000 for an average rate of $0.063 per kilowatt hour. The 800,000 kilowatt hours of cogenerated power would reduce purchased power costs by $135,000 for an average value of over $.168 per kilowatt hour. When operated in a peak shaving mode the value of cogenerated power can be doubled.

Ratcheted Demand Charges

Many demand rates include a ratchet provision whereby the customer is billed the greater of the actual peak demand that occurs during the month or some fraction of the historic maximum demand. The cost impacts of this type of rate design can be significant for those facilities with seasonally varying demands.

Figure 7-4 illustrates this type of rate for two different months. The rate itself consists of the simple demand energy rate described above; however, in this case, the billing demand is assumed to be the maximum of the actual demand or 100% of the peak which occurred in the last 11 months. Two curves are shown on Figure 7-4; the lower (less expensive) curve being identical to that of Figure 7-1. The second, more expensive rate, is based on ratcheted demands where the actual winter demand is only 75% of the summer demand, but is billed at 100% of that summer demand.

During a winter month, the cost of conventional power based on 10,000 kilowatts and 3,750,000 kilowatt hours would be $337,500 or $.090 per kilowatt hour. Without the ratchet, the cost would be $300,000 or $.080 per kilowatt hour. With the cogeneration system in place, the actual demand during the winter months would be 2,500 kilowatts; however, the billing demand would be 5,000 kilowatts (equal to the summer peak of 5,000 kilowatts). The monthly usage would be only 330,000 kilowatt hours and the cost would be $91,500 for an average rate of $.277 per kilowatt hour. The

FIGURE 7-4
DEMAND RATCHET

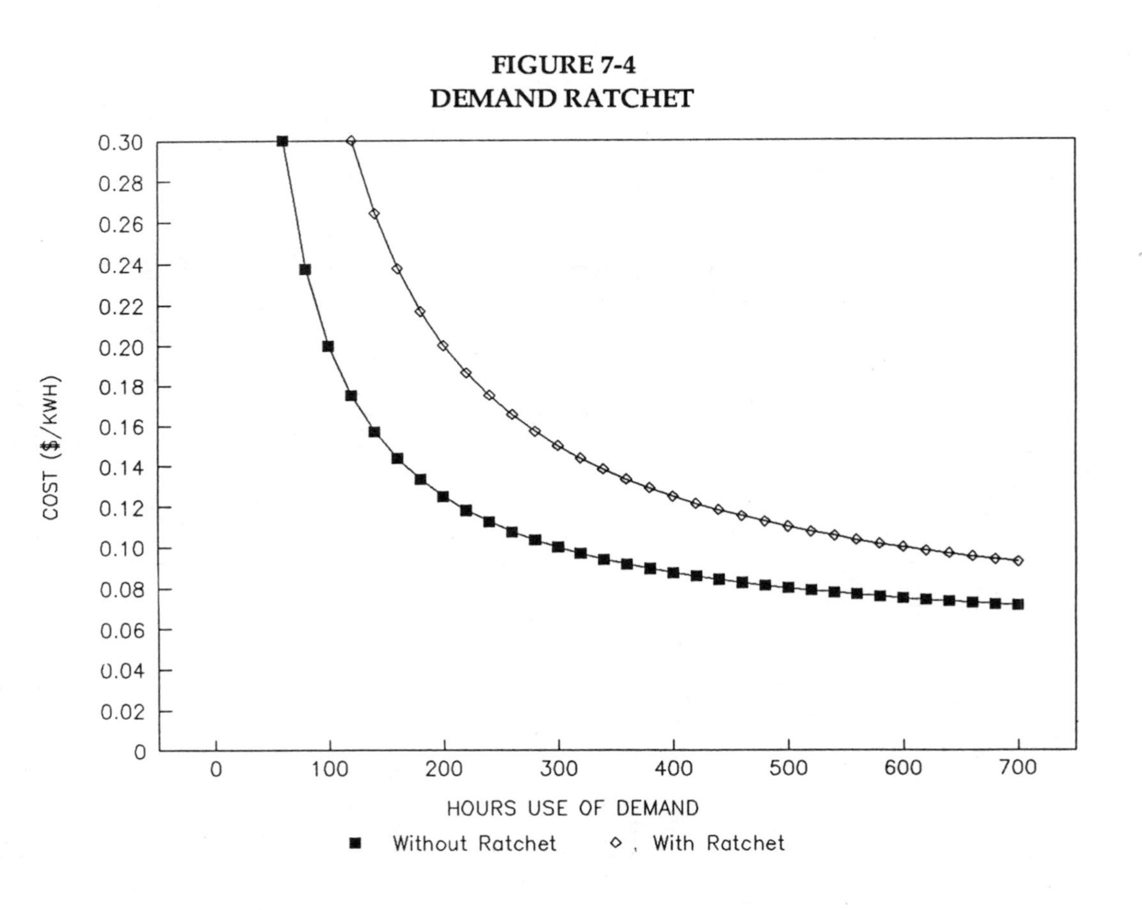

3,420,000 kilowatt hours of cogenerated power reduces the purchased power cost by $246,000 for an average of $.072 per kilowatt hour.

Energy Only Rate Design

This type of rate is the simplest and is illustrated by Figure 7-5. In this case, the value of cogenerated power does not significantly differ from the average retail cost of power.

VALUE OF COGENERATED POWER SOLD TO A UTILITY

The cogenerator also has the option of selling the cogenerated power to the local electric distribution company or, if that utility will provide wheeling, to another electric utility. The value of the cogenerated power is the utility's "avoided cost" which is defined as the cost that the utility would incur if the power delivered by the cogenerator were produced by the utility or purchased from another source. The avoided cost can include both a capacity component, which is related to the fixed cost of the capacity that the utility would not require, and an energy component which is related to the variable costs that the utility avoids.

Historically, either very large or very small cogeneration projects have elected to sell power to an electric utility. Most PCS sites have used power on-site to displace power purchased from the utility. The most notable exception to this rule has been for small systems of less than 100 kilowatts, where some state regulatory commissions have allowed the cogenerator to "run the meter backward," thus receiving an avoided cost equal to the electric utility's retail rate.

There has been considerable controversy in the determination of the electric utility's avoided cost. Some state commissions have required utilities to develop a plan which identifies the next central power plant which would be developed, including the capital and operating costs of that plant. If cogenerated power is sold to the utility under a long-term contract which allows the utility to defer the construction of that plant, the cogenerator then receives both the value of the energy cost and the capacity charges associated with that plant. Much of the dispute between cogenerators and utilities

FIGURE 7-5
ENERGY-ONLY RATE

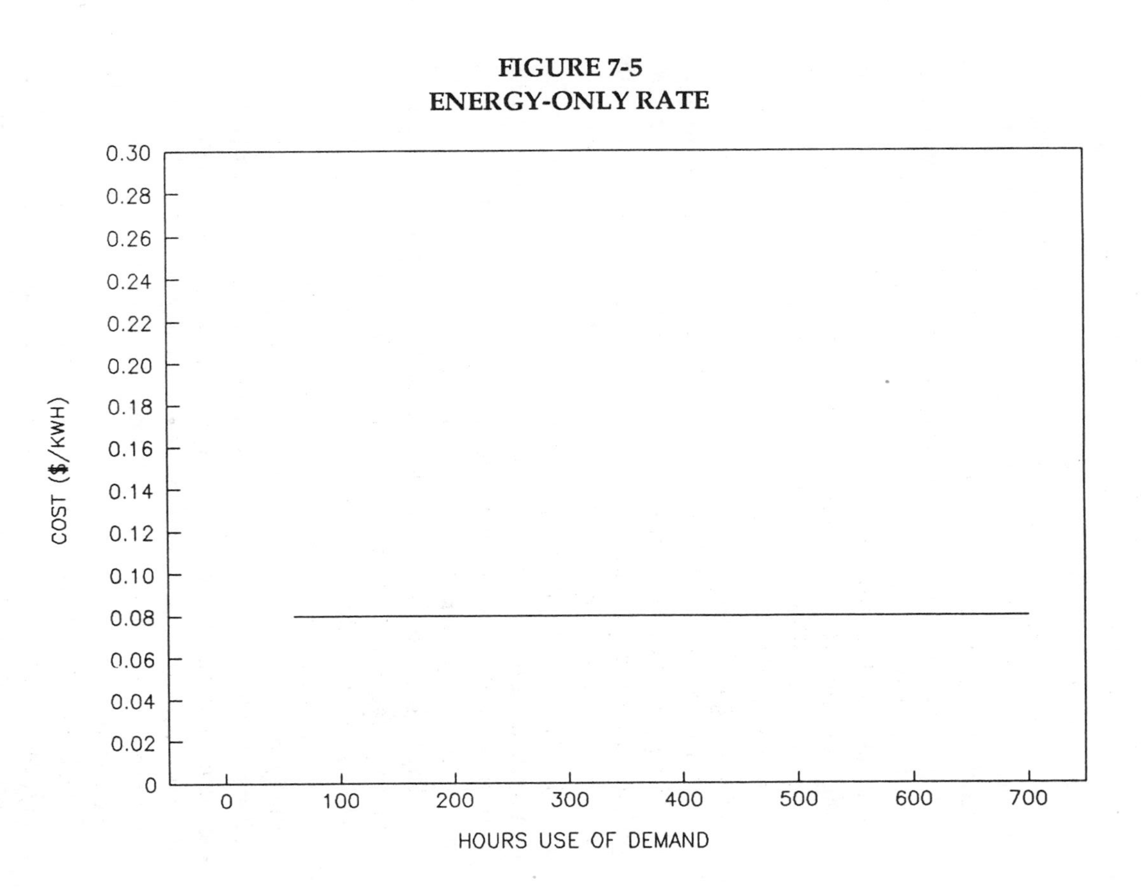

has focused on the type of plant which would be built, the fuel it would use, the time when the plant would be required to be in service, and the capital cost of that plant.

In some states, in order to facilitate project financing, the electric utility has provided additional capacity payments during the early years of the power purchase contract, while receiving lower payments during the contract's latter years. A discount factor is used to insure that the total present value of the capacity payments is not increased. In some cases, the capacity payments during the contract's latter years are heavily discounted in order to compensate the utility for the risk of higher front end payments. In other states, contracts are sometimes backend loaded to insure that funds are available for increasing fuel and maintenance costs.

More recently, many electric utilities and states have adopted a bidding procedure as a mechanism for determining avoided costs and for allocating capacity credits. The electric utility typically initiates the process by identifying a need for additional capacity and the type of capacity (peaking or baseloaded) that will be required. In some bidding schemes, the utility also identifies the plant that would be constructed to meet this need, and the plant is then used as a benchmark in the bidding procedure.

The utility then solicits bids from non-utility power producers, including cogenerators. While many evaluation factors are considered in these bidding procedures, cost is usually the dominant criterion, and the lowest responsible bids are selected to meet the capacity need. This approach allows the market to determine the utility's avoided cost, since all incremental power requirements are met from non-utility sources and the highest accepted bid is the utility's avoided cost. When properly employed, a bidding procedure can provide the lowest cost power to all customers purchasing power from the utility.

Determination of the value of cogenerated power, as sold to the utility, can be a more uncertain task when the utility uses a bidding process. In addition, the high cost of preparing a bid may eliminate small cogeneration projects from the process. One alternative being considered by some utilities is the use of the highest winning bid as the avoided cost for all small projects.

VALUE OF RECOVERED HEAT

The value of the useful, recoverable heat is based on the cost of the conventional fuel which is displaced; either natural gas, oil, purchased steam, coal or, in some cases, purchased power. Smaller cogeneration systems will generally displace either natural gas, fuel oil or power as the alternative fuel source. However, several points are significant in establishing a value for recoverable heat.

First, thermal energy recovered from a prime mover is available as heat, while fuel must be purchased as fuel and then converted to heat in a boiler. If the end user boiler operates at an annual efficiency of 75%, then each Btu of recovered heat is worth 1.33 Btus of fuel. Alternatively, if the recovered heat displaces purchased power used with a 100% efficiency, as for example in a water heater, then each Btu of heat displaces a single Btu of power.

Second, the value of the recovered heat is also dependent on the cost of the fuel that is actually displaced. For purposes of this example, assume that natural gas is priced at $3.60 per million Btu (MMBtu). If the end user's boilers operate at an efficiency of 75%, the heat would be valued at $4.80 per MMBtu. If the recovered heat were to displace off-peak electricity purchased at $.04 per kilowatt hour for a 100% efficient water heater, the heat would be valued at $11.72 per MMBtu. Alternatively, if water heating were provided from an energy only rate of $.080 per kilowatt hour as discussed earlier, the value of recovered heat would be $23.44 per MMBtu.

COGENERATION SYSTEM OPERATING COSTS

The actual cost reduction resulting from the operation of a cogeneration system will be dependent on the facility's energy requirements, the mechanical/electrical systems, purchased power rates, fuel costs, space requirements, staffing, insurance, taxes and other operational requirements. Each of the key operating cost parameters is reviewed below.

Purchased Power Costs

The rates for power purchases from an investor owned electric utility are specified in a tariff which is usually subject to approval by a state regulatory commission. Unlike privately owned systems, rates for municipally owned systems or coops are not usually regulated by a state commission. A utility rate consists of one or more of the following components.

Customer Charge – This item is a fixed monthly charge applicable to all customers. It is intended to recover the fixed overhead costs (e.g., meter reading, billing, etc.) associated with each account.

Demand Charge – This item is the amount charged by the utility for the capacity to deliver power. A *ratchet* refers to a demand billing which is based on some historic peak demand rather than the peak incurred in the current month. The billing demand, as contrasted to the actual or measured demand, consists of some fraction (usually from 50% to 100%) of the peak demand incurred during the ratchet period–usually 11 months.

Energy Charges – This item is the amount charged by the utility for actual energy delivered. It is based on the variable costs of power production including fuel.

Taxes – This item is the amount charged by the utility on behalf of some governmental body such as the state or local municipality. These taxes may be included in the rate itself as a Gross Receipts Tax or Franchise Fee, or they may be listed as a sales tax.

Generally, a utility bill will contain one or more of the above items.

Cogenerator Power Purchase Rates

Cogenerators who meet the FERC requirements for status as a Qualified Facility (QF) can purchase power from the electric utility and/or sell power to it. Sales to a QF can fall into any of the following categories.

Standby – Sales of power to the cogenerator which are required due to an unscheduled shutdown of the cogenerator's power plant.

Supplemental – Sales of power to a cogenerator which are required to make up the difference between the power produced by the cogenerator and the site demand.

Maintenance – Sales of power to a cogenerator that are made according to a time schedule developed by the utility and the cogenerator.

These sales may be on either a firm or an *interruptible* basis. This latter case refers to sales which are scheduled at the convenience of the electric utility. These sales can be suspended unilaterally by the utility in accordance with a pre-established set of conditions specified by the utility tariff or by contract.

Cogenerator Power Sales Rates

As discussed earlier, QF's may also sell power to the electric utility and, according to regulations developed by FERC, these sales are to be at 100% of the utility's avoided cost. The billing for a power sale to a utility may take the same form as a sale by the utility with demand and energy charges.

Many utilities have developed avoided cost rates, with some utilities differentiating in the rate depending on the cogenerator's size. In addition, utilities have established different values for the capacity component depending on the date at which the cogeneration system becomes operational, the degree of control the utility has over operation of the facility, and the term of the contract between the cogenerator and the utility. To the extent that the cogenerator can provide long-term technical and economic guarantees of power availability to the utility, the capacity payment for this power is usually increased. Smaller, factory-assembled modules of less than a few hundred kilowatts are not likely to receive significant capacity payments.

In some areas where avoided costs exceed purchased power costs the cogenerator can, at its option, sell all cogenerated power to the utility while simultaneously purchasing all required power from that same utility.

Fuel Costs

Small and medium-sized cogeneration systems are fueled primarily by gaseous and liquid fuels, with liquid fuel usage limited to lighter and midweight oils. Larger systems may also use heavy or residual fuels and solid fuels, including municipal, industrial and agricultural refuse.

Gas distribution utilities generally offer *firm gas* which is available on a year-around basis and *interruptible gas* with deliveries suspended at the convenience of the distribution system. While the interruptible gas may be offered at a significant discount off the firm cost of gas, that gas may not be available anywhere from a few days to 4 months per year. The availability of interruptible gas will depend on the supplier, the interstate gas pipelines, and the local distribution system. Use of interruptible gas for cogeneration requires the cogeneration system to use an alternate fuel and/or purchased power and conventional boilers during periods of fuel interruption.

Smaller purchases of firm natural gas are usually based on a declining cost block rate structure with smaller, high cost initial blocks and larger tail end blocks which approach the utility's own variable costs for gas. If natural gas is used for heating or process use,then the power plant fuel is an incremental load which should be priced accordingly. This pricing, which allows an end user to purchase all fuel at a lower average cost, may not be available when the cogeneration system is owned by a third party, which will be a separate utility customer.

Baseloaded cogeneration systems operating at rated capacities exhibit fairly uniform monthly fuel requirements, therefore operating at a high annual load factor. In these cases, a gas rate based on a *demand-commodity* structure may provide significant cost savings. These rates, which are similar to the electric utility's demand-energy rate structure, may provide natural gas at rates that are significantly lower than heating gas rates.

A cogeneration system may utilize a fuel such as sewer or landfill gas which might have previously been flared or otherwise disposed. Such fuels may have economic value in the current energy market and it should not be assumed that a by-product fuel has no value because it has not been previously used.

Insurance

Insurance can be obtained to protect against *catastrophic failure* of the cogeneration plant, *loss of savings* or revenue due to the inability to operate the plant, and *liability* insurance against personal injury or property damage caused by the operation of the cogeneration system. In some cases, cogeneration equipment may be included under the site's overall insurance coverage. However, the more likely case is that in which the cogeneration equipment will be either itemized on that policy or separately insured. The cost of such insurance will vary depending on equipment performance history, site and design characteristics; however, the annual charge may range from .25% to 2% of the equipment capital cost.

Taxes

It is also necessary to examine cogeneration economics on an after-tax basis including consideration of income, property and sales taxes.

A cogeneration system will produce reduced utility costs for the end user. Depending on the end user's accounting policies, this cost reduction may contribute to increased profit through lower production costs or it may generate a direct profit through internal energy sales. In either case, this increased profit will generally be subject to federal, state and local income taxes. In addition, the cogeneration system itself will incur local property tax liabilities. In any evaluation of cogeneration economics it is important to utilize tax rates applicable to the particular end user in the specific locality.

When a third party owns and operates the cogeneration property, then energy sales from the cogenerator to the end user may be subject to local sales or utility gross receipts taxes. This cost element should be explored in any third-party project.

Many facilities, such as hospitals and nursing homes, which have characteristics favorable to cogeneration, also enjoy tax-exempt status. Third-party ownership of a cogeneration system serving a tax-exempt end user it not itself a tax-exempt entity and tax liability must be considered.

Staffing

Smaller cogeneration systems are usually capable of unattended operation and their economic success frequently depends on the ability to operate without a full-time staff dedicated to any individual installation. Larger systems and particularly those with steam heat recovery systems are usually staffed by operating engineers. In some areas, the staffing may be dictated by local boiler codes; however, most systems delivering hot water or low pressure steam do not require an operating engineer. It must be stressed that the cogenerator should review local codes to determine staffing requirements.

Maintenance Costs

Maintenance costs are another key cost element that can be somewhat ambiguous. All prime movers require periodic inspection, servicing and replacement of parts; however, the time period for each scheduled maintenance procedure will depend on the type of prime mover, engine operating conditions and site conditions. The design and the cost of a maintenance program represents a trade off between more frequent, less costly scheduled service calls and routine procedures and more costly unscheduled maintenance calls caused by outages or engine failures. Maintenance requirements and costs for various prime movers were reviewed in Chapter 2.

Administrative and Management Costs

All business activities incur clerical and management costs.. When a cogeneration system is owned and operated by a third party, these costs can be as high as 5% of total operating costs.

COGENERATION SYSTEM CAPITAL COSTS

The cogeneration system's projected budget includes equipment costs including all ancillary gear, installation costs including fuel supply, engine pad, electrical interconnect, exhaust stack, etc. and both professional and service fees such as engineering, legal and financing services. The more significant fees are reviewed below.

A/E design fees for small PCS systems may range from 8% to 15% (for small systems) of the cost of the equipment plus installation. *Construction management* services can add another 6% to 8% to the cost of a project.

Legal Fees can be quite significant for project financed systems or in cases where a third party develops the plant. While these fees are less dependent on the size of the project, they can range up to 10% of the project budget for mid- and large-sized installations.

Permitting costs including building construction fees and licenses will vary with local conditions. Where environmental issues are a primary concern emission permits may require regional air quality and water modelling and can be quite costly.

In third-party developments, it may be necessary to retain *specialized consultants* to review and monitor development work. As a percentage of the project budget, these consulting fees will generally be in the range of 1% or less to as high as 6% for small PCS installations.

Project management costs, whether incurred as a professional service or internalized by the project developer, can be another 2% to 8% of the total installation cost.

Tables 7-1 and 7-2 list both operating and capital cost components.

ECONOMIC MEASURES

Some frequently used economic performance measures such as *simple payback* and *return on investment* are approximations that are easy to apply and understand. These two measures do not consider the time value of money and this deficiency can be significant. *Internal rate of return* (IRR), *net present value* (NPV), annual *cost-benefit ratio* (C/B) and *return on equity* are exact measures which consider the discount rate. Very importantly, each of these measures is capable of accurately reflecting factors such as taxes whose impact will vary over time.

TABLE 7-1
COGENERATION CAPITAL COST CHECKLIST

Equipment and Installation
- Hardware costs
- General trades work
- Construction contingency
- Land acquisition costs

Ancillary Costs
- Insurance
- Permits
- System startup and testing

Professional Fees
- A/E fees
- Legal fees
- Special consultants
- Construction testing and inspection services
- Soils and subsurface investigations
- Construction Manager fee
- Environmental studies
- Project Management

Financing Costs
- Bond Consultant
- Legal opinion
- Governmental fees
- Rating agency fees
- Interest paid during construction
- Interest earned during construction
- Bond insurance
- Bank fees
- Loan insurance

General Contingency

TABLE 7-2
EQUIPMENT COSTING CHECKLIST

- Engine Generator Set
- Heat Recovery System
- Controls
- Exhaust System including Stack
- Gas Compressor
- Thermal Storage
- Water Treatment
- Shipping charges
- Site handling and rigging
- Base or concrete pad
- Electric Utility interconnection
- Fuel Supply
- Modifications to building structure
- Mechanical system interface
- Electrical system interface
- Credit for emergency generator
- Supplemental duct burner
- Fuel storage and processing system
- Heat rejection radiators
- Diverter valve

Most investors who explicitly consider the opportunity cost use IRR as an economic measure.

The measures that consider the time value of money also include some measure of opportunity cost. That is, they recognize that the value of a dollar available for investment today is greater than the value of a dollar one, five or ten years from now. Today's dollar can be invested at some interest rate to yield more than one dollar in the future. The *discount rate* is the factor which is used to calculate the value of money available at some future time.

Present value can be calculated using the following formula:

$$PV = VALUE/(1 + i)^n$$

where

PV = Present value of some future dollar amount,
VALUE = Amount of the future dollars,
i = Discount rate, and
n = Number of years.

Organizations that develop a cogeneration project with no equity, either through 100% debt or third-party relationships, have no invested capital and, therefore, measures such as simple payback and internal rate of return are not relevant. Net present value of future utility costs over the economic life of the project is an appropriate comparative economic measure.

The IRR is the discount rate, expressed as a percentage, which will produce zero as the present value of the sum of all project costs and revenues. The calculation of the internal rate of return, therefore, requires the development of cash flow projections for the economic life of the project.

REVIEW

A decision as to whether or not a cogeneration system is viable requires a detailed analysis of all incremental capital and operating costs, both for the conventional and the cogeneration system. For an internal use system, determination of the value of cogenerated power will require consideration of the electric utility's rates both for full and partial service and the cogeneration system's operating mode. The valuation of cogenerated heat will be dependent on the cost of the fuel that the recovered heat displaces and not the cost of the cogeneration system fuel.

Cogeneration projects that sell most or all of their electrical output to an electric utility must consider the costs that the utility avoids through the purchase of the cogeneration energy and capacity. This last factor introduces considerable complexity into the valuation of cogeneration power introducing concerns regarding the timing of the availability of the capacity, system reliability and availability, and the ability of the electric utility to dispatch the cogeneration plant. The need for levelized rates from the utility as a condition of financing can further complicate the evaluation of the cogenerated power.

The preparation of a capital budget for a cogeneration system also requires careful analysis in order to include all actual costs. While most equipment costs and engineering fees are easily estimated, installation costs which can include ancillary equipment and modifications to the existing facility and its energy systems are frequently overlooked. In many cases, soft costs, which include professional fees for investment counselors, attorneys, outside engineering consultants and even the cogeneration developer's internal staff time, are not fully considered.

Chapter 8

PROJECT DEVELOPMENT PROCESS

The final decision as to whether or not a cogeneration system is an appropriate investment is usually made after a detailed, thorough engineering study. The engineering study is usually proceded by one or two less comprehensive technical and economic analyses. Once the engineering study is completed and a decision is made to proceed, that decision in and of itself, represents just one of the initial steps required to develop a project. Other activities, each of which can bring development to a stop, include:

- Environmental permitting,
- Zoning approvals,
- Utility interconnection approval,
- Detailed final design,
- Negotiation of power sales, steam sales and fuel supply agreements,
- Project financing,
- Construction, and
- Start-up

The development activities can be categorized into three groups according to project uncertainty or risk. The first category consists of *initial analytical activities* which are rather inexpensive. These activities are the preliminary analyses that are required to determine whether or not the project is potentially viable. The second category includes all *permitting, negotiations and approval activities* and preliminary detailed design. These activities can be quite lengthy and expensive, and are usually completed before any funds are expended for equipment purchases or construction. The level of design that occurs at this stage of development is only to the detail required to support permitting or interconnection approval. The final category includes *final design and construction activities* and these activities are typically undertaken only after all necessary negotiations are completed and both construction and long-term financing are available. This stage of development is the most costly; however, the risk is limited to more conventional and manageable design, construction and equipment performance risks. These activities are illustrated in Figure 8-1.

FIGURE 8-1
PROJECT DEVELOPMENT PROCESS

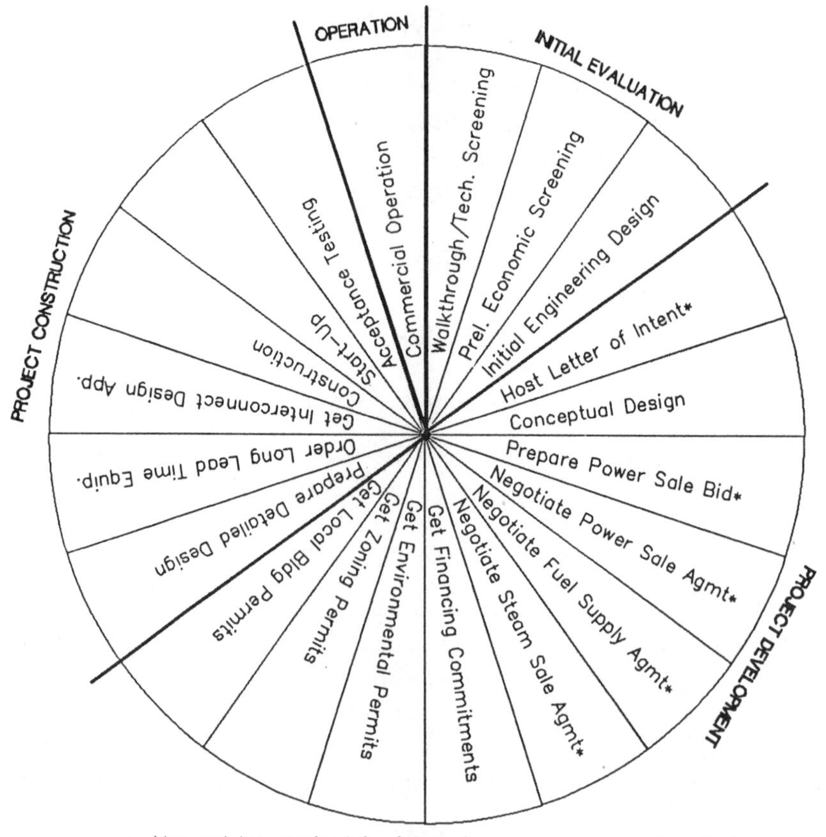

* May not be required for internal use or user owned projects.

As discussed above, the development of a cogeneration system can be considered in stages, with each stage including one or more steps. The first three steps typically require the end user to invest in analytic efforts with no assurance that a project with operating cost savings will result. These steps allow a potential cogenerator to explore the economics of cogeneration with a series of resource investments, each greater than the previous and each producing the information required to determine whether the costs of the next step are warranted. The three initial steps described below are:

- Site Walkthrough and Technical Screening;
- Preliminary Economic Screening; and
- Detailed Engineering Study.

These processes are illustrated by Figure 8-2, and are described below:

WALKTHROUGH

An initial site inspection, or walkthrough, has two objectives or goals. The first is to determine whether the site is technically compatible with cogeneration, with the second being to determine whether a cogeneration system has the potential for economic viability.

Technical feasibility is based on compatibility between the cogeneration system and the user's mechanical and electrical systems; a determination as to whether there is adequate space for a cogeneration plant, including fuel processing equipment; and an analysis of whether the existing systems are adequate. The economic analysis is frequently based on the average cost of power and fuel and often assumes 100% use of cogenerated power and heat. This type of analysis can result in significant errors in the estimate of operating cost reductions. Chapter 9 includes a methodology for conducting a walkthrough analysis.

ENGINEERING AND ECONOMIC SCREENING

The walkthrough economic analysis usually makes many simplifying assumptions and, as one proceeds to the more thorough

FIGURE 8-2
OVERALL DECISION MAKING PROCESS

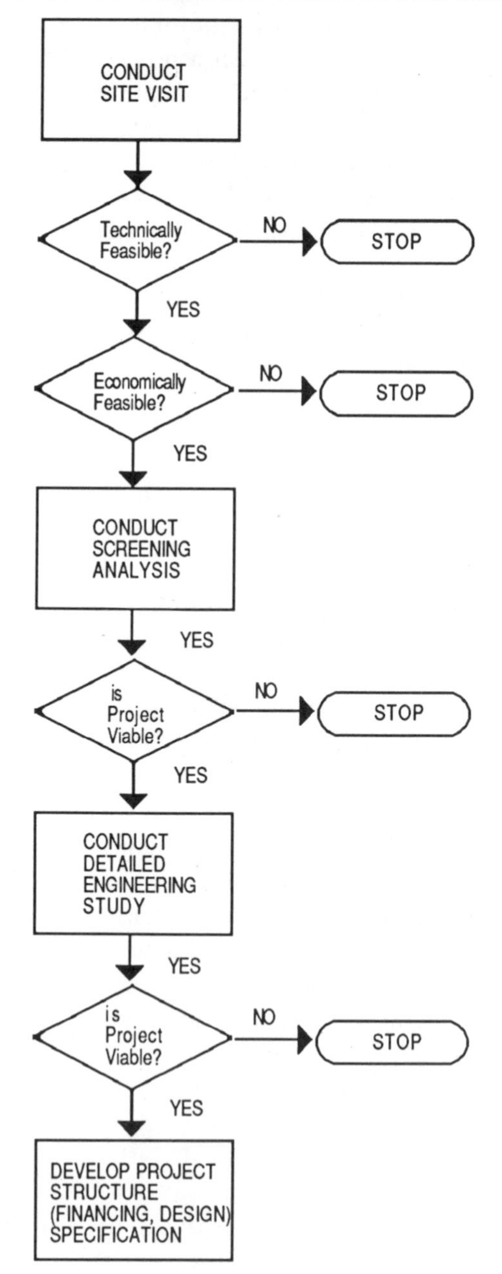

analyses, these assumptions are replaced by more realistic approximations for the specific site.

The most significant difference between the screening analysis and the more general approach used in the walkthrough is the development and use of monthly and hourly energy use data, with historic or anticipated monthly conventional energy characteristics used as the basis for estimating operating costs. At this step, it is more important to understand the purposes for which fuel is used in order to determine the quality of the required thermal energy and to estimate the fraction of the end user's fuel use that can be displaced with recovered heat. Among the additional factors that must be considered in this analysis are the end user's mechanical and electrical systems, purchased power rates for both full and partial service, conventional and cogeneration system fuel costs and availability, space requirements, staffing, insurance and various operational requirements.

This type of study utilizes historic energy use data as a basis for the modelling of energy needs, potential variations in those needs, and the amount of purchased energy that can be displaced with cogenerated heat and power. As described in Chapter 9, the hourly energy data can be used to develop load-duration curves as a basis for modelling system performance. Hour-by-hour modelling of system performance is usually performed in later design stages. Finally, it includes a capital and operating cost analysis of both conventional and cogeneration systems, both as a basis for the investment and for system optimization.

The output of a screening analysis includes a number of parameters, including an estimate of the approximate size of the cogeneration system and its capital costs, the type of prime mover to be used and the design and operating philosophy. Some measure of economic performance such as total project savings, payback period, Internal Rate of Return, Return on Equity or Net Present Value is also computed.

The next set of calculations is a cash flow analysis wherein financing costs, tax implications and inflation are considered. At this point, the total installed cost of the system must be estimated and a simple payback is computed. In addition, anticipated cost escalators and the owner's financing details and tax position can be input to produce projected cash flows and an IRR. At this stage of the analysis it is often possible to examine alternative ownership and financing options. One important decision that results from this step is a choice between internal funding or third-party ownership.

The screening analysis provides the information required to allow the decision maker to determine whether or not they should proceed to the next step–the detailed engineering study.

PRELIMINARY ENGINEERING STUDY

It is during the preliminary engineering analysis where conceptual designs are more fully detailed, systems optimized, and costs fully developed. Data requirements for an existing facility include specific expansion, construction, renovation, and demolition plans; as-built plans and drawings; and information on recently implemented or anticipated energy conservation plans. This latter information is required to assure that the feasibility analysis considers probable future conditions rather than historic patterns which may no longer be relevant.

This level of engineering evaluation often includes hour-by-hour analyses of the end user's energy requirements and the modelling of equipment performance for each hour. As a consequence, it is important to consider the impact of recent or planned energy conservation activities because an effective conservation program can significantly change cogeneration economics.

While the cost of the engineering analysis can be quite meaningful, that cost should be viewed as the final expenditure in a multistep process leading to the selection of the most cost-effective utility system. In most projects, this step is undertaken only after preliminary, less comprehensive investigations have shown that the cost of the detailed analysis is justified.

The objective of the preliminary engineering analysis is to provide the information required to undertake more costly and time consuming activities including negotiations of power, steam and fuel supply contracts as appropriate, environmental permitting, zoning approval and the development of project financing.

Chapter 9

COGENERATION PERFORMANCE MODELLING

The cogeneration development process is an iterative design procedure in which a preliminary concept is developed based on limited data and then continuously refined as additional information is acquired. The very first step, the walkthrough, uses a minimum of data to develop a preliminary concept. That concept is refined and developed during the next screening step as additional and more comprehensive site data and local requirements are incorporated into the process. Detailed design, environmental permitting and zoning approvals, electrical interconnection approvals and fuel supply negotiations usually result in additional design input. Finally, once financing is secured, the project moves into detailed design where construction documents are developed based on more detailed equipment and site modelling. Performance modelling is required after the project is operational in order to determine if it is being properly operated and maintained and providing the anticipated benefits. This chapter reviews the initial data collection and modelling activities.

SITE DATA COLLECTION

A feasibility analysis attempts to determine whether any cogeneration alternative is compatible with the end user's technical, economic and financial requirements. It usually requires information about specific energy use patterns and utility costs at each facility.

Typical site specific data needs include a description of the end user's mechanical and electrical systems, the existing and projected uses of energy, the efficiency with which the conventional systems convert energy, seasonal and diurnal variations in energy use, and both current and projected costs of that energy. Other information such as published utility rates for power and fuel, the electric utility's interconnection requirements and equipment cost and performance are not as site specific. In this chapter, preliminary site specific data collection needs and techniques are reviewed.

While most cogeneration modelling is based on a 12-month period, 3 years of site data should be collected to determine whether there are any long-term trends which should be considered and if the 12 months being used for the cogeneration analysis is typical.

Electrical Data

In general, the most reliable source of purchased power data is the local electric utility. Utilities can provide long-term billing histories and, for larger accounts where recording meters may be required by time-of-day rates, they can usually provide more detailed demand profiles. As with all data, including both energy use characteristics and costs (even that which is supplied by a utility), one should not assume that all data are correct. One objective of the initial walkthrough is to determine what data are available and whether such data are likely to be correct. A secondary objective is to identify any on-site meters and sub-meters that might be used in a supplemental data collection effort. Many newer facilities include submetering and, in some cases, utility meters can be read on a periodic basis by on-site personnel to provide electrical demand profiles. The most troublesome problem resulting from the use of the utility billing meter is that it is not intended for recording hour-by-hour variations in energy use and may be inaccurate when so used. In some cases, moving averages can be used to smooth some of the variation caused by larger scale meters, while preserving the required detail.

In general, an electric utility bill will provide both the billing demand and the total energy requirements for the facility. Depending on the specific utility, that bill may also include additional details such as the actual or measured demand, the site average or peak demand, power factor, the time at which the peak demand occurred and the energy used during the various time-of-day periods.

Figure 9-1 is an illustration of a utility electric bill. As can be seen, it sometimes includes helpful supplemental data such as whether or not the transformers are owned by the utility or the end user, delivery voltage levels and the location of the utility metering.

In an effort to control the patterns of energy use, many utilities have instituted time-of-day rates and have installed recording demand and energy meters. In general, the demand interval is either 15, 30 or 60 minutes and varies from utility to utility. Figure 9-2 is a sample printout from a recording meter. While these detailed printouts are not typically provided with the monthly utility bill, they are usually available from the utility for a nominal charge. Detailed demand profiles may not be as readily available for smaller commercial customers.

Data such as that shown in Figure 9-2 are extremely useful in sizing the cogeneration system and, in particular, in determining how much purchased power can be displaced by a baseloaded cogeneration system. These data are used to develop load-duration curves (Figure 9-3) which illustrate the frequency with which electrical demands exceed specified levels. In practice, the curve is developed using demand increments or bins, the size of which may range from a few percent to 5% of the peak annual demand. For example, the data in Figure 9-2 can be sorted into 50-kilowatt bins (Table 9-1). It is then possible to calculate the percentage of time the site demand falls within each bin and represent the cumulative percentage as a load-duration curve. Curves can be developed on a monthly, seasonal or annual basis depending on both data availability and analytic needs.

In a period when utility rates can increase dramatically, there is no assurance that historic utility costs are useful as predictors of future costs. The basic charges, the fuel adjustment charges or the structure of the rate may have changed. Therefore, the local utility should be contacted to obtain a copy of the utility's complete tariff and any pending requests for rate increases and for long-term forecasts of rates. These forecasts must be independently confirmed.

FIGURE 9-1
SAMPLE ELECTRIC BILL

Please Make Checks Payable To / Return This Portion With Your Payment

Account Number	Statement Date

Service Used At

Account Status

PREVIOUS BILL	NOV 2	$73,451.39
PAYMENT - THANK YOU -	NOV 19	$73,451.39 CR

BALANCE $0.00

ELECTRIC RATE 037

METER #	BILLING PERIOD FROM	TO	DAYS	METER READING PREVIOUS	CURRENT	CONSTANT	DEMAND	KILOWATT HOURS USED
76215840	OCT 31	DEC 3	33	01064	01169	4800.00	2227.2	504000 OFFPK
	OCT 31	DEC 3	33	01238	01360	4800.00	2404.8	585600 ONPK

THIS IS YOUR ELECTRIC BILL CALCULATION:

CUSTOMER SERVICE CHARGE		$202.700000	
BILLING DEMAND	2404.8 KW		
DEMAND CHARGE	2404.8 KW X $6.00	$14,428.800000	
OWNERSHIP DISCOUNT	1000.0 KW X $0.20 CR	$200.000000 CR	
	1404.8 KW X $0.10 CR	$140.480000 CR	
TOTAL DEMAND CHARGE RATE 037		$14,088.320000	
ENERGY CHARGE	493920 KWH X $0.050830	$25,105.953600	
	573888 KWH X $0.060830	$34,909.607040	
FUEL ADJUSTMENT	1067808 KWH X $0.002170 CR	$2,317.143360 CR	
GU ADJUSTMENT	1067808 KWH X $0.001470	$1,569.677760	
OC ADJUSTMENT	1067808 KWH X $0.000376	$401.495808	
TOTAL ENERGY CHARGE RATE 037		$59,669.590848	
* TOTAL CHARGE RATE 037		$73,960.610848	$73,960.61
		AMOUNT NOW DUE	$73,960.61

- IF WE RECEIVE "AMOUNT NOW DUE" BY JAN 02, 1985, YOU WILL AVOID A 1 1/4% LATE PAYMENT CHARGE.
- * TOTAL RATE AMOUNT WOULD HAVE BEEN $74,782.43 ON RATE 038
- 2% PRIMARY METER DISCOUNT 1089600 KWH. 1067808 KWH BILLED.

FIGURE 9-2
METERED DEMAND DATA

09/01/84 - 09/30/84 DEMANDS

MONTHLY ONE HALF HOUR DEMANDS IN KILOWATTS
(NOTE: - 1 INDICATES MISSING DATA)

TIME	1230	0100	0130	0200	0230	0300	0330	0400	0430	0500	0530	0600	0630	0700	0730	0800		
	0830	0900	0930	1000	1030	1100	1130	1200	1230	0100	0130	0200	0230	0300	0330	0400		
	0430	0500	0530	0600	0630	0700	0730	0800	0830	0900	0930	1000	1030	1100	1130	1200		
DATE																	MAX KWO	TOTAL KWHRS
09/25/84	838	839	844	852	828	831	828	830	848	888	937	981	1065	1136	1198	1234		
	1259	1250	1298	1322	1298	1308	1317	1318	1281	1265	1255	1260	1224	1184	1119	1078		
	1057	1034	1036	1074	1077	1047	1020	1008	998	985	919	860	851	830	827	802	1322	25218
09/26/84	799	794	805	805	792	792	792	789	810	845	894	950	1032	1119	1196	1264		
	1206	1278	1272	1259	1272	1258	1235	1230	1165	1160	1145	1133	1104	1066	1026	996		
	963	961	962	991	951	831	808	796	761	742	703	669	658	641	632	633	1286	23033
09/27/84	620	618	623	624	626	614	613	609	627	647	690	761	843	917	1001	1014		
	1004	1016	1020	1031	1022	1020	1028	1021	1014	1018	1002	996	968	953	915	876		
	846	829	847	887	864	852	823	812	779	752	701	680	662	653	648	620	1031	19801
09/28/84	619	625	625	640	620	622	614	612	636	749	706	763	830	923	984	1003		
	1015	1027	1024	1039	1033	1024	1044	1023	1028	1040	1019	1007	988	963	936	897		
	859	844	869	879	852	822	795	781	754	734	708	668	663	646	636	627	1044	19840
09/29/84	636	613	616	602	613	601	620	625	636	634	677	705	745	793	827	823		
	830	827	829	831	846	821	808	802	787	788	788	775	758	743	728	723		
	722	730	728	712	706	703	687	679	680	680	670	650	648	620	625	611	846	17151
09/30/84	631	607	604	596	600	595	624	618	612	624	655	688	713	728	746	742		
	742	744	794	800	792	808	814	775	771	763	763	756	770	748	720	709		
	705	695	701	682	703	686	670	673	683	658	660	645	632	623	625	625	814	16659

MAXIMUM HALF HOUR NON COINCIDENT DEMANDS IN KILOWATTS

TIME	1230	0100	0130	0200	0230	0300	0330	0400	0430	0500	0530	0600	0630	0700	0730	0800	
	0830	0900	0930	1000	1030	1100	1130	1200	1230	0100	0130	0200	0230	0300	0330	0400	
	0430	0500	0530	0600	0630	0700	0730	0800	0830	0900	0930	1000	1030	1100	1130	1200	
MAX KWO	838	839	844	852	828	831	828	830	848	888	937	981	1065	1136	1198	1264	
	1286	1278	1298	1322	1445	1572	1534	1545	1606	1547	1551	1547	1507	1481	1433	1249	
	1300	1287	1280	1317	1299	1128	1020	1008	998	985	919	898	891	861	856	841	1606
TOT KWO	10074	9998	10013	9945	9880	9843	9866	9862	10184	10434	11157	11941	12826	13651	14674	15058	
	15220	15369	15653	15796	16170	16284	16178	16086	16331	16533	16488	16308	16006	15653	14972	14465	
	14208	14087	14139	14046	13642	13244	12662	12345	12091	11797	11342	10876	10683	10394	10226	10069	628770

MONTHLY SUMMARY;

MAXIMUM 30 MIN. DEMAND	1606 09/24/84 1230
TOTAL CONSUMPTION RECORDED	628770 KWH
RECORDING START TIME	09/01/84 00:00 (115 MIN. INTERVAL START TIME)
RECORDING STOP TIME	09/30/84 24:00

USAGE CHARACTERISTICS

MAXIMUM 30 MINUTE DEMAND	1606 kW 09/24/84 1230
AVERAGE MONTHLY DEMAND	873.29 kW
LOAD FACATOR	0.543

FIGURE 9-3
LOAD DURATION

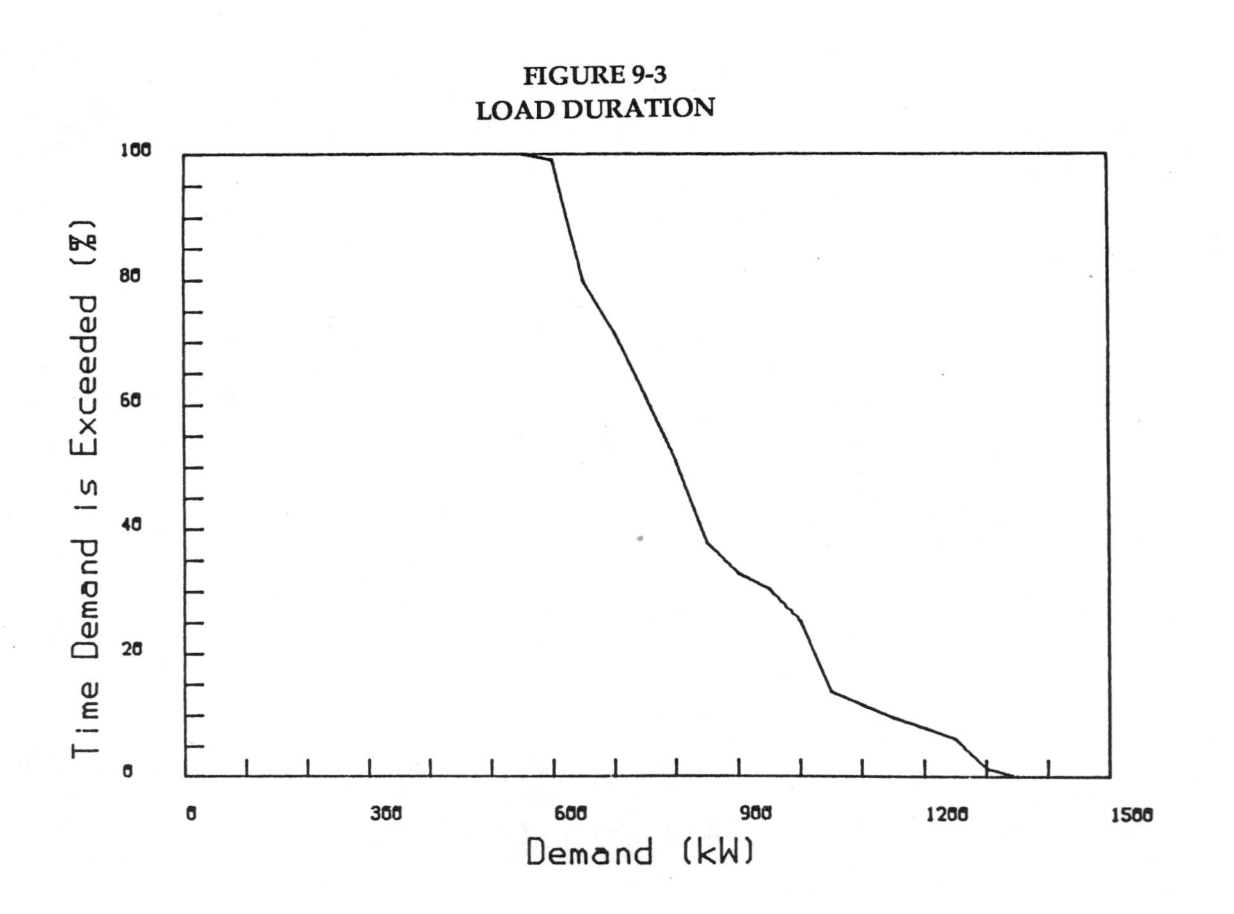

TABLE 9-1
DEMAND DATA

Demand Interval (kW)	No. of Readings	Percent of Total Readings	Cumulative Total (%)
551-600	3	1.0	100.0
601-650	55	19.1	99.0
651-700	24	8.3	79.9
701-750	29	10.1	71.6
751-800	30	10.4	61.5
801-850	38	13.2	51.1
851-900	14	4.9	37.9
901-950	7	2.4	33.0
951-1000	15	5.2	30.6
1001-1050	33	11.5	25.4
1051-1100	6	2.1	13.9
1101-1150	6	2.1	11.8
1151-1200	5	1.7	9.7
1201-1250	5	1.7	8.0
1251-1300	14	4.9	6.3
1301-1350	4	1.4	1.4
TOTAL	288	100.0	

Estimating purchased power requirements and costs for new construction, and particularly for commercial structures, is a more significant problem. For more complex structures and industrial processes, the design engineer should be able to provide anticipated energy requirements and projected costs. In other cases, and particularly for smaller commercial facilities, the data may not be available, even during construction.

Fuel Use Data

The analysis of fuel use data is usually more troublesome and should be approached with caution. First, the fuel data should cover the same 12-month period used for the electrical data. If this is not readily possible, then site (e.g., occupancy, production, etc.) and weather conditions for the two 12-month periods should be compared to determine if they were equivalent. If there are significant differences the fuel and/or electric data should be adjusted accordingly.

Second, it is then necessary to determine how much of the fuel use is displaceable by recovered heat. For example, natural gas used for cooking cannot be replaced by heat recovered from an engine. In addition, if both high and low pressure steam are required, the cogeneration system's recovered heat may only be able to satisfy the low pressure needs, and the high pressure steam load should be excluded in analyzing the specific option. During the initial site visit, it is important to identify all such non-displaceable fuel uses.

Several approximations can be used to estimate the amount of displaceable fuel. In a central power plant, where boiler steam output is monitored, it may be assumed that all the fuel used to produce steam is replaceable. When steam data are available, it is helpful to verify the accuracy of the steam meters as they can be out of calibration, or located such that not all steam is metered. This latter case frequently occurs when steam is used within the power plant.

A second approach is to inventory all functions, such as cooking and drying, that require fuel rather than heat and to estimate the amount of fuel required by each piece of equipment. This total is then subtracted from the billed fuel use.

A third approach, which is useful when fuel use is weather sensitive, is to determine if there is a mathematical relationship between fuel use and weather (Figure 9-4) where heating degree days are used as a measure of weather. The fuel use corresponding to zero degree days is the base usage, which should then be further divided into replaceable uses, such as for water heating, and non-displaceable uses, such as for process and cooking. When natural gas is purchased on an interruptible rate, with oil or propane as an alternate fuel, it is likely that the oil and natural gas is replaceable with recovered heat.

Sites using fuel oil or other fuels delivered in bulk, present one additional problem since the billing data are based on deliveries and not actual usage. If the end user maintains fuel inventory data

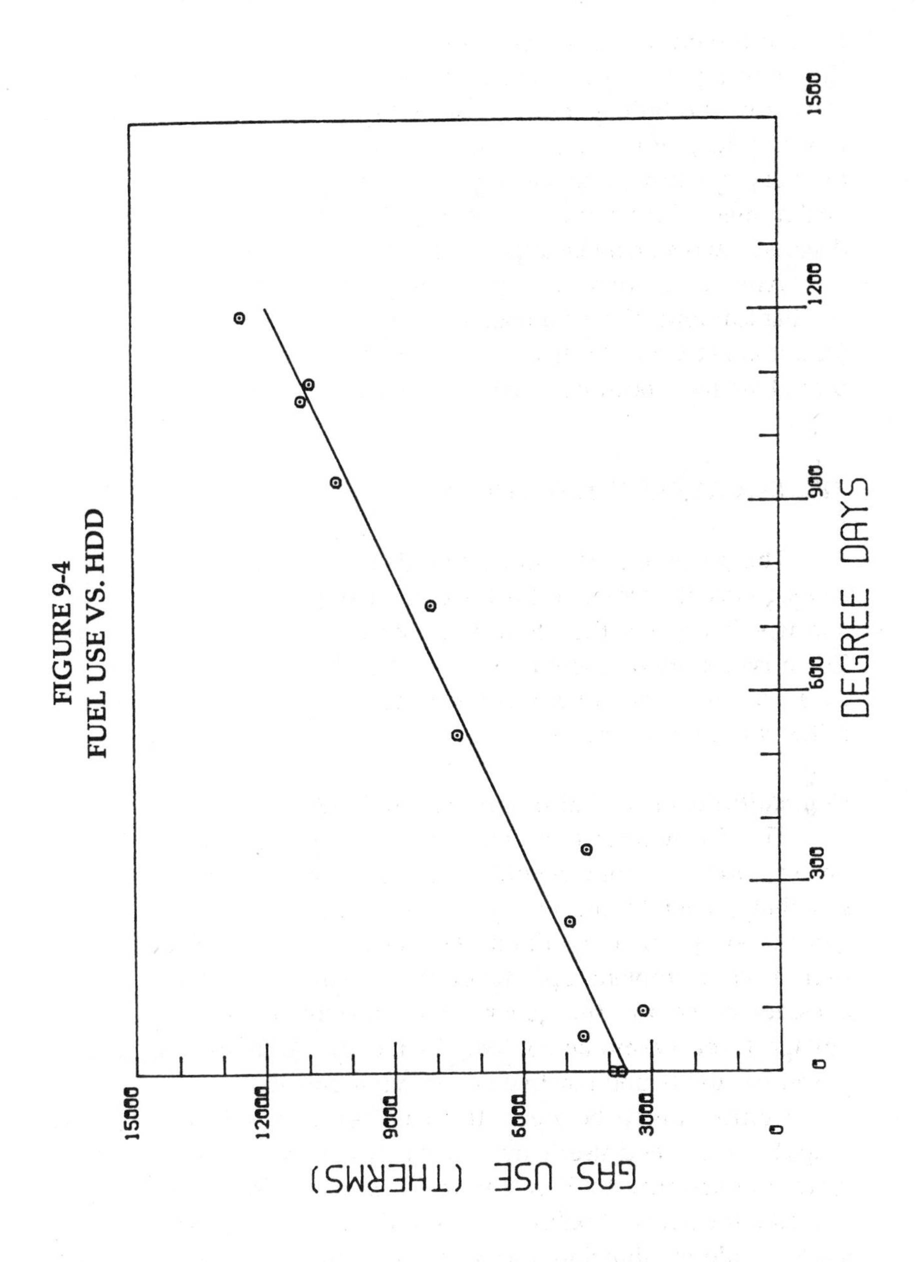

FIGURE 9-4
FUEL USE VS. HDD

the calculation of fuel usage is rather straightforward. However, in the more typical case where inventories are not maintained and a review of the billing information indicates that deliveries may be out of phase with usage, it may be necessary to allocate actual delivery quantities to calender months to approximate usage. Techniques relating fuel use to weather or output, such as those discussed above, may be appropriate.

Any feasibility analysis must consider the electrical, mechanical and plot characteristics, as they currently exist. Therefore, it is necessary to determine whether electrical, mechanical and plot plans have been updated to reflect the actual facility.

TECHNICAL CONSIDERATIONS

The purpose of the initial site visit is to determine if there are any significant barriers to the development of a cogeneration system and whether or not the site has the potential for economic viability. These issues are discussed below, and it should be noted that each consideration or potential issue does not need to be addressed in the initial site inspection.

Compatibility of Available and Required Energy

The cogenerator must be able to use the electricity to reduce power purchases from the utility or, alternatively, it must be able to sell that power to the utility for an assured revenue stream. In general, electrical compatibility between a cogenerator and an end user is not a problem. Specific electrical parameters that should be considered include the generator's transient overload capability, voltage level, harmonic content, frequency variation, regenerative power capability and reactive power requirements.

Compatibility between the cogeneration system's thermal output and the end user's mechanical system requirements may be more troublesome. As discussed in Chapter 2, reciprocating engines provide the least flexibility in satisfying high quality thermal loads, while combustion and steam turbines provide the greatest capability. The site's thermal needs should be catalogued to determine which prime movers are acceptable.

The following guidelines can be used in evaluating building cogeneration applications. Hot water, readily available from reciprocating engines at temperatures of up to 220°F, is acceptable in the following building mechanical systems:

Two-Pipe Fan-Coil Heating or Cooling Systems – These systems usually require heat sources at temperatures as low as 95°F to 140°F.

Four-Pipe Fan-Coil Heating and Cooling Systems – These systems usually require heat sources at temperatures ranging from 160°F to 220°F.

Water Source Heat Pump – These systems are usually capable of operation with heat sources at temperatures as low as 60°F.

How Water Baseboard Heating System – These systems usually require heat sources at temperatures ranging from 180°F to 220°F.

Radiant Panel Hot Water Heating Systems – These systems usually are capable of operation with heat sources at temperatures ranging from 100°F to 140°F.

Service Water Heating Systems – These systems usually require heat sources at temperatures ranging from 100°F to 120°F for new construction and at approximately 140°F in existing buildings.

Single Effect (Single Stage) Absorption Cooling Systems – These systems usually require heat sources at temperatures ranging from 220°F to 250°F.

Double Effect (Two Stage) Absorption Chillers – These systems require steam at pressures of 115 psig or more.

Building mechanical systems such as gas or electric furnaces, electric baseboard heating, incremental air source heat pump heating and cooling systems and electric radiant panel heating systems are generally not compatible with the recovered heat. Other systems such as hot water or steam unit heater systems may be compatible depending on the temperature used in the system and the costs of any necessary modifications.

Finally, it should be noted that buildings with combinations of compatible and incompatible systems may be excellent. Electric heating and a four-pipe heating and cooling system provide a balanced electrical and thermal requirement. Electric hot water heating systems in an all electric building may be quite feasible.

In order to fully utilize recoverable heat, the building mechanical systems must also provide an adequate temperature difference and flow in order to remove the heat available from the cogeneration system. A recirculating service water system may have a temperature difference of 20°F and, therefore, the cogeneration system heat recovery system must be capable of rejecting the heat to a 20°F difference or, alternatively, be capable of increasing flow to provide adequate cooling of the engine.

Adequacy of Existing Systems

Cogeneration is not a cure for most existing mechanical or electrical system problems. A cogeneration system can provide additional heating capacity supplementing inadequate site heating equipment, or it can provide more reliable power or better voltage regulation at weaker portions of the electric utility grid. It cannot however, of itself, remedy an inadequate hot water distribution system, improve the forced air circulation system, or correct deficiencies in the electrical distribution system. Finally, a cogeneration system cannot restore deteriorated mechanical system performance caused by inadequate maintenance.

Fuel Supply

Cogeneration systems are capable of being fired by a variety of fuels. If oil or a solid fuel is used, there must be additional space for both long-term and short-term (daily) storage of fuel. Space will also be required for liquid fuel pumps and filters, and for solid fuel handling systems. This space requirement may not be significant if the cogeneration system is using the same fuel as the site's boiler.

Naturally aspirated, natural gas, reciprocating engines are capable of operation on gas at low, distribution system pressures of a few inches of water. Turbocharged engines may require gas pressures of approximately 30 psig. When this pressure is not available from the utility, it may be necessary to install a pressure booster. Combustion turbines require high pressure gas sometimes as high as 400 psig. This pressure is not usually available from the gas distribution utility and space must be provided for one or two gas compressors. Dual compressors are usually installed to improve system reliability.

Prudent Energy Management or Conservation

Cogeneration is not a substitute for prudent energy management. If an end user is wasteful, the cogeneration system is likely to provide a better economic return because the site will use greater quantities of heat. While a building with a high thermal energy budget and around-the-clock heating requirements is a good cogeneration candidate, an investment in less expensive conservation measures may be more cost efficient.

In investigating cogeneration economics it is important to examine the potential for energy usage reductions due to conservation. A system may be viable based on the thermal loads that exist prior to initiating a conservation program, and may result in poor economic performance due to the rejection of recoverable heat after the site thermal loads have been reduced.

ECONOMIC CONSIDERATIONS

The primary reasons for installing a cogeneration system are economic performance and the reduction of the uncertainty regarding future energy costs. In this section, a simple analytic approach is used to develop a preliminary estimate of cogeneration system economic performance. It uses average utility costs and many simplifying and optimistic assumptions and is, therefore, likely to overstate savings. If a project is marginal based on this type of approach, a more detailed analysis will likely result in a decision that the project is not acceptable.

The worksheet (Table 9-2) and supporting data are included. However, these data should be used only when data for an actual engine are not available. The worksheet is based on an average conventional boiler efficiency of 80%.

The initial step consists of determining whether the site is compatible with cogeneration. Table 9-3 can be used for most building mechanical systems. If HVAC retrofit costs are anticipated they should be estimated and included in the worksheet as Item A.

The second set of questions deals with historic utility costs and rates. The average purchased power costs, Item C, and heating or boiler fuel costs, Item E, can be determined from historic utility bills. The standby charge, Item D, and the cogeneration fuel cost, Item F, may require inquiries to local electric and gas utilities respectively.

The minimum electric billing demand, Item B, and the lowest monthly fuel consumption, Item G, can also be determined from historic utility bills or fuel records.

Table 9-4 provides factors which can be used to develop an initial size estimate for the system. As can be seen, it is necessary to select a prime mover type, and this selection is usually based on the quality of the required heat. In addition, it is necessary to select an approximate system size, and this size can be estimated from the electric billing data or, alternatively, a trial and error procedure can be used. The minimum monthly fuel consumption, which is converted to a heat load using an 80% efficient boiler, is used to estimate the size of a thermally baseloaded cogeneration system. The Heat Recovery Factor is selected from Table 9-4 and, when divided into the site's minimum thermal requirements, result, in the total amount of power, expressed in kilowatt hours, that a thermally baseloaded system will produce per month. In order to estimate the required capacity, in kilowatts, the system's electrical output is divided by the number of operating hours (see Table 9-5) in the month.

The cogeneration system cost can be estimated using Figure 9-5. Mid-range values should be used unless the installation is exceptionally simple and the cogeneration plant is at an industrial facility or located in a building mechanical room. The upper half of the cost range should be used for any atypical installations. The HVAC costs must be included in the total project budget.

The calculation of the system's power output and fuel requirements are based on the assumption that the system is baseloaded and that the engine output does not vary over time. Typical engine heat

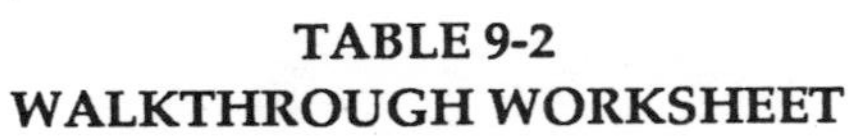

TABLE 9-2
WALKTHROUGH WORKSHEET

SITE: ____________

ELECTRIC UTILITY: ____________

FUEL SUPPLIER: ____________

TECHNICAL FEASIBILITY

HVAC System ____________

Compatibility with Cogeneration (Table 9-3) ____________

A. HVAC retrofit costs (Add to "M") $ ____________

B. Minimum Demand ____________ kW

ECONOMIC FEASIBILITY

C. Average Purchased Power Cost $ ____________ /kWh

D. Standby Charge $ ____________ /kW/Month

E. Heating Fuel Cost $ ____________ /MMBtu

F. Cogeneration System Fuel Cost $ ____________ /MMBtu

G. Lowest Monthly Fuel Requirement ____________ MMBtu/Month

H. Monthly Electric Output
(See Table 9-4 for value of HRF)
.8 x "G" / HRF = ____________ kWh/Month

I. Maximum Installed Capacity
(See Table 9-5 for value of HOURS)
"H" / HOURS = ____________ kW

J. System Size
(Less than or equal to "B" or "I") ____________ kW

K. System Unit Cost
(See Figure 9-5 for COSTS) $ ____________ /kW

L. Cogeneration System Cost
"J" x "K" = $ ____________

M. PROJECT INVESTMENT
"A" / "L" = [$]

N. Annual Electric Output
12 x "J" x HOURS = ____________ kWh/Year

O. Annual Fuel Use
(See Table 9-4 for value for FUEL)
"N" x FUEL = ____________ MMBtu

P. Annual Cogeneration Fuel Cost
"F" x "O" = $ ____________

Q. Annual O & M Cost
(See Table 9-6 for value for OM)
"N" x OM = $ ____________

R. Annual Standby Costs
"D" x "J" x 12 = $ ____________

S. TOTAL OPERATING COSTS
"P" + "Q" + "R" = [$]

T. Recovered Thermal Energy
"N" x HRF = ____________ MMBtu/Year

U. Annual Conventional Fuel Cost $ ____________

V. Saved Heating Fuel Costs
"E" x "T" / .8 = $ ____________

W. Saved Purchased Power Costs
"C" x "N" = $ ____________

X. TOTAL OPERATING COST REDUCTION
"V" + "W" = [$]

Y. ANNUAL SAVINGS
"X" − "S" = [$]

Z. SIMPLE PAYBACK
"M" / "Y" = [ ] Years

TABLE 9-3
BUILDING HVAC

System	Feasibility
Central heating and cooling plant	Good match
Two-pipe heating or cooling system	Good Match
Water source heat pump	Good Match
Radiant hot water panels	Good Match
Gas or electric furnaces – both interior and rooftop	Possible, depending on costs
Hot water baseboard	Possible, depending on costs
Unit heaters – steam	Possible, depending on costs
Unit heaters – hot water	Possible, depending on costs
Air handling or built-up air conditioning system with heating coils and electric cooling	Possible, depending on costs
Electric baseboard	May require modifications
Unitary air source heat pumps	May require modifications
Unit heaters – gas or electric	May require modifications
Unitary electric heating-cooling units	May require modifications
Radiant panel, electric heating	May require modifications

TABLE 9-4
ENGINE CHARACTERISTICS – "FUEL" AND "HRF"
(HHV)

Prime Mover	Fuel (MMBTU/KWH)	HRF (MMBTU/KWH)
Reciprocating Engine		
Less than 100 kW	.0130	.0060
100 kW to 500 kW	.0125	.0055
More than 500 kW	.0110	.0050
Gas Turbine		
Less than 1,000 kW	.0167	.0095
1,000 kW to 5,000 kW	.0133	.0070
5,000 kW to 15,000 kW	.0111	.0050
Larger than 15,000 kW	.0100	.0045

TABLE 9-5
ENGINE OPERATING HOURS - "HOURS"

Prime Mover	Hours (Hours/Mo)
Reciprocating Engine	667
Gas Turbine	717

TABLE 9-6
O & M COSTS - "OM"

Prime Mover	Cost ($/kWh)
Reciprocating Engine	.0150
Gas Turbine	.0050

rates are found in Table 9-4. Typical unit costs are included in Table 9-6 and these can be used to estimate maintenance costs. The values should be increased slightly for reciprocating engines of 75 kilowatts or less and decreased for reciprocating engines of more than 500 kilowatts and combustion turbines of more than 5,000 kilowatts. No labor costs are included in these values.

The total amount of recovered heat can be estimated using the Heat Recovery Factors of Table 9-4, and the value of the cogenerated heat is computed assuming that all the recoverable heat is economically applied. A similar assumption is made for the cogenerated power, which is valued at the average retail rate. As discussed previously, cogenerated power may be valued at a price which is substantially above or below the average retail rate. The procedures outlined in the worksheet are also based on an assumption that no power is exported, or that if power must be sold off site, the exported power is also valued at the average retail rate.

The procedure results in an estimate of the size and cost of the project, the type of prime mover, projected operating cost savings and a simple payback. Because of the many optimistic assumptions and the use of average costs, the payback estimate is likely to be reduced as a result of more comprehensive and detailed analyses. Further action should only be undertaken if the payback is clearly acceptable.

The next step in the analysis, if warranted, usually requires hourly load data, is based on incremental rates and is usually performed using a computerized model.

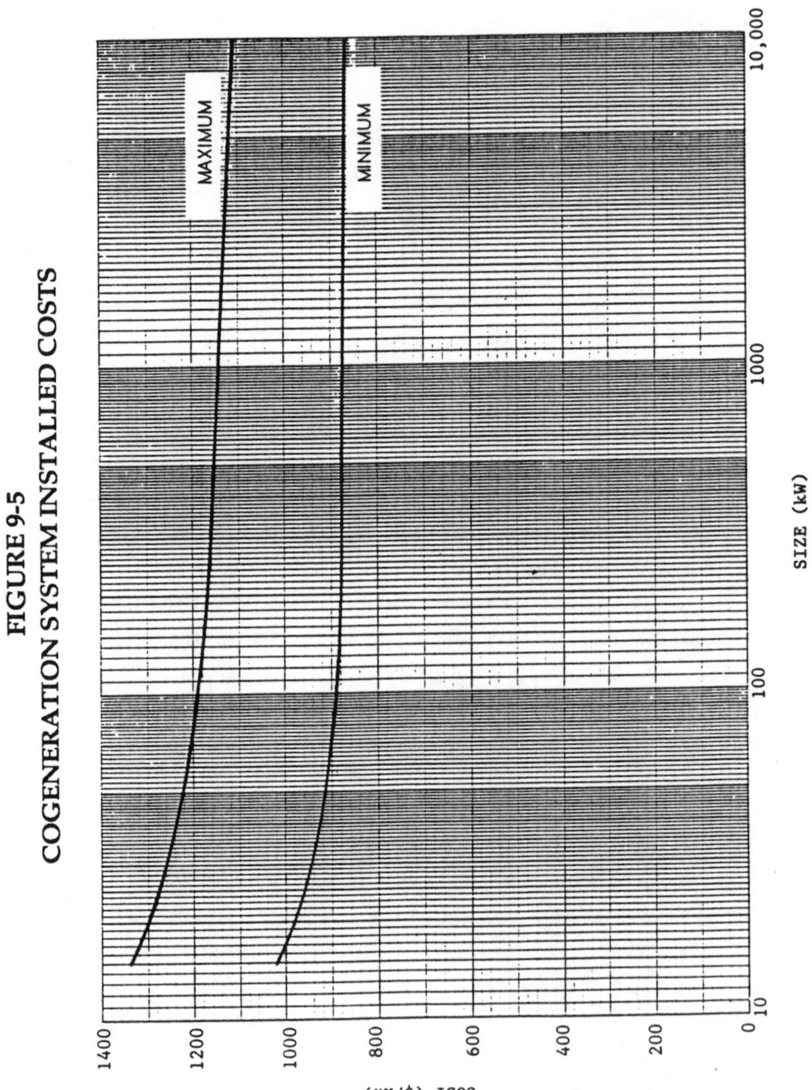

FIGURE 9-5
COGENERATION SYSTEM INSTALLED COSTS

COST ESCALATION

Over the past decade energy costs have shown considerable variability. Natural gas prices first increased dramatically and then leveled off; purchased power cost increases have been less rapid and more regional in nature. With many utilities about to complete or cancel nuclear power plants, there remains the potential for dramatic increases in the rates of selected utilities. Escalation in purchased power and fuel costs is an important consideration in cogeneration system economic performance.

Developing projections for the costs of both purchased power and fuel is perhaps the most difficult task required in the economic analysis. While different projections may be found, they are usually developed by different sources and frequently by sources who use such projections to promote specific fuel use policies. Thus, there are few sources of independent, comprehensive energy cost projections available to the potential cogenerator.

Rather than spend the resources required to develop independent cost projections, it is suggested that the potential cogenerator investigate a number of projections as available from alternative sources including electric utilities, natural gas utilities, fuel oil dealers, state energy offices and state regulatory commissions. In this manner, the cogenerator can examine economic performance over a range of projections and determine the economic risks associated with the cogeneration investment.

A second, more detailed approach is based on developing a model of the utility's cost structure and projecting changes in each major cost component. The major elements in an electric utility's costs are fuel, labor, depreciation, return on rate base, taxes, retained earning and dividends. These data are readily available from both state and federal level utility reports and multi-year trends can be developed. In addition, in those cases where major investments, such as a nuclear power plant, are scheduled for completion and inclusion in the rate base, the effect can be directly determined. Major elements of gas utility costs are similar to those of the electric utility; however, fuel costs will represent well over 50% of the total cost.

This analysis should not be limited to energy costs. Many maintenance and operating contracts include adjustments for inflation, and the potential impact of these adjustments should be examined.

REVIEW

Modelling the technical and economic performance of an internal use cogeneration project requires extensive information on the end user's electrical and thermal requirements. In general, the collection of all of the data that are required will be time consuming and expensive, with no assurance that the project will ultimately be viable. One approach which minimizes risk is a staged analytic procedure, where readily available information is used to determine whether the expense of more detailed data collection is justified.

This chapter provides a simple model which uses monthly energy billing data and a number of equipment cost and performance assumptions to determine the potential for cogeneration feasibility at a specific site. The model is based on the use of average costs and is likely to result in significant errors. Its purpose is limited to use in a decision as to whether or not more detailed analyses and data collection should be undertaken.

The final decision regarding a specific investment for an internal use cogeneration project will likely require a detailed hour-by-hour analysis of system performance for a typical year. In comparison, a cogeneration project where most of the revenues are derived from the sale of power to a utility may require less extensive performance modelling, but more comprehensive analyses of utility and fuel costs.

Chapter 10

PROJECT OWNERSHIP AND FINANCING

Once it has been determined that a cogeneration project is viable, the next issue that must be decided is the financing and ownership structure. Several alternative structures are possible, and the final choice will be dependent on many factors, including the availability and cost of capital and a willingness and ability to accept both technical and financial risks.

As discussed above, in considering whether or not a cogeneration system should be developed, the first question to be answered is whether or not the project produces operating cost savings. Is the cogeneration project itself based on viable economics? Once the viability question has been answered, the decision maker must address two important issues.

- What are the project risks? What factors can effect the project's viability and how can the impact of these factors be quantified? How much financial risk can the end user accept? How much of the operating cost savings or return can be given up to reduce risk?

- What financing structures are available? What are the pros and cons of each structure? Which structure best fits the end user's goals and available resources, including staff capabilities and availability, money and development time?

It should be noted and emphasized that the starting point for any examination of financing is a determination that the project itself is economically viable.

USER OWNERSHIP

For smaller cogeneration systems straightforward end user financing, wherein the energy end user raises all the required capital, assumes responsibility for the cogeneration system's design, construction and operation, and retains all the financial benefits, is among the more common choices. The most notable exception to this trend is small PCS applications where most commercial end users have chosen third-party ownership as a mechanism for reducing risk or day-to-day operating responsibilities.

For self ownership the source of funds may be 100% equity (with the total investment taken from a capital or operating budget) or from traditional debt financing. Among the advantages of this approach are simplicity, familiarity for all parties and relatively low financing charges. Disadvantages include the use of limited cash or debt for activities not primary to the end user's basic business, requirements for staff time and 100% of the project risk. While tax benefits are not generally a key factor in the decision to develop a cogeneration system, in some cases the investor may not be able to fully use all the tax benefits that the cogeneration project produces. Third-party ownership may provide one mechanism for using these benefits.

One key consideration in the evaluation of self ownership is the source of the funds for debt repayment. Will the cogeneration project's operating savings cover both principal and interest? Will it be necessary to use the business's overall assets or revenues to support debt repayment? A project may have negative cash flows during the term of the debt and positive cash flows once that debt is retired. A project that can support itself may be more attractive for self ownership.

Leases provide another source of funds for cogenerators. Under a true lease, the lessor or developer may design, build, own and operate the cogeneration facility which is then leased to a lessee in exchange for rental payments. One advantage of this type of structure is that the tax benefits are retained by the lessor who may be better able to use them. While the lessee retains much of the project's risk, a true lease does provide access to "off balance sheet" capital. In addition, a lease may provide more favorable financing costs than would be possible through direct ownership.

One variation of the true lease is the finance lease/installment purchase contract wherein the developer may design, build and operate the cogeneration system and then sell the project to the end user. This approach reduces the end user's risks for project design, construction and start-up; however, off balance sheet financing is not possible.

Sources of funding for installment purchases are banks, financial institutions and, more recently, governmental agencies through the use of Industrial Development Bonds (IDB), Municipal Bonds and other tax-exempt instruments.

THIRD-PARTY OWNERSHIP

Project financing is also available for those cogeneration projects where the end user is unable to raise the required capital, unwilling to accept the project risks, or unwilling to accept the project rate of return. In these cases, the end user gives up some fraction of the project's benefits in exchange for third-party responsibility and financing.

One type of third-party structure takes the form of an energy service contract (ESC). In this case, the third-party agrees to finance, own and operate the cogeneration facility selling power to the end user or to an electric utility or both, and selling thermal energy to the host end user. The energy end user may receive a share of the savings in the form of a rate for heat and/or power that is some fraction of conventional costs, a fixed rate for heat and/or power that is less than the cost of conventional power, or a fixed annual payment such as a ground lease.

In general, in order to develop project financing, the end user usually must commit to take a specified quantity of electricity and/or heat, if available, or to pay a minimum annual charge. In the case where the cogenerated power is sold to an electric utility, the utility is asked to agree to similar conditions including payments for power that are based on a fixed capacity charge and a variable energy charge.

Most ESCs place the project's technical and economic risk with the third-party developer and some will even specify some guaranteed minimum savings level to the end user. The capital is considered off balance sheet; however, in order to qualify for an ESC structure, the end user must be a viable organization capable of fulfilling the "take-or-pay" obligations. The most significant disadvantage of an ESC is that the end user must give up a significant fraction of the project's economic benefits, and sometimes up to 90% of the operating cost savings, in return for shifting significant risk to a third party.

In the past, many end users elected some form of third-party ownership and the tax status of third-party owners became uncertain. In 1983, legislation was passed which established four criteria that had to be met if a third-party project was to be considered an Energy Service Contract for tax purposes. These criteria are:

- The service recipient cannot operate the cogeneration facility;
- The end user cannot incur any significant financial burden if there is nonperformance by the cogeneration system;
- The end user cannot receive any significant financial benefit if the operating costs of the facility are less than those contemplated in the contract; and,
- The energy customer cannot have an option or an obligation to purchase all or part of the project at a price other than the fair market value.

The effect of satisfying these requirements is to ensure that any third-party ESC will not be treated as a lease or installment sale.

EVALUATING THIRD-PARTY OWNERSHIP

In determining whether third-party ownership is a viable option, it is necessary to answer two questions. First, does this particular project qualify for third-party financing? Will the operating cost savings support the third party's required return on investment and provide the end user with some financial benefit?

To determine if the project qualifies for third-party financing, it is necessary to develop a cash flow model of the third party's economics. This model will require the following assumptions and modifications to the cash flow model that were developed to determine project viability.

1. Assume that the third party will be able to use all tax benefits;

2. Assume that the third party will structure the project so that deductions are used to minimize operating profits and tax consequences;

3. Assume that project financing fees will add from 15% to 35% to the project budget;

4. Assume that operating costs will be increased by up to 5% to cover management expenses;

5. Assume that operating savings are reduced by the end user's share of the savings; and,

6. Assume that the third party will be able to secure financing at rates that are 1% to 3% above the long-term prime rate.

Given these assumptions, the end user can determine whether or not the project can support both the third party and the end user's financial requirements. In addition, it should be recognized that third-party requirements will vary considerably, with some developers seeking simple paybacks of 2 years or less while others will accept paybacks of up to 4-1/2 years.

DEVELOPING A THIRD-PARTY AGREEMENT

The development of a third-party ownership agreement requires professional engineering, legal and financial advice. One prerequisite to any agreement is the end user's complete understanding of its own operating costs, including both the cost of purchased power and thermal energy. Hidden costs such as insurance, staffing, parts, etc. should all be considered.

Many of the ESCs provide power and heat to the site end user with charges based on a discounted published utility tariff or posted fuel prices or on a sharing of project "profits." As was discussed in Chapter 7 it is important to note that cogeneration systems generally displace the least expensive blocks of power, and cogenerated power may be worth less than the host's average cost of conventional power. In an equitable ESC, the site owner should usually pay a rate for cogenerated power that is a fraction of the rate that would have been paid had that power been purchased from the utility.

Other issues that should be addressed in an ESC are:

- Responsibility for providing power and thermal energy during outages of the cogeneration system.

- Responsibility for the costs of supplemental, standby and maintenance power.

- Rights of access to the site.

- The host's rights to take over plant operations under specified default conditions by the third party.

- Termination costs should either the host or the third party terminate the agreement prior to the scheduled term.

- Disposition of the system upon expiration of the agreement with cost responsibilities.

- Responsibility for maintenance of the conventional systems and the building environment.

- Identification of the cost responsibility for any ancillary loads; e.g., engine room fans, fuel pumps, etc., that result from the installation and operation of the cogeneration system.

- Identification of modifications to the existing building and processes, including fuel supply and storage and electrical switchgear that are required by the installation of the cogeneration system, and assignment of costs.

- Detailed descriptions of the measurements that are to be used for billing power and thermal energy, the computation of the appropriate bills, any late charges and mechanisms for the resolution of billing disputes.

REVIEW

In today's marketplace, cogeneration projects that produce simple paybacks of less than 4 years, or an Internal Rate of Return of over 20%, can generally be "project financed." That is, the cogeneration project itself can be the sole basis and security for the plant investment. Various off-balance sheet financing alternatives are available which will allow the energy end user to retain ownership of the system.

In many cases a combination of technical or financial factors may make third-party ownership more attractive. For example, the small commercial or institutional end user may not have the expertise to develop and operate a cogeneration system. Alternatively, the end user may not wish to engage in an activity, power generation, which is far removed from their day-to-day operations, or wish to undertake the risk associated with the project. In some cases, the end user may not wish to take on any incremental staffing requirements. In other cases, public and other tax-exempt facilities may not be able to fully use tax benefits. In any event, there are a number of circumstances under which independent, third-party development, ownership and operation are desirable. In these cases, the energy end user should prepare an *independent and comprehensive* evaluation of the site's cogeneration potential as a basis for negotiating a third-party contract.

Chapter 11

TYPICAL PROJECT ANALYSES

Cogeneration system performance and economic viability are site specific, dependent both on the facility's energy usage characteristics and on utility rates and fuel costs. This chapter presents analyses of three different type applications; an industrial facility, a major health care facility and a small commercial facility. The energy characteristics, utility rates and cogeneration systems evaluated for each were selected to illustrate a variety of applications. In addition, two economic analyses were performed for each facility. The first consisted of the simpler "walkthrough economic analysis" based on average costs and requiring a minimum of site data. The worksheets and supporting data for this approach are included at the end of this chapter. The second analysis is the more comprehensive incremental analysis used in the preliminary or screening economic modelling.

While fictitious names are used in these analyses, the data are based on actual facilities.

INDUSTRIAL FACILITY

The first system being considered consists of a chemical plant, CHEMCO, located in the Midwest. The facility has a boiler plant which produces 125 psig steam which is used for space and water heating and in different processes. The facility is sited in an area where there are adequate transmission and distribution facilities and, therefore, gas is available 12 months a year.

Conventional electric and fuel requirements and costs are shown in Table 11-1 and illustrated by Figures 11-1 through 11-4.

Walkthrough Analysis

The initial step in the development process is to "qualify" the site; that is, to determine whether or not the system has the potential to be an economically viable cogeneration application. In this example, the procedures outlined in Chapter 9 were used for the preliminary analysis.

Technical Compatibility

The site has a central boiler plant which produces 125 psig steam for use in various processes. The need for higher pressure steam suggests the use of a combustion turbine in the initial analysis. Since the site currently has natural gas and fuel oil boilers, with oil storage tanks, and since the steam available from the cogeneration system HRSG could be fed to the boiler plant header, the site is technically compatible with cogeneration.

Economic Potential

The analysis of cogeneration economics is illustrated in Table 11-2, and each entry is reviewed below.

Because the conventional system is a boiler plant, mechanical connection of the cogeneration system steam header with the plant header poses no unique or special problems and these costs were considered to be within the range of a typical installation. No unique retrofit costs were anticipated and Item "A," *HVAC retrofit costs,* was assumed to be zero. Table 11-1 was used to determine the value for Item "B," the *minimum billing demand,* which is 8,398 kilowatts.

The *average purchased power cost,* Item "C," was computed directly from the site's actual bills for a 12-month period. These costs have been summarized in Table 11-1, and the average cost is $.059 per kilowatt hour. It was necessary to contact the local electric utility to obtain a copy of the utility's partial service rates in order to determine *standby charges,* Item "D." In this case, the reservation charge was determined to be $4 per kilowatt per month.

The average *heating fuel cost,* Item "E," was determined by totaling fuel usage and costs for a 12-month period. This facility only uses natural gas, and the average annual cost is $2.80 per million Btu (MMBtu). It was also necessary to contact the local gas utility to determine if the *cogeneration gas cost,* Item "F," differs from the cost of conventional boiler gas. In this case, the cogeneration gas load, which is a non-seasonal application, resulted in a lower annual rate of $2.60 per MMBtu for all gas used at the site. Where fuel cost data are not readily available, as might be the case for direct purchases of natural gas, a conservative approach would be to set the price of cogeneration fuel equal to the price of heating fuel.

Development of the above information was required prior to the actual system analysis. The first step in that analysis was to determine the cogeneration system size, and the approach used in this worksheet is modelled on a thermally baseloaded system. Item "G," the *lowest monthly displaceable fuel requirement,* is used to size the cogeneration plant and the required data are found on Table 11-1.

The minimum monthly usage is 70,400 MMBtu, and the next step in the analysis was to determine how many kilowatt hours of cogenerated power must be produced to provide an adequate amount of recoverable steam. This computation was performed as Item "H," where the *monthly electrical output* was computed. Table 9-4 shows typical fuel requirements and *heat recovery factors* (HRF) for different prime movers. As discussed above, a combustion turbine was chosen and, in order to select the appropriate HRF, it was necessary to determine the approximate size of the cogeneration system. Again, Table 11-1 was useful. The minimum billing demand is 8,398 kilowatts and therefore, it was assumed that the cogeneration system would be between 5,000 kilowatts and 15,000 kilowatts. As will be shown later, this approach to sizing can result in excess capacity.

The cogeneration system's minimum monthly thermal output was computed by multiplying the minimum monthly fuel requirement by 80%, the efficiency assumed for the boiler. The heat requirement was divided by the HRF to determine the amount of power that must be produced by a thermally baseloaded cogeneration system. In this example, the system would be required to produce 11.26 million kilowatt hours.

In order to estimate the size of the cogeneration system, it was then necessary to determine the engine's *monthly availability*. Table 9-5 was used for this purpose and, for a gas turbine, the appropriate availability is 717 hours per month. The total monthly output was then divided by the hours of availability to compute the *maximum installed capacity*, Item "I." This value, 15,700 kilowatts, was compared to the minimum peak demand of 8,398 kilowatts, Item "B," and the lesser of the two values was then used as the *system size*, Item "J." The selection of the lower value results in a higher probability that the system will be able to use all cogenerated power on-site with no need for external sales.

Once the system size was estimated, Figure 9-5 was used to estimate the cogeneration *system unit cost*, Item "K." Because this facility was judged to be average or typical, a mid value of $950 per kilowatt was used. The *cogeneration system cost*, Item "L," the product of the unit cost and the installed capacity, was estimated to be $8.0 million. Because there are no HVAC retrofit costs, the *project investment*, Item "M," was also equal to $8.0 million.

The above calculations resulted in an initial estimate of the cogeneration system size and budget. The next step was the calculation of the operating costs and the resulting energy cost savings.

The system's *annual electric output*, Item "N," was computed based on the assumption that the turbine operates at a constant output for 12 months. It was the product of the installed capacity, the monthly hours of availability and the number of months. In this case, the total annual output was computed to be 72.2 million kilowatt hours. The cogeneration system's *annual fuel use*, Item "O," was computed using the fuel requirements of Table 9-4 (.0111 MMBtu per kilowatt hour) and the electric output. The *annual fuel cost*, Item "P," was the product of the fuel use and the average annual cost, or $2,085,000. *Annual O & M costs*, Item "Q," were estimated using the appropriate value from Table 9-6 ($.005 per kilowatt hour) and the system's electrical output, for a total of $361,000 per year. The *annual standby cost*, Item "R," was the product of the monthly reservation charge and the installed capacity, and totaled $403,000 per year. The *total operating cost*, Item "S," the sum of these components, was projected to be $2,849,000 per year.

The next step was the calculation of the reduction in conventional operating costs. The total *recovered thermal energy,* Item "T," was determined by multiplying the total system electrical output by the HRF, and was projected to be 361,000 MMBtu. The *annual conventional fuel cost,* Item "U," was taken from Table 11-1, and was projected to be $3.84 million. The *saved heating fuel costs,* Item "V," was the minimum of the actual annual fuel costs, Item "U," and the value of recovered heat. This latter factor was computed by dividing the recovered thermal energy by the boiler efficiency and multiplying by the conventional fuel cost. The reduction in heating fuel costs was estimated to total $1,264,000 annually. The *saved purchased power costs,* Item "W," was the product of the system's electrical output and the average value of purchased power and was computed to be $4,263,000 annually. It should be noted that it was assumed that the excess power could be sold to another end user or the utility at a price equal to the retail rate. The *total operating cost reduction,* Item "X," was computed as the sum of the value of the recovered heat and the displaced power and was projected to be $5,527,000 per year.

The *annual saving,* Item "Y," was computed as the difference between the operating cost reduction and the cogeneration system operating costs and was projected to be $2,678,000. The *simple payback,* Item "Z," was computed as the project budget divided by the annual savings, or 3.0 years. With a simple payback of 3 years, this project was considered as having the potential for economic viability.

This approach can also be applied to the evaluation of alternative systems, and Table 11-3 illustrates the analysis of a cogeneration system consisting of two turbines of 4,200 kilowatts each. In this case, because the turbine falls between 1,000 kilowatts and 5,000 kilowatts, an HRF value of .0070 MMBtu per kilowatt hour and a fuel value of .0133 MMBtu per kilowatt hour were used. The turbine unit cost was $960 per kilowatt for a required investment of $8.06 million and produced a simple payback of 2.9 years.

Engineering and Economic Screening Analysis

Based on the results of the walkthrough analysis, a decision was made to conduct a more comprehensive engineering and economic screening analysis. That analysis is described in this chapter, noting the differences between it and the walkthrough analysis. The screening analysis consists of the following five steps.

1. Projection of the facility's *energy requirements* both with and without the cogeneration system in place.

2. Estimation of *supplemental costs* of purchased power and fuel.

3. Estimation of cogeneration *system operating costs.*

4. Preparation of life cycle *cash flows.*

5. *Sensitivity analyses* to identify key parameters and risks.

As with the walkthrough analysis, the screening analysis is reviewed in detail.

Energy Analysis

One year of "baseline" energy data were required for the analysis of cogeneration economics at CHEMCO. The various sources of information were discussed in Chapter 9.

The local electric utility provides power to CHEMCO at two services which are *conjunctively billed*. That is, the peak coincident demand is determined and is used as the billing demand, rather than the sum of the individual peak demands.

The maximum demands, energy usage and load factor for a 12-month period are shown in Figures 11-1 through 11-3. There was little month-to-month variation in usage and both demand and usage peaked during the winter months. Figure 11-3 presents monthly load factors which are defined as the total kilowatt hours used during the month divided by the maximum measured demand. The high load factor and the limited range in the seasonal variations in demand and usage are characteristic of good cogeneration applications.

Monthly billing data such as those shown in Figures 11-1 to 11-3, while readily available, were of limited use in analyzing the performance of a potential cogeneration system. Hourly electrical and thermal data were required to more realistically size the system and to determine how much of the site's energy requirements could be satisfied by various size cogeneration systems. The local electric utility provided electrical load data at 30-minute intervals, and these data were used for the more comprehensive analysis.

Figure 11-5 presents the total requirement for each hour for typical winter weekdays at CHEMCO. There was little variation in the hourly demands with the actual hour-to-hour variations attributed to slight variations in site occupancy. Figure 11-6 illustrates the demand profiles for Saturday and Sunday as compared to the average weekday demands. The weekend usage was within the range of weekday usage variations.

Detailed profile data for a summer month and spring/fall, or transition months were also required. The seasonality of the CHEMCO diurnal demand variations is shown by Figure 11-7 where average weekday demands are shown for each of the three types of months, and Figure 11-8, where weekend demands are shown.

Figure 11-9 is another way of presenting the detailed demand data. In this case, the hourly demands for a winter time period were sorted into intervals or bins, and the frequency with which each interval occurred is indicated in the demand histogram. These demand data were then represented as a load-duration curve as shown in Figure 11-10. This curve shows the percentage of time that specific demand levels were exceeded. Figure 11-11 presents load-duration curves for all three types of months considered in this analysis.

Finally, the profile data illustrated above were used to construct a curve showing the relationship between the amount of on-site capacity and the on-site load that that capacity can supply (see Figure 11-12). In this case, a cogeneration system sized at 50% of the peak demand is capable of supplying about 63% of the plant's monthly winter requirements. Increasing the size of the cogeneration system to 100% of the annual peak would only increase the amount of site load which could be supplied by 37%.

Hourly demand data were used to develop the load capacity curves for each season as shown in Figure 11-13.

The central steam plant consists of three boilers, each of which has the capability to fire on either natural gas or fuel oil. Steam is required 12 months a year and is used primarily for process, space conditioning and service water.

Figure 11-4 presented monthly fuel usage as taken from gas utility billings. The seasonality of the thermal requirements is clearly evident with winter consumption significantly exceeding that of the spring, summer and fall months. Plant logs and an inventory of plant equipment were used to estimate how much of the gas was used directly for process and, therefore, not replaceable with recovered heat.

As was the case with electrical data, it was necessary to examine daily and hourly variations in steam requirements in order to prepare a more complete and comprehensive analysis of cogeneration options. In this analysis, thermal data consisting of boiler charts maintained by the staff were used.

Hourly data for a winter period, as shown in Figure 11-14, were used to develop a steam demand histogram, a steam load duration curve, and a steam capacity vs. load curve (Figure 11-15). The thermal capacity vs. load curve was then used in a manner identical to the electrical capacity vs. load curve. A summer and transition months capacity vs. load curve were also developed.

As was the case with the walkthrough analysis, a combustion turbine was chosen as the basis for a cogeneration analysis at CHEMCO. The turbine can provide all the recoverable heat as higher pressure steam and has high availability.

Sizing of the turbine was initially based on both the thermal and electrical requirements. With a minimal site electrical requirement of approximately 6,400 kilowatts during the winter, two turbines each rated at approximately 3,500 kilowatts to 4,000 kilowatts, depending on site conditions, were chosen.

The selected turbine had a continuous duty, ISO electrical rating of 3,925 kilowatts when burning a gaseous fuel. The turbine was derated by 79 kilowatts to compensate for the site altitude, and 134 kilowatts to compensate for inlet and exhaust pressure losses. Actual pressure loss calculations must be repeated during detailed design. These adjustments resulted in a site rating of 3,712 kilowatts for the turbine at 59°F.

This gas turbine requires 250 psig natural gas and since the local distribution company was only capable of providing 150 psig gas at this location, fuel gas compressors must be included in the cogeneration plant. As a result, the turbine's site maximum of 3,712 kilowatts of power was further derated by 35 kilowatts to compensate for the compressor's parasitic energy requirements. The turbine's thermal output was not derated because the amount of recoverable heat only depends on the total turbine power output.

The turbine has a fuel requirement of 11,960 Btus of natural gas per kilowatt hour of output. However, the heat rate was reported in the fuel's Lower Heating Value (LHV), the typical basis for engine specifications, while billings are typically based on Higher Heating Value (HHV) as measured in million Btu or therms. The heat rate was increased to 13,290 Btu per kilowatt hour (HHV) for natural gas.

If all the heat that is recoverable from the exhaust is used, the turbine will provide approximately 4,600 Btus of heat per kilowatt hour of turbine output, with that heat available as steam.

The turbine's output was also modified to compensate for climatic conditions. The on-peak electrical capacity of the turbine, as measured in kilowatts, was derated based on the *average maximum* local daily temperature; the average output, measured in kilowatt hours, was derated based on the local *average* daily temperature.

Table 11-4 provides a summary of the site's projected monthly energy requirements. As shown there, any of the site's thermal requirements were assumed to be met with recovered heat from a heat recovery steam generator (HRSG). If there was an additional thermal requirements it was assumed that that load would be met with steam from the same HRSG, with the output increased by burning fuel in an exhaust duct burner. Finally, it was assumed that any remaining steam need was met by firing the existing conventional boilers. The cogeneration system's duct burner capacity was set at 7,700 Btu per kilowatt hour of turbine output, with an efficiency of 90%. The average conventional boiler efficiency of 80% was used.

It should be noted that although supplemental firing will be required in all months, the cogeneration system will not be able to use all recoverable heat and will be required to reject some heat. The reason for the need, both to fire supplemental burners and to reject heat in the same month, is that the turbine itself will not be able to meet the plant peak steam load while exceeding the site minimum steam load.

Finally, it was assumed that the turbine generator set would have an availability of 8,500 hours per year, and that all planned turbine outages would be scheduled during off-peak hours. The amount of power produced in each time period was based on the number of hours of turbine availability and the turbine average capacity during those hours.

Cost Analysis

Under current conditions the site's seasonal requirements for boiler fuel would be met with natural gas priced at $2.80 per MMBtu. This rate was estimated based on the gas utility's actual tariff. No. 2 fuel oil for the boilers was projected to cost $.70 per gallon.

Table 11-5 shows the projected monthly costs of supplemental power and fuel with the cogeneration system in place. The electric utility rate consisted of a demand charge of $12 per kilowatt for the first 100 kilowatts plus $10 per kilowatt for all excess demand. The energy charge was based on $.060 per kilowatt hour for the first 50,000 kilowatt hours then $.045 per kilowatt hour for the next 100,000 kilowatt hours then $.040 per kilowatt hour for the next energy block (equal to 300 hours use of demand) plus a charge of $.0375 per kilowatt hour for all additional kilowatt hours. The utility's fuel adjustment charge was $.005 per kilowatt hour. The same rate (see Figure 11-16) was used for all utility supplied power (i.e., the power costs without cogeneration and for supplemental power costs with cogeneration).

The two turbines would be capable of supplying most of the site's requirements; however, because of the low load factor of the resulting supplemental load, the average cost of utility supplied power was computed to be $.094 per kilowatt hour as compared to $.059 for conventional power.

It was assumed that standby service would also be taken from the local utility based on the provisions of their Auxiliary Service Rate. Under this rate CHEMCO would be billed $4.00 per kilowatt of standby capacity per month as a "reservation charge" for the availability of standby service. This charge of $29,516 per month (see Table 11-6) would be paid regardless of whether standby is used or not. If standby is used, the facility will then be billed for the use of standby. In this analysis, the standby was computed based on the conventional demand charge, less the reservation fee. Based on two assumed outages per year the cost for the use of standby power was estimated at $88,448.

Based on rate negotiations with the local gas supplier, it was determined that cogeneration gas could be available for $2.60 per million Btu (MMBtu). The total cost of supplemental heating gas was projected to be $2,491,000 per year. Fuel costs for the turbine (see Table 11-6) were also computed based on the above natural gas rate and were projected to total $3,186,000 annually.

Based on quotations from turbine vendors who would provide a maintenance contract, the general maintenance for the turbine was priced at a rate of $.005 per kilowatt hour with no incremental labor costs. It was also assumed that there would be no incremental maintenance costs for the heat recovery boiler, since this boiler and the duct burner would displace the use of the conventional boilers. This cost was estimated at $324,000 for the current year.

It is anticipated that there will be minor sales of power to the utility, and these sales were valued at the average avoided cost rate of $.020 per kilowatt hour for total annual revenues of $15,000 per year. Owning and administrative costs were estimated as a percentage of the equipment cost and were projected at $29,500 per year.

Table 11-7 is a summary of projected utility costs with and without the cogeneration system. Those projections are based on the assumption that the plant is in place and operating today. Total conventional costs would be $7.85 million, while costs with cogeneration would total $6.80 million with $14,600 in power sales revenue. The net reduction in operating costs would be $1.07 million.

Project Budget

CHEMCO has space available for the cogeneration system at the site of the existing boiler plant and the cogeneration plant would be conveniently sited with regard to the existing steam distribution system, and within a short distance of the electrical service.

The cost of the turbine generator set, including switchgear, controls, interconnection, fuel compressor and minor building modifications was estimated to be $2,950,000. Installation costs were projected at $2,450,000 for a total budget of $5,400,000. Fees were estimated at 12% for professional services ($648,000) and 10% ($540,000) was included for contingencies. The preliminary cost estimates for the system totaled $6,588,000.

Cash Flow Analysis

Under current rates and costs, the baseloaded cogeneration system could produce annual savings of $1.07 million. However, it was anticipated that if CHEMCO moved forward with the project, 1992 would be the first full year during which the system would be operational. The next step in the analysis was to adjust the projected savings for cost changes during the development period.

Table 11-8 provides detailed cash flow projections on a year-by-year basis. It is based on a general rate of inflation of 5% per year, with natural gas increasing at a rate of 2% above inflation and purchased power at 1% above inflation.

The operating cost savings for 1992 were projected to increase to $1,436,000.

Economic Analysis

The simple question to be answered is whether or not the cost savings that can result from the cogeneration plant justify the capital investment. Over the 20-year life, the project could produce total savings of almost $27.7 million. Deducting project construction and development costs resulted in a net saving of $21.1 million.

If the plant were put into operation in 1992 with savings of $1,436,000 per year, the simple payback period would be 4.6 years. However, payback is not a precise measure of economic viability in that it does not consider relative changes in costs or the time value of money. Based on a 100% equity investment, the project's pretax Internal Rate of Return (IRR) was projected to be 24.2% over its 20-year economic life. Based on a discount rate of 13%, the NPV of the project was computed to be $3.95 million.

Sensitivity Analysis

In this section, the relationship between the projected cost reductions and different factors is examined. Internal Rate of Return (IRR) was chosen as the economic measure to be used as the basis for comparisons.

Figure 11-17 illustrates the relationship between the IRR and *relative changes* in power and fuel costs with all other assumptions unchanged. The alternative rates of increase used in this analysis were only applicable to the first 5 years, and it was assumed that purchased power rates would change relative to fuel costs. At one extreme, if power costs were to increase at a rate of 5% per year less than fuel costs, the cogeneration system would produce an IRR of approximately 10.6%. At the other extreme, if power prices increase 5% faster than fuel the IRR would increase to 31.6%.

The effect of absolute changes in gas prices, both for boiler fuel and cogeneration use were also considered. The baseline analysis was based on a conventional gas price of $2.80 per MMBtu with a cogeneration rate of $2.60 per MMBtu. Figure 11-18 shows the impact of alternative cogeneration gas rates.

Figure 11-19 illustrates the impact of changes in project budget on the IRR of the project. It should be pointed out that a 100% overrun is highly unlikely. The single most expensive piece of equipment are the turbine-generator sets at approximately $2.95 million. This price is not likely to change significantly, and certainly not by 100%. Additionally, the project budget includes a contingency of $540,000 which would partially offset any installation cost increases.

Conclusions

Two approaches to the analysis of cogeneration system performance were documented. The first used average costs and made many optimistic assumptions regarding the ability to use cogenerated power and steam, and resulted in the conclusion that an 8,400-kilowatt gas turbine cogeneration system would produce a simple payback of 2.9 years. The more comprehensive analysis, based on incremental power and fuel costs, the use of load-duration curves to estimate the amount of power and recovered heat that can be used, and a fuller listing of operating costs, resulted in a simple payback of 4.6 years and an IRR of 24.2%.

The next step in the development process would require the use of hour-by-hour performance modelling, part-load equipment performance data, and detailed engineering as a basis for costing. This more detailed engineering would further refine the economic analysis, and is required prior to any decision to commit to construction.

HEALTH CARE FACILITY

The second example considered here consists of a full-service hospital, located in an urban area. The facility is a multibuilding, medical campus, purchasing all power from the local electric utility through two services at a primary voltage. Steam is raised to 100 psig in three dual fuel (natural gas or oil) boilers, each rated at 40,000 pounds of steam per hour. Chilled water is produced using various combinations of four 850-ton electric and absorption chillers.

The hospital's utility requirements, as taken from bills, are summarized in Table 11-9 and illustrated in Figures 11-20 through 11-23.

Walkthrough Analysis

The hospital currently produces steam from a central plant; therefore, interconnecting the existing mechanical systems with a cogeneration system should present no problem. While power is delivered through two services, they are both at a single point and electrical interconnection should be straightforward. The site is rather crowded, and the actual installation of the plant may incur additional costs.

Table 11-10 presents the simple calculation of cogeneration system size, operating cost savings and payback. This site is interesting in that the use of absorption chillers results in summer gas usage that far exceeds winter loads. In fact, the two months with the lowest gas use are May and November. The electric usage also peaks in summer, based in part on the use of electric chillers. Metered data indicate that the boilers operate at an annual efficiency of approximately 84%.

A 2,270-kilowatt combustion turbine was considered. The choice of a turbine was dictated by the requirement for high pressure steam, and the size by the electric billing demands. Since the thermal load will support a 2,434-kilowatt system, which is greater than the minimum peak electrical load, even at the reduced size this plant will produce more power than can be used at the hospital.

Based on this analysis, the turbine would produce annual savings of $852,000 and require an investment of $1,910,000 for a simple payback of 2.2 years. Based on the results of this analysis, a decision was made to proceed with the next level of evaluation.

Engineering and Economic Screening Analysis

The engineering analysis required the development of detailed energy use data and refinements of budget estimates as described below.

Energy Analysis

A baseline year of data were required as the starting point of the cogeneration analysis, and the needed information was developed from electric and gas utility billing records, boiler plant logs and steam charts.

The maximum demands for the year are shown in Figures 11-20. & 11-21 presents monthly energy usage allocated to both on- and off-peak periods; Figure 11-22 presents monthly load factors. As can be seen, the hospital demands were relatively constant at about 2,300 kilowatts during winter months, increasing to over 3,000 kilowatts during the cooling season. Power usage shows the same pattern with winter requirements of approximately 1,400,000 kilo-watt hours per month and summer usage increasing to almost 1,900,000 kilowatt hours per month. Load factor was relatively high, averaging over 570 hours use of demand per month, with a range of 520 hours during the spring to slightly more than 650 hours during the winter.

Hourly electrical demand data were provided by the electric utility. Figure 11-24 presents the requirement for each hour for typical winter weekdays at the hospital. The demand pattern is very repetitive being relatively flat at 1,700 to 1,800 kilowatts during the early morning hours, starting at 6:00 a.m. and flattening out at 2,200 kilowatts at 9:00 a.m. During the afternoon, the demands drop reaching a level of 1,800 kilowatts at midnight. Figure 11-25 presents the average, maximum and minimum demand for weekdays for a 2-week period. Figure 11-26 illustrates the demand profiles for Saturday and Sunday as compared to the average weekday demands. The difference between weekday and weekend requirements is not as significant as is found in most buildings.

The seasonality of the demand variations is shown by Figure 11-27 where average weekday demands are shown for each of the three types of months. Figure 11-28 presents weekend demands. The effect of building occupancy is clear (Figure 11-27), while the cooling load is evidenced in both Figures 11-27 and 11-28.

Figure 11-29 presents the demand histogram, Figure 11-30 the load duration curve and Figure 11-31 the capacity-load curve for a winter month. Load capacity curves were developed for each season and are shown in Figure 11-32.

The central steam plant consists of three boilers, each of which has the capacity to fire either natural gas or fuel oil. Steam is required 12 months a year and is used primarily for space conditioning, both heating and cooling, service water and sterilizers.

Figure 11-23 presented monthly fuel usage as taken from daily boiler logs. While all the boilers are capable of firing on oil, only natural gas was used.

Daily and hourly data for the winter period, as shown in Figures 11-33 through 11-35, were used to develop a steam load duration curve (Figure 11-36) and a steam capacity vs. load curve (Figure 11-37). A summer and transition month capacity vs. load curve as shown in Figure 11-38 were also developed.

Sizing of the turbine was initially based on the hospital's thermal requirements. With a minimum steam requirement of approximately 12,200 pounds per hour during spring and fall months and with a typical heat recovery factor of 6,000 to 7,000 Btus per kilowatt hour for smaller combustion turbines, it was determined that the largest cogeneration system that could be sited without excessive heat rejection was approximately 1,000 to 1,500 kilowatts. Because of the limited availability of combustion turbines in this size range, a single 1,080-kilowatt (ISO) turbine was examined.

The turbine has a continuous duty, ISO electrical rating (including generator losses) of approximately 1,080 kilowatts when operated on gaseous fuels. This turbine can be operated at higher output levels in a standby mode; however, this operating mode is limited to short periods of time.

The turbine was derated by 20 kilowatts to compensate for altitude and 40 kilowatts to compensate for inlet (3 inches of loss) and exhaust (7 inches of loss) pressure losses. Pressure loss calculations must be revised during final design based on actual inlet filters, ducting, boiler selection and stack design. These capacity adjustments resulted in a site rating of 1,020 kilowatts for the turbine at 59°F. The local gas utility could deliver natural gas at 100 psig while the turbine would require 150 psig gas and, therefore, an additional loss of 10 kilowatts was required for the compressor. Finally, the turbine's output was adjusted based on both the average ambient temperature and the average maximum ambient temperature.

The turbine's fuel requirement of 14,785 Btus (LHV) of natural gas per kilowatt hour was increased to 16,430 Btu per kilowatt hour based on the HHV. If all the heat that is recoverable from the exhaust is used, the turbine can provide approximately 7,770 Btus per kilowatt hour of turbine output. Assuming a nominal 1,000 Btus per pound, the thermal output is equivalent to 7.77 pounds of 100 psig steam per kilowatt hour of engine output.

Table 11-11 provides a summary of the hospital's projected monthly energy requirements based on the operation of the cogeneration system. The system's duct burner capacity was set at 8.2 MMBtu per hour (equivalent to 8,200 Btus per kilowatt hour of turbine output), with an efficiency of 90%. As can be seen in Table 11-11, the cogeneration system, including the duct burner, would be capable of satisfying the hospital's steam requirements for 4 months per year, thus allowing the shutdown of the back-up boilers.

Cost Analysis

Supplemental power costs, which are defined as the cost of power which is purchased from the utility on a regular basis, and supplemental fuel costs, which is the cost of fuel required for the conventional boiler and the duct burner, were computed and are detailed in Table 11-12. The incremental costs of operating the cogeneration system are detailed in Table 11-13. All purchases are subject to a 5% sales tax.

The cogeneration system would only supply a portion of the hospital's electrical requirements, and the local utility would still supply approximately 50% of the load. The cost of these power purchases were estimated to be $701,000 per year. The average rate for purchased power increased somewhat from $.070 per kilowatt hour, based on an average load factor of 574 hours use per month for the conventional case, to approximately $.077 per kilowatt hour for supplemental power.

It was assumed that standby service would also be taken from the local utility, with a "reservation charge" of $2.75 per kilowatt per month plus a $210 customer charge. In this case, it was found that the reservation charge is a minimum bill. When utility supplied standby is actually used, there will be a charge of $1.00 per kilowatt per day of use. Based on an analysis of the standby rate and projected supplemental power purchases, it was determined that there would be no additional charges for the use of standby.

The local gas utility provided burner tip natural gas costs for fuel use both in the conventional boilers and in the cogeneration system, including the duct burner. The cost for boiler fuel would be $3.10 per MMBtu plus a $1,500 per month customer charge. The cogeneration gas would be billed at $2.75 per MMBtu; however, there would be no additional customer charge. All costs would then be subject to the 5% sales tax. The total cost of supplemental fuel was computed to be $545,000 and resulted in an average rate of $3.37 per MMBtu. Fuel costs for the turbine were projected at $426,000 per year.

The general maintenance for the engine was priced at a rate of $.005 per kilowatt hour with no incremental labor costs for a charge of $44,500 per year. Owning and administrative costs were estimated as a percentage of the equipment cost and were projected at $9,900 per year.

Table 11-14 is a summary of projected 1990 utility costs both with and without the cogeneration system. It is based on the assumption that the plant is in place and operating today. Total conventional costs would be almost $2,060,000 while the cogeneration system would result in operating costs of $1,764,000 for a net reduction in utility operating costs of over $293,000 per year.

Project Budget

The cost of the turbine generator was estimated at $590,000 for the complete factory assembled package. The HRSG, rated at 16,000 pounds per hour, was estimated to cost $250,000; switchgear was estimated at $90,000 and the fuel compressor at $60,000, for a total of $990,000. An additional allowance of $25,000 was made for required building construction which was then added to installation costs which were estimated at $250,000. Fees were estimated at 12% for professional services and 10% was included for contingencies. The budget for the baseline system was projected to total $1,540,000.

Cash Flow Analysis

Under current rates and costs, the baseloaded cogeneration system could produce annual savings of $293,000. However, it is anticipated that if the hospital moves forward with the project, 1992 would be the first full year during which the system would be operational. Table 11-15 provides cash flow projections on a year-by-year basis. In 1992, the operating cost savings were projected to increase to $330,000.

Economic Analysis

The simple payback based on the 1992 savings was computed to be 4.7 years. The turbine based cogeneration system would produce a pretax IRR of 24.6%, and based on a 20-year life and a discount rate of 9%, the NPV was projected to be $2.65 million. The undiscounted savings were projected to total over $9.7 million.

Sensitivity Analyses

Figures 11-39 and 11-40 show the relationship between the project capital budget and the resulting IRR and NPV respectively. Figures 11-41 and 11-42 illustrate the relationship between the IRR and the NPV and *relative changes* in purchased power and fuel costs.

Conclusion

The project described above would likely meet the investment requirements for most health care facilities. A second analysis was conducted based on two turbines, and the resulting IRR was estimated to be less than 22%. Thus, the incremental investment in the second turbine, while producing an incremental IRR of approximately 20%, could also be viable.

As was the case with the industrial facility, the use of incremental costs and hourly profiles resulted in a payback period which was somewhat lengthier than the payback which was computed based on average costs and the assumption that all heat and power can be used on-site.

COMMERCIAL FACILITY

The development of packaged cogeneration systems has opened up the smaller industrial and commercial market to cogeneration. This example is based on a commercial laundry which operates five days per week. The laundry produces steam at 60 psig; however, only a small portion of that steam is required for process. Most of the steam is used in the mechanical room where it generates hot water. In addition, the laundry also uses considerable quantities of natural gas directly in dryers.

Conventional electric and gas requirements and costs are shown in Table 11-16 and Figures 11-43 and 11-44.

Walkthrough Analysis

The industrial facility is well suited for cogeneration, with ample space in the existing mechanical room. At present, 60 psig steam is produced using either of two natural gas fired steam generators. Most of the steam is used to produce hot water in either of the two hot water generators. Natural gas is already available at the site.

The analysis of cogeneration economics is shown in Table 11-17. The first important difference between the laundry and the previous examples is the lack of demand data. The laundry is served on an "energy only" rate and, therefore, only energy usage data were available. As a commercial customer, this power is somewhat expensive at $.107 per kilowatt hour, which includes a 10% sales tax.

A second difference is that no standby charges were entered on the worksheet. While the utility offers standby service, with a reservation charge of $4.50 per kilowatt per month, a decision was made that the laundry would schedule all maintenance during non-operating hours and would shut down operations during an unscheduled outage. The laundry's cogeneration system will not be interconnected with the utility grid.

Only the displaceable fuel use was used in Item G, the lowest monthly fuel requirement. Therefore, the 1,000 MMBtu of estimated process use was subtracted from the total gas billing and resulted in displaceable fuel use of 1,124 MMBtu during a summer month. Inasmuch as most of the steam is used to produce hot water, a reciprocating engine can be used for this application. In this size range, the reciprocating engine will be more efficient than the combustion turbine.

Since no demand data are available, it was necessary to develop an initial estimate of the engine size in order to select the appropriate Heat Recovery Factor, HRF. The facility operates for 50 hours per week based on a typical 10-hour operating day for 5 days a week. Total monthly usage was estimated to be 250 hours. Dividing these hours of use into the minimum electric usage of 72,000 kilowatt hours resulted in an estimated demand of approximately 288 kilowatts in winter. The estimated demand increased to 520 kilowatts in summer. Based on these estimates an HRF was selected for an engine size between 100 kilowatts to 500 kilowatts.

Based on the monthly thermal loads, a 245-kilowatt system with a budget of $1,000 per kilowatt was selected.

The calculation of the annual power output required an additional departure from the previous analyses. A baseloaded 260 kilowatt system would produce in excess of 2,000,000 kilowatt hours, and would require selling over one-half of the cogenerated power to the utility. In this case, no interconnection with the utility was anticipated and therefore no sales would be possible. The system output was capped at the actual annual usage.

Clearly, this assumption resulted in an error, since it was previously discovered that the demand may be as high as 520 kilowatts. A 245-kilowatt cogeneration system could not satisfy all site electrical requirements. If the site demand should exceed the capacity of the cogeneration system, it would then be necessary to purchase some supplemental power from the utility, and as a result, to interconnect with it. Should supplemental power be required, the actual amount of cogenerated power would be further decreased.

The total operating costs for the system were estimated at $59,000 per year. Annual savings were estimated at $139,000 and resulted in a net operating cost reduction of $80,000 and a simple payback of 3.1 years. Based on this preliminary analysis it was decided to proceed to the next analytic step.

Engineering and Economic Screening Analysis

The engineering analysis required the development of much more detailed data than were readily available at the laundry.

Energy Analysis

In order to better understand the laundry's energy use patterns, a recording demand meter was temporarily installed at the site. The results are shown in Figures 11-45 and 11-46. The laundry load picks up starting at 4:00 a.m. and levels out at 300 kilowatts during the day. The decrease which occurs during the luncheon break is clearly visible, and the load starts to again decrease during the afternoon hours. More importantly, there exists a base load of approximately 25 kilowatts during all hours.

Weekend demands remain low, except for a slight increase which occurs on Saturday morning (see Figure 11-47). Therefore, as shown in Figure 11-43, the electric power use was split into two categories: process and base. Because of the significant difference in the relative demands, it was decided to isolate the base load and to continue to serve it from the electric utility. Since this base requirement would be a normal utility load, no standby charge was anticipated. A second reason for isolating the load is that it would be inefficient for a 300-kilowatt engine-generator set to operate at 25 kilowatts.

Since the recovered heat would be used for hot water heating, a condensate meter was installed in the return from the hot water generators. The meter was read during the laundry's working hours and the results are shown in Figure 11-48. The steam requirements were relatively constant during any given working day.

Table 11-18 shows the energy performance of the cogeneration system. A 325-kilowatt engine with a heat rate of 13,025 Btus (HHV) per kilowatt hour and fueled by natural gas was selected. The system would produce 6,000 Btus of hot water per kilowatt hour and would be available 8,000 hours a year. As shown in Table 11-18, it would only be operated during the laundry's occupied hours, and the laundry would continue to purchase the base power requirements from the utility. Since the engine would produce only a small fraction, a maximum of 57% during the summer, of the laundry's heat requirements, all recovered heat could be usable. The conventional boiler would still be required to produce the 60 psig steam required for selected processes.

Cost Analysis

The supplemental costs are shown in Table 11-19. Since the power is purchased on an energy only rate, the rate for supplemental power is essentially equal to the rate for power as a full service customer. Power costs were reduced significantly from $116,000 to $23,000 per year. Boiler fuel costs were reduced from $110,000 to $90,000.

The cogeneration system's projected operating costs were computed and are shown in Table 11-20. Maintenance charges for the reciprocating engine were estimated to be $.015 per kilowatt hour, as compared to $.005 per kilowatt hour for a turbine.

The system would reduce utility costs from $226,000 to less than $164,000 (see Table 11-21) for a net reduction of slightly more than $62,000 per year.

Project Budget

The cogeneration engine cost was estimated at $200,000 for a 325-kilowatt reciprocating engine with heat recovery. Installation costs were somewhat less than average since there was room for the engine in the existing building, and mechanical costs would be low. The isolation from the electric utility would also result in reduced switchgear costs; however, this saving would be partially offset by costs incurred in rewiring the base load so that it would remain on the utility grid. The total projected cost including fees and contingency was estimated at $354,000, or slightly more than the $1,000 per kilowatt used in the walkthrough analysis.

Project Economics

Table 11-22 provides detailed cash flow projections. Based on a 1-year period for project implementation, the cost savings for the first year of operation were projected at $65,000 for a simple payback of 5.4 years. This payback was judged unacceptable.

Conclusions

In this last example, a system which seemed to produce an excellent return based on average costs and optimistic assumptions was found to produce an unacceptable return when incremental costs and actual load data were used in the project analysis.

REVIEW

Cogeneration viability is dependent on the particular energy use characteristics, purchased power rates and fuel costs for each specific application. Developing the data required to perform a comprehensive analysis of cogeneration economics is costly, as is the analysis of that data. In this chapter, three different potential applications were evaluated, with each evaluation taken to two different analytic levels. In each case, the more comprehensive analysis, based on detailed energy use data and site specific engine and system modelling, provided a less attractive return on investment.

Any decision to invest in a cogeneration system is a complex one that should only be made after thorough and independent analysis of the project performance.

TABLE 11-1
CHEMCO CONVENTIONAL SYSTEM COST

FAC 0.005 $/kWh AVE COST 0.059 $/kWh

MONTH	DAYS	MEASURED DEMAND (kW)			ON PEAK ENERGY (kWh)			TOTAL ENERGY (kWh)	CONV'L LD FACT (HOURS)	COST ($)
JAN	31	8,536.2			5,384,800			5,384,800	631	322,693
FEB	28	8,876.0			5,241,600			5,241,600	591	320,260
MAR	31	8,795.4			5,849,800			5,849,800	665	345,242
APR	30	8,530.4			5,683,200			5,683,200	666	335,313
MAY	31	8,496.0			5,728,800			5,728,800	674	336,881
JUN	30	8,398.0			5,337,000			5,337,000	636	319,176
JUL	31	8,593.8			5,548,800			5,548,800	646	330,282
AUG	31	8,530.4			5,760,000			5,760,000	675	338,577
SEP	30	8,455.6			5,444,200			5,444,200	644	324,351
OCT	31	8,576.6			5,716,000			5,716,000	666	337,203
NOV	30	9,043.2			5,821,800			5,821,800	644	346,716
DEC	31	9,008.6			5,913,600			5,913,600	656	350,245
TOTALS	365				67,429,600	0	0	67,429,600	649	4,006,940
MAX		9,043.2	ERR	0.0					675	
MIN		8,398.0	ERR	0.0					591	
DATE	02/15									
Revision		1.4								

AVE COST ALT HTG FUEL 2.80 $/MMBtu 140,000 BTU/GAL

MAX STEAM DEMAND 0 M Lb/HR
BOILER EFFICIENCY 80.0%

MONTH	CONV'L GAS USE (THERMS)	NON-DISPL'B GAS USE (THERMS)	DISPL'BLE GAS USE (THERMS)	FUEL OIL USE (GAL)	STEAM USE (M LB)	TOTAL DISP FUEL (MMBTU)	TOTAL FUEL COST ($)	BOILER EFFIC. (%)
JAN	1,487,178	170,080	1,317,098		103,834	131,710	416,410	78.8
FEB	1,352,919	178,193	1,174,725		94,565	117,473	378,817	80.5
MAR	1,401,271	202,134	1,199,137		97,093	119,914	392,356	81.0
APR	1,194,714	134,809	1,059,905		82,461	105,991	334,520	77.8
MAY	1,067,214	171,173	896,041		71,135	89,604	298,820	79.4
JUN	896,041	142,049	753,992		59,163	75,399	250,892	78.5
JUL	863,190	131,454	731,736		57,347	73,174	241,693	78.4
AUG	820,733	115,838	704,895		57,163	70,489	229,805	81.1
SEP	996,531	216,270	780,260		62,722	78,026	279,029	80.4
OCT	1,133,747	217,541	916,206		73,936	91,621	317,449	80.7
NOV	1,179,418	175,679	1,003,739		82,745	100,374	330,237	82.4
DEC	1,342,040	162,749	1,179,291		95,431	117,929	375,771	80.9
TOTALS	13,734,997	2,017,971	11,717,026	0	937,595	1,171,703	3,845,799	80.0
MAX								82.4
MIN								77.8
DATE	02/15							

FIGURE 11-1
MONTHLY DEMAND – CHEMCO

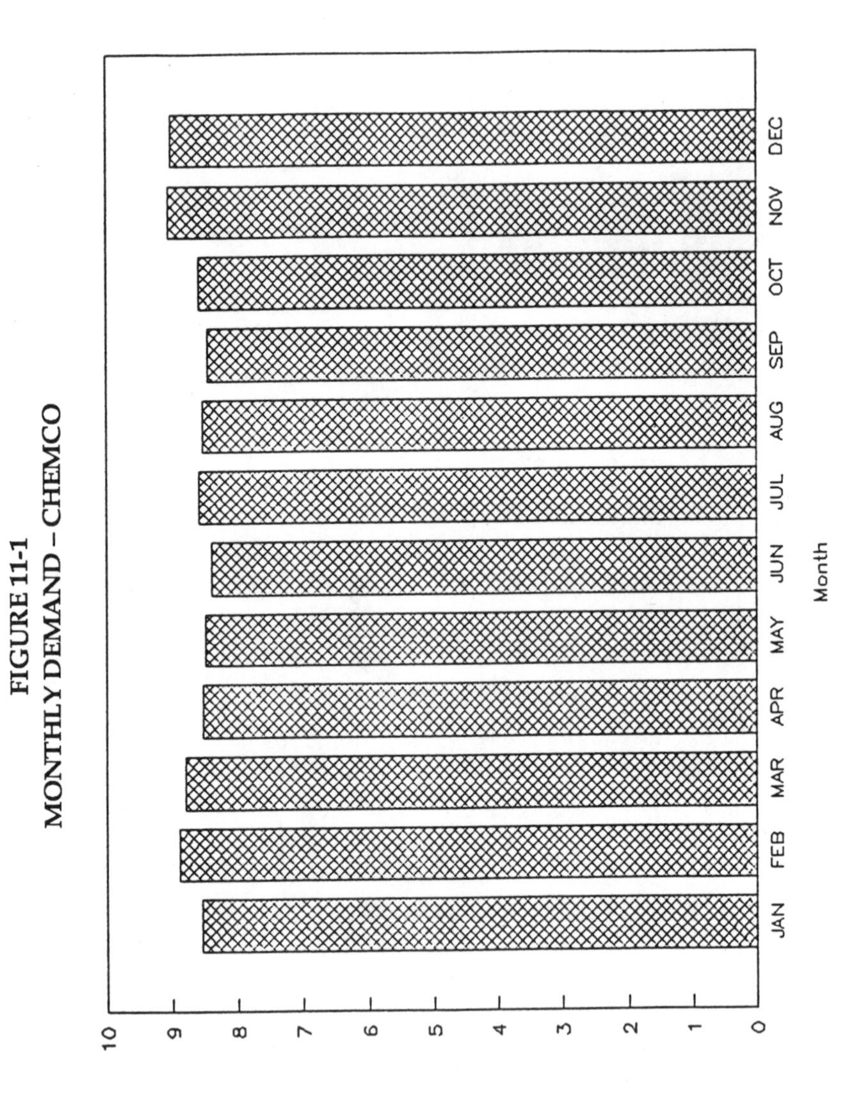

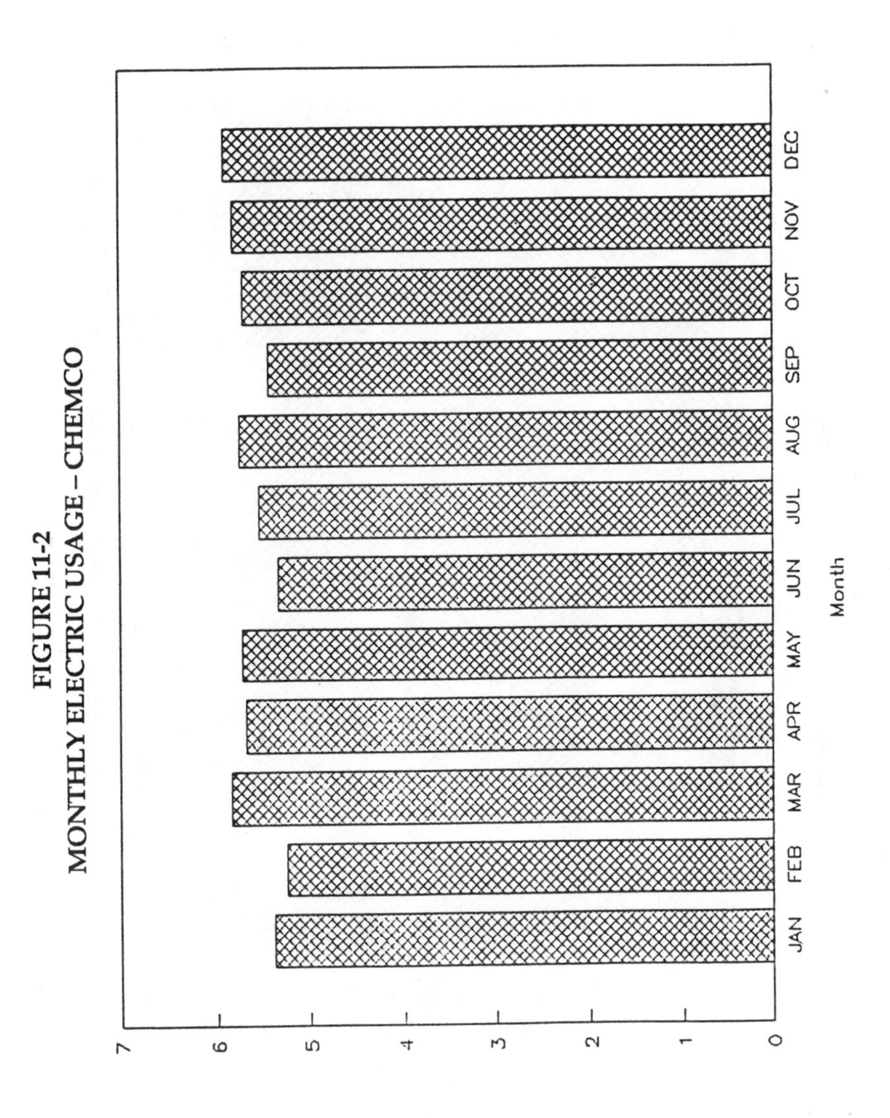

FIGURE 11-2
MONTHLY ELECTRIC USAGE – CHEMCO

FIGURE 11-3
MONTHLY LOAD FACTORS – CHEMCO

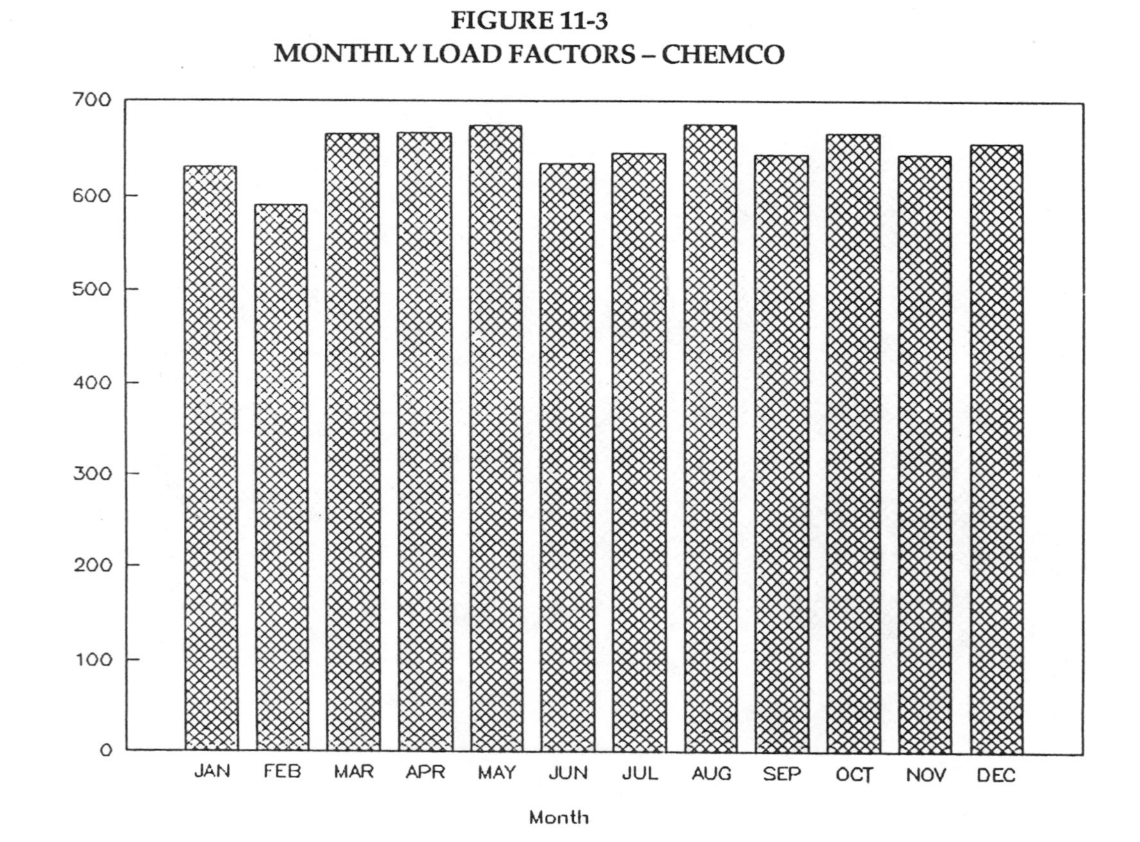

FIGURE 11-4
MONTHLY GAS USAGE – CHEMCO

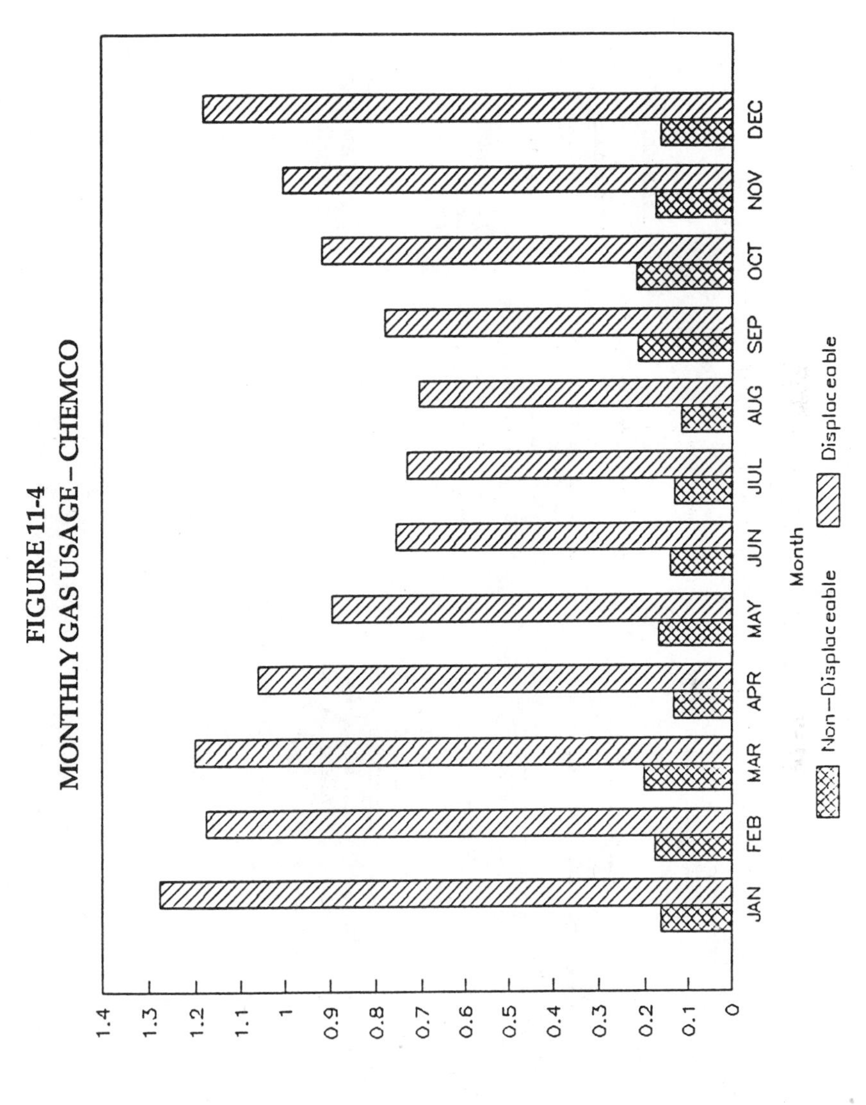

TABLE 11-2
WALKTHROUGH WORKSHEET

SITE: CHEMCO

ELECTRIC UTILITY: ____

FUEL SUPPLIER: ____

TECHNICAL FEASIBILITY

	HVAC System	Boiler Plant	
	Compatibility with Cogeneration (Table 9-3)	Excellent	
A.	HVAC retrofit costs (Add to "M")	$ 0	
B.	Minimum Demand	8,398	kW

ECONOMIC FEASIBILITY

C.	Average Purchased Power Cost	$ .0590	/kWh
D.	Standby Charge	$ 4.00	/kW/Month
E.	Heating Fuel Cost	$ 2.80	/MMBtu
F.	Cogeneration System Fuel Cost	$ 2.60	/MMBtu
G.	Lowest Monthly Fuel Requirement	70,400	MMBtu/Month
H.	Monthly Electric Output (See Table 9-4 for value of HRF) .8 x "G" / HRF =	11.26×10^6	kWh/Month
I.	Maximum Installed Capacity (See Table 9-5 for value of HOURS) "H" / HOURS =	15,700	kW
J.	System Size (Less than or equal to "B" or "I")	8,398	kW
K.	System Unit Cost (See Figure 9-5 for COSTS)	$ 950	/kW
L.	Cogeneration System Cost "J" x "K" =	$ 7,978,000	
M.	PROJECT INVESTMENT "A" + "L" =	$ 7,978,000	

N.	Annual Electric Output 12 x "J" x HOURS =	72.26×10^6	kWh/Year
O.	Annual Fuel Use (See Table 9-4 for value for FUEL) "N" x FUEL =	802,000	MMBtu
P.	Annual Cogeneration Fuel Cost "F" x "O" =	$ 2,085,000	
Q.	Annual O & M Cost (See Table 9-6 for value for OM) "N" x OM =	$ 361,000	
R.	Annual Standby Costs "D" x "J" x 12 =	$ 403,000	
S.	TOTAL OPERATING COSTS "P" + "Q" + "R" =	$ 2,849,000	
T.	Recovered Thermal Energy "N" x HRF =	361,000	MMBtu/Year
U.	Annual Conventional Fuel Cost	$ 3,845,000	
V.	Saved Heating Fuel Costs "E" x "T" / .8 =	$ 1,264,000	
W.	Saved Purchased Power Costs "C" x "N" =	$ 4,263,000	
X.	TOTAL OPERATING COST REDUCTION "V" + "W" =	$ 5,527,000	
Y.	ANNUAL SAVINGS "X" − "S" =	$ 2,678,000	
Z.	SIMPLE PAYBACK "M" / "Y" =	3.0	Years

TABLE 11-3
WALKTHROUGH WORKSHEET

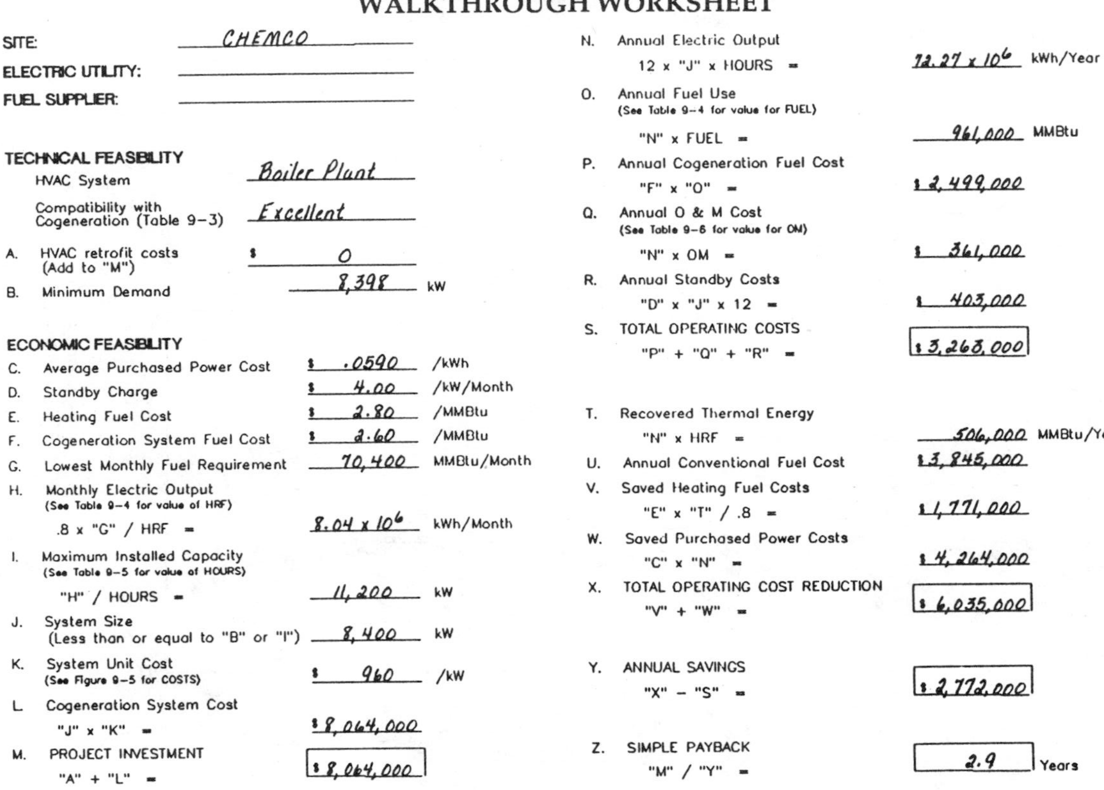

SITE: *CHEMCO*

ELECTRIC UTILITY: ______

FUEL SUPPLIER: ______

TECHNICAL FEASIBILITY

	Item	Value	Units
	HVAC System	*Boiler Plant*	
	Compatibility with Cogeneration (Table 9-3)	*Excellent*	
A.	HVAC retrofit costs (Add to "M")	$ *0*	
B.	Minimum Demand	*8,398*	kW

ECONOMIC FEASIBILITY

	Item	Value	Units
C.	Average Purchased Power Cost	$ *.0590*	/kWh
D.	Standby Charge	$ *4.00*	/kW/Month
E.	Heating Fuel Cost	$ *2.80*	/MMBtu
F.	Cogeneration System Fuel Cost	$ *2.60*	/MMBtu
G.	Lowest Monthly Fuel Requirement	*70,400*	MMBtu/Month
H.	Monthly Electric Output (See Table 9-4 for value of HRF) .8 x "G" / HRF =	*8.04 x 10⁶*	kWh/Month
I.	Maximum Installed Capacity (See Table 9-5 for value of HOURS) "H" / HOURS =	*11,200*	kW
J.	System Size (Less than or equal to "B" or "I")	*8,400*	kW
K.	System Unit Cost (See Figure 9-5 for COSTS)	$ *960*	/kW
L.	Cogeneration System Cost "J" x "K" =	$ *8,064,000*	
M.	PROJECT INVESTMENT "A" + "L" =	$ *8,064,000*	

	Item	Value	Units
N.	Annual Electric Output 12 x "J" x HOURS =	*72.27 x 10⁶*	kWh/Year
O.	Annual Fuel Use (See Table 9-4 for value for FUEL) "N" x FUEL =	*961,000*	MMBtu
P.	Annual Cogeneration Fuel Cost "F" x "O" =	$ *2,499,000*	
Q.	Annual O & M Cost (See Table 9-6 for value for OM) "N" x OM =	$ *361,000*	
R.	Annual Standby Costs "D" x "J" x 12 =	$ *403,000*	
S.	TOTAL OPERATING COSTS "P" + "Q" + "R" =	$ *3,263,000*	
T.	Recovered Thermal Energy "N" x HRF =	*506,000*	MMBtu/Year
U.	Annual Conventional Fuel Cost	$ *3,845,000*	
V.	Saved Heating Fuel Costs "E" x "T" / .8 =	$ *1,771,000*	
W.	Saved Purchased Power Costs "C" x "N" =	$ *4,264,000*	
X.	TOTAL OPERATING COST REDUCTION "V" + "W" =	$ *6,035,000*	
Y.	ANNUAL SAVINGS "X" – "S" =	$ *2,772,000*	
Z.	SIMPLE PAYBACK "M" / "Y" =	*2.9*	Years

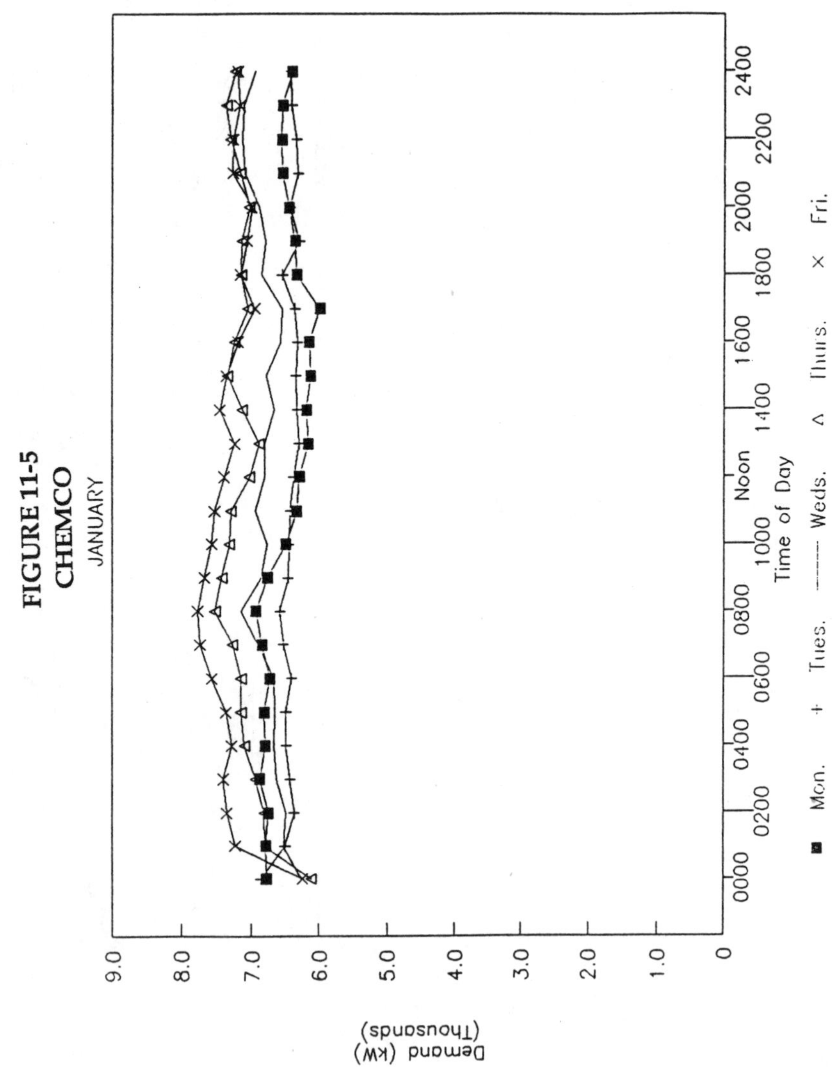

FIGURE 11-5
CHEMCO
JANUARY

FIGURE 11-6
CHEMCO
JANUARY

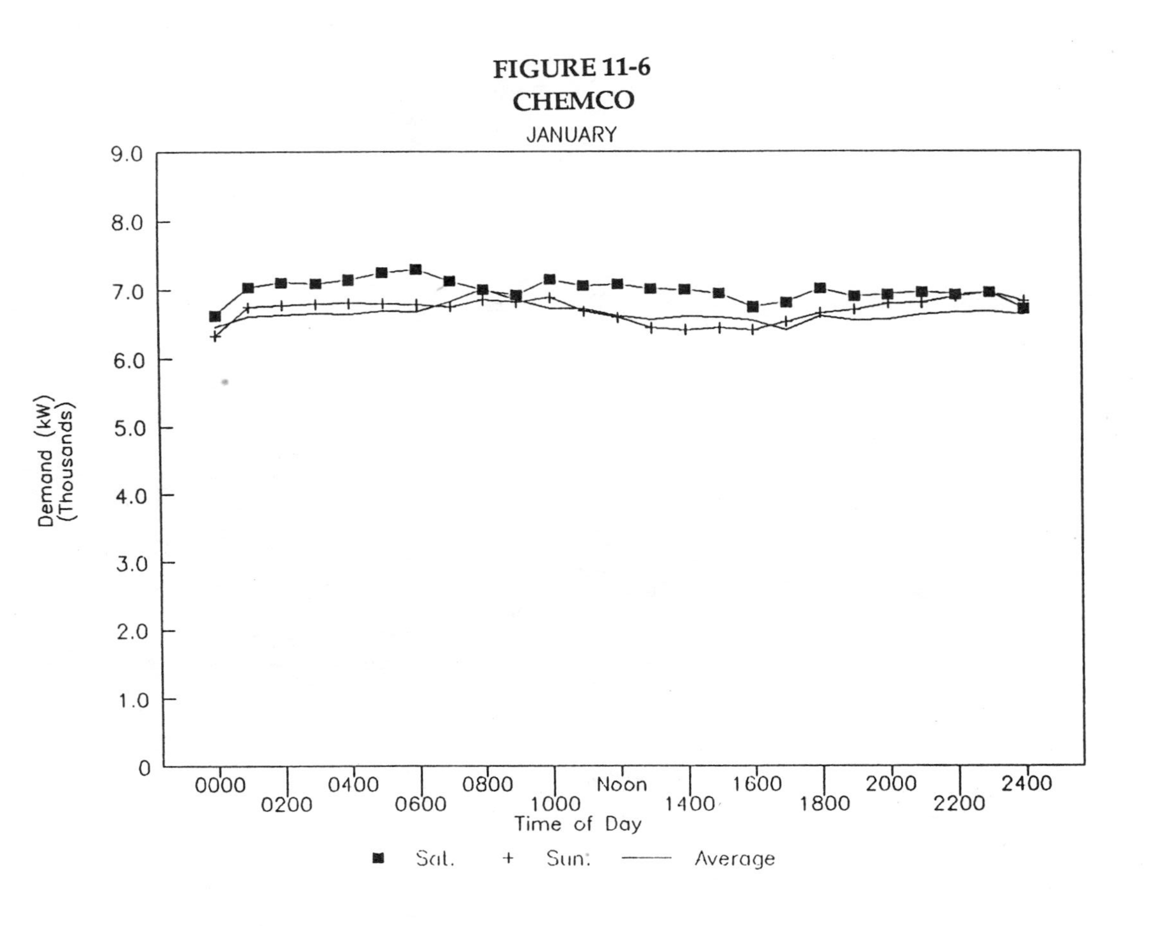

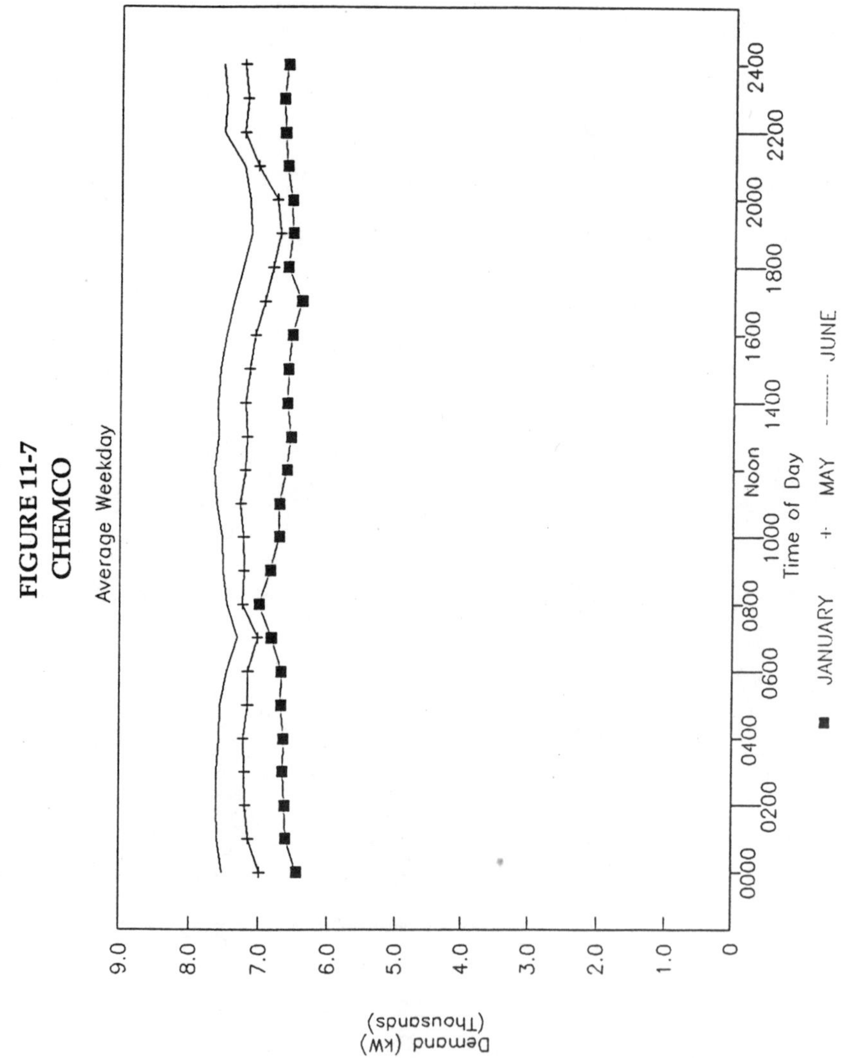

FIGURE 11-7
CHEMCO

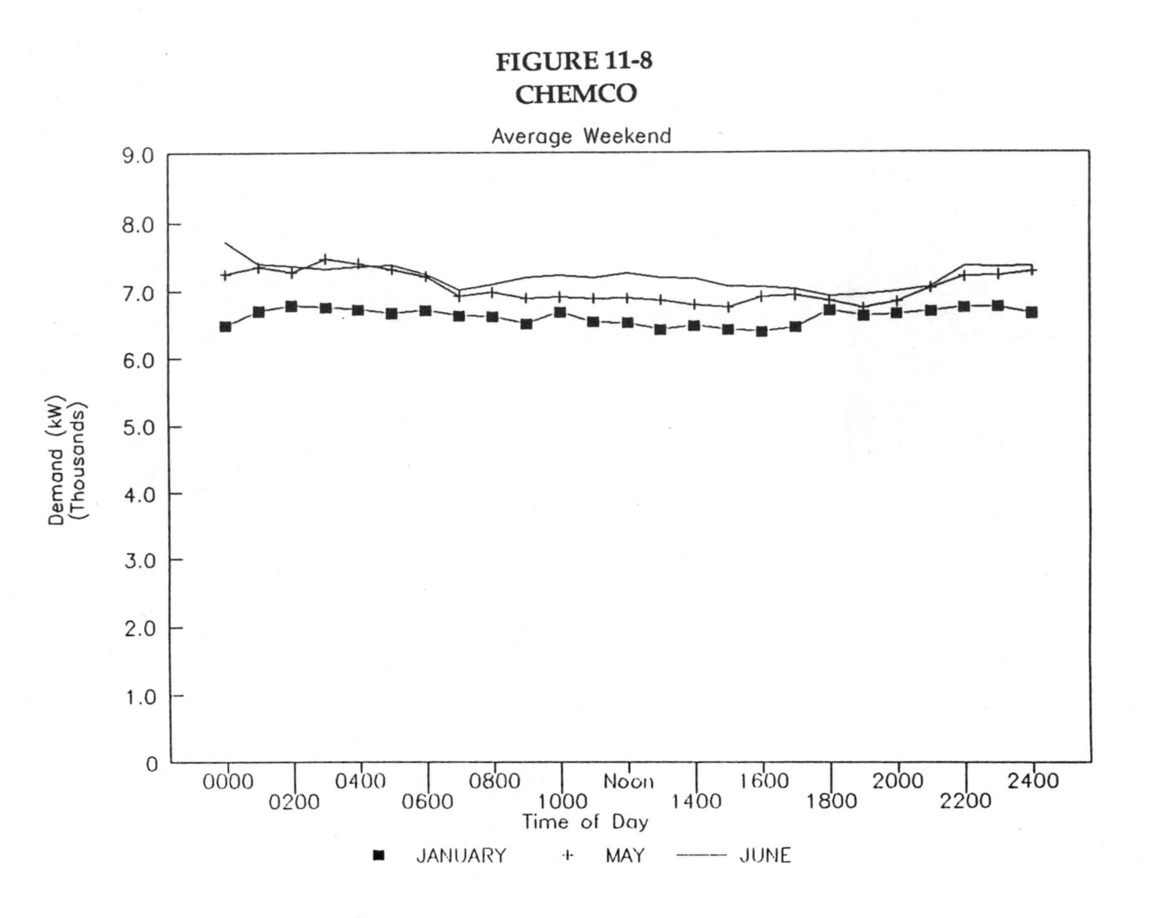

FIGURE 11-8
CHEMCO

FIGURE 11-9
DEMAND HISTOGRAM

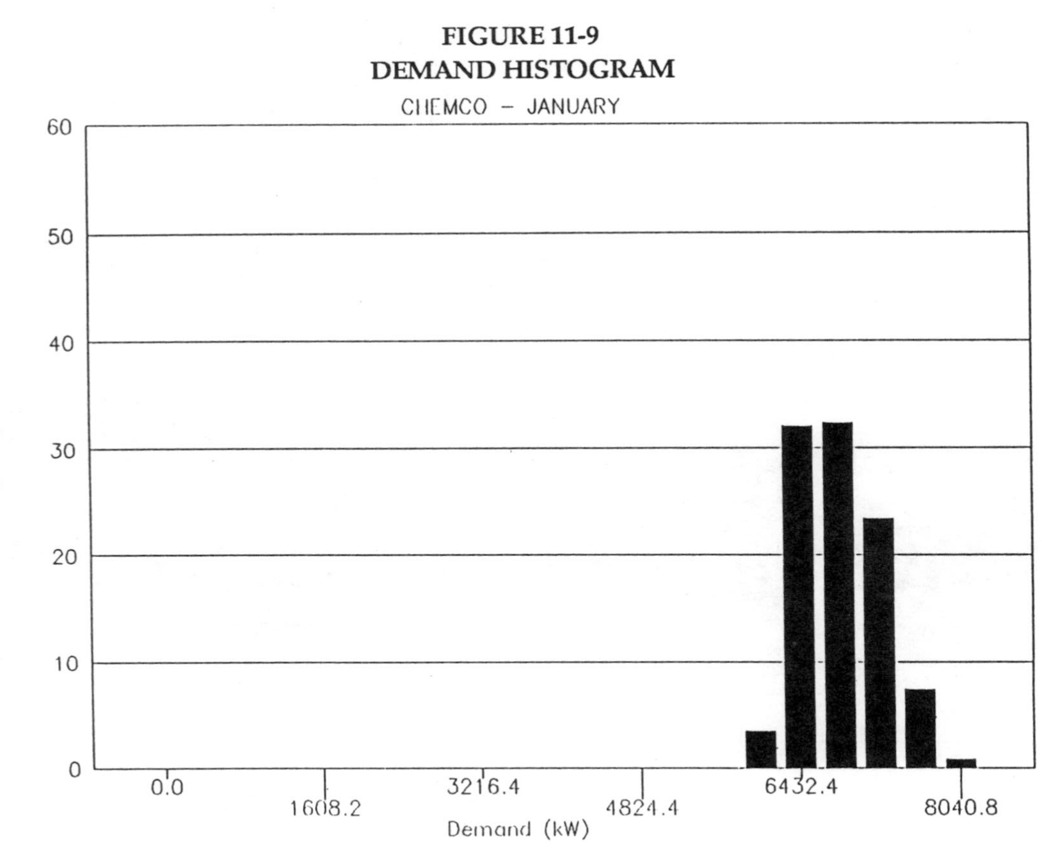

FIGURE 11-10
LOAD VS. DURATION

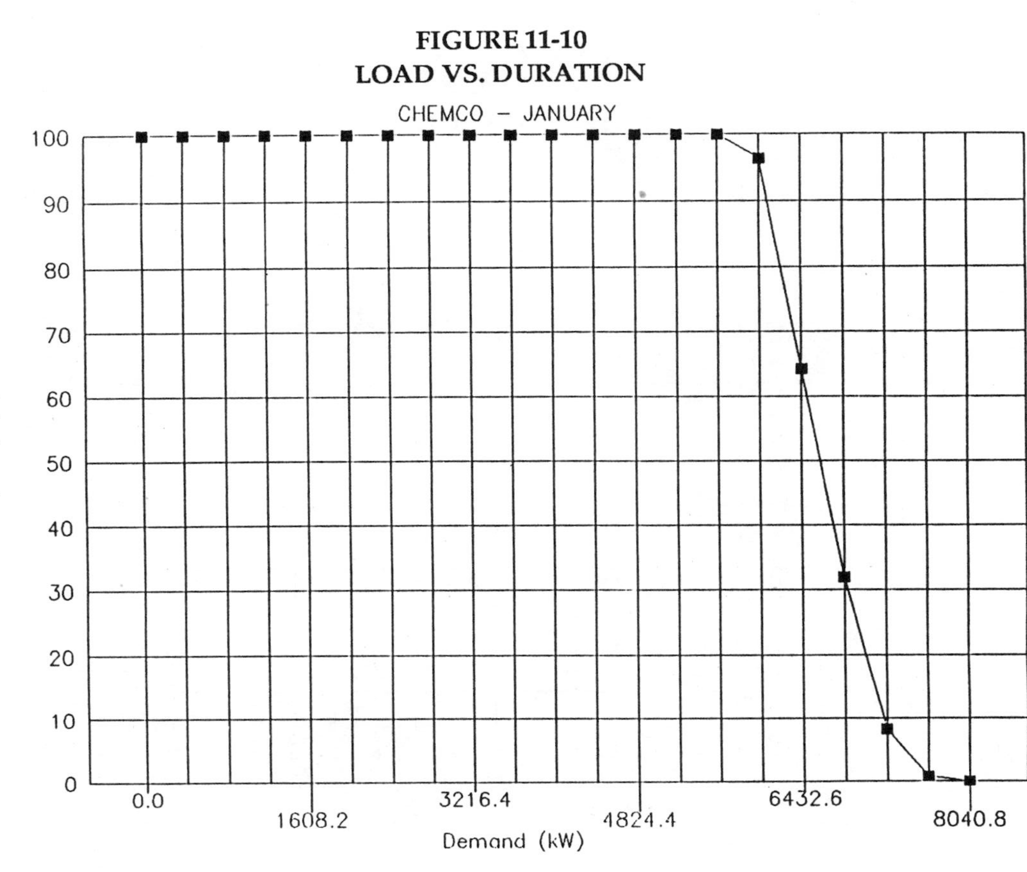

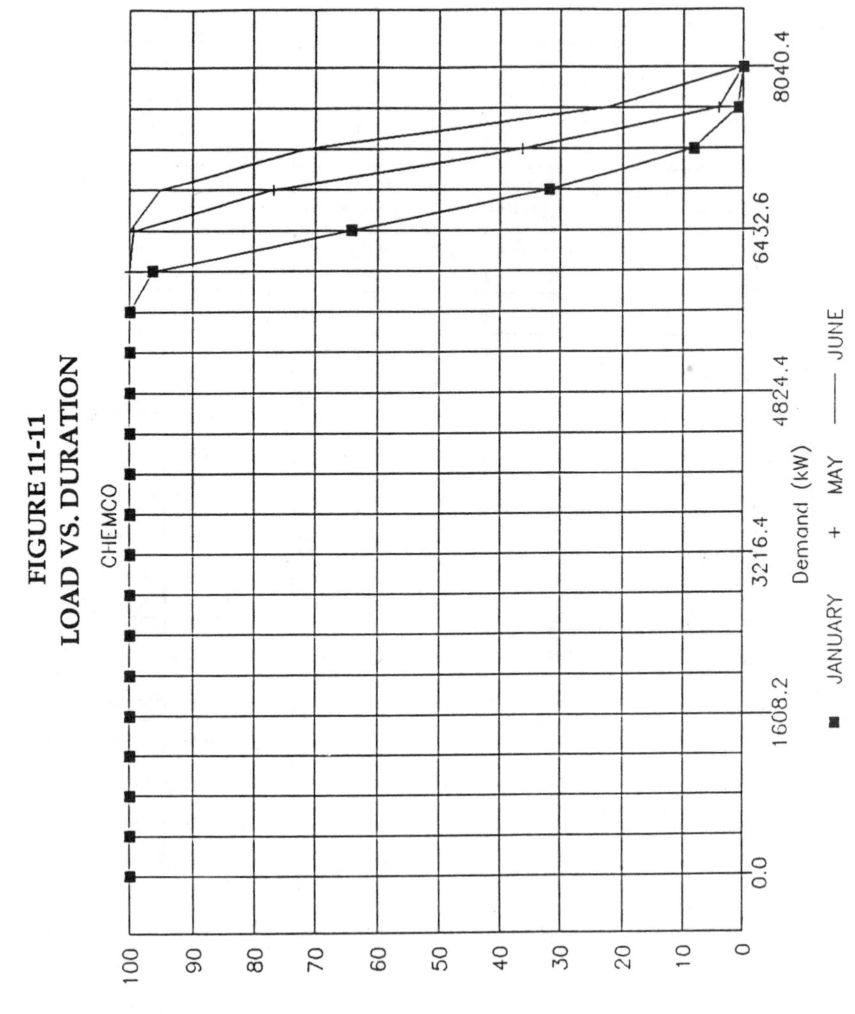

FIGURE 11-11
LOAD VS. DURATION

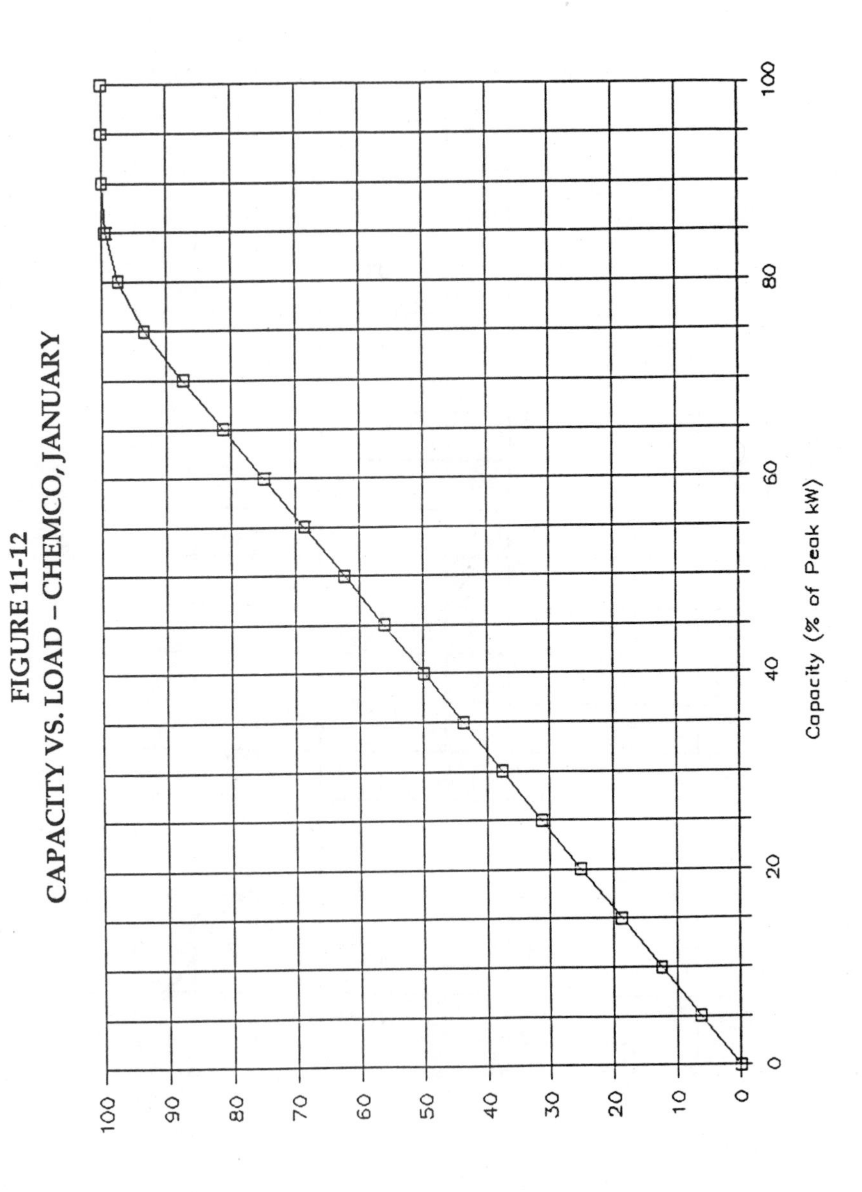

FIGURE 11-12
CAPACITY VS. LOAD - CHEMCO, JANUARY

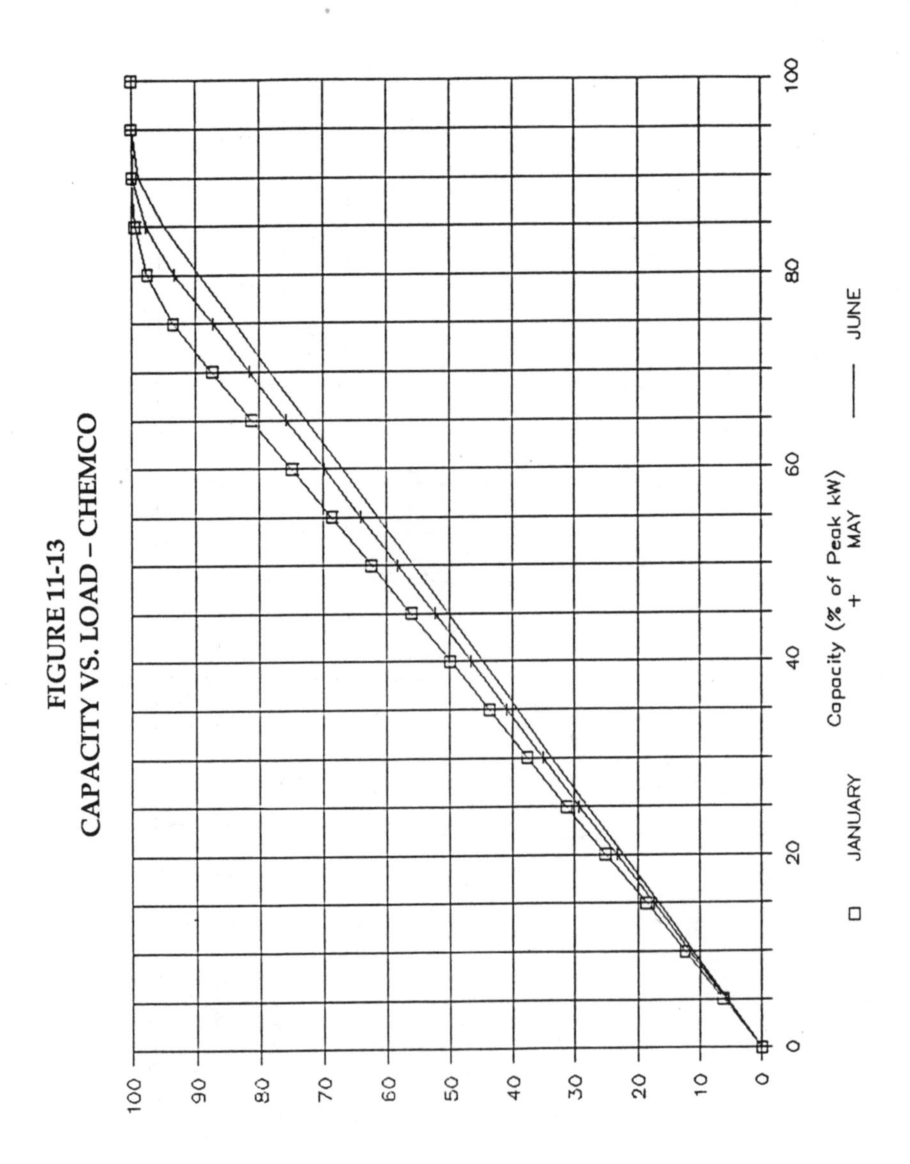

FIGURE 11-13
CAPACITY VS. LOAD – CHEMCO

FIGURE 11-14
CHEMCO

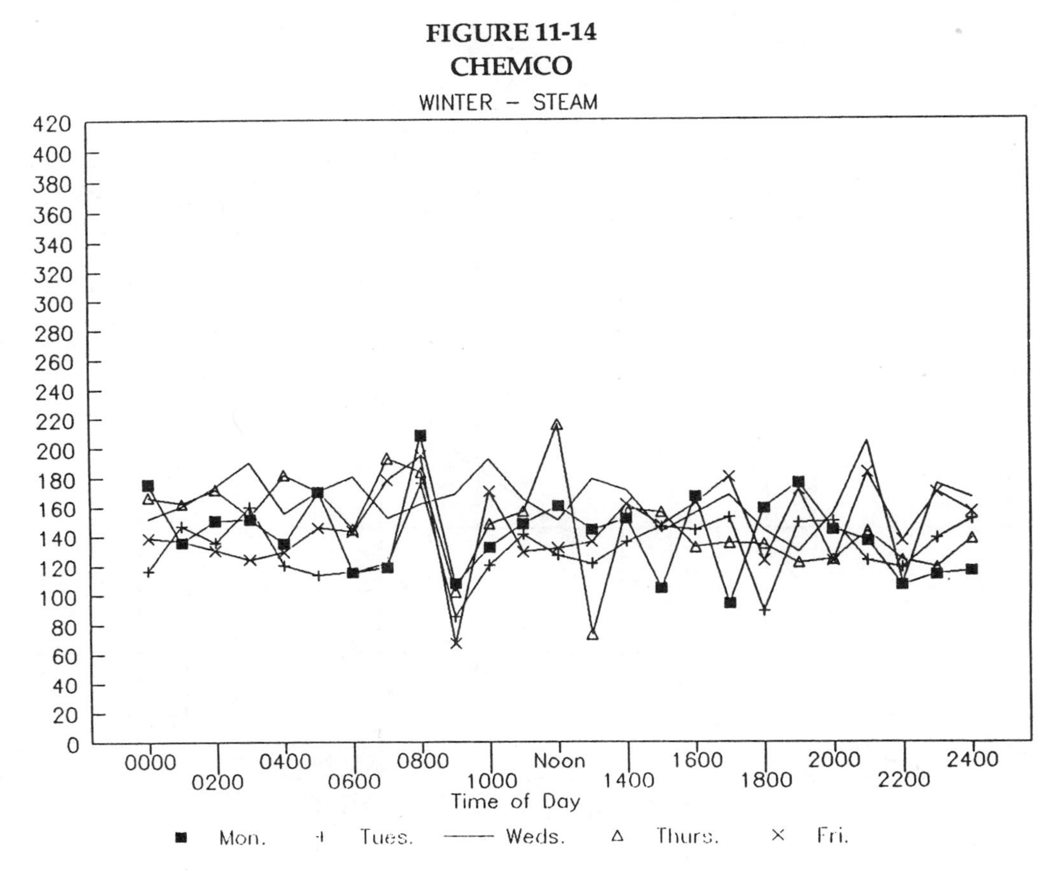

FIGURE 11-15
CAPACITY VS. LOAD – CHEMCO

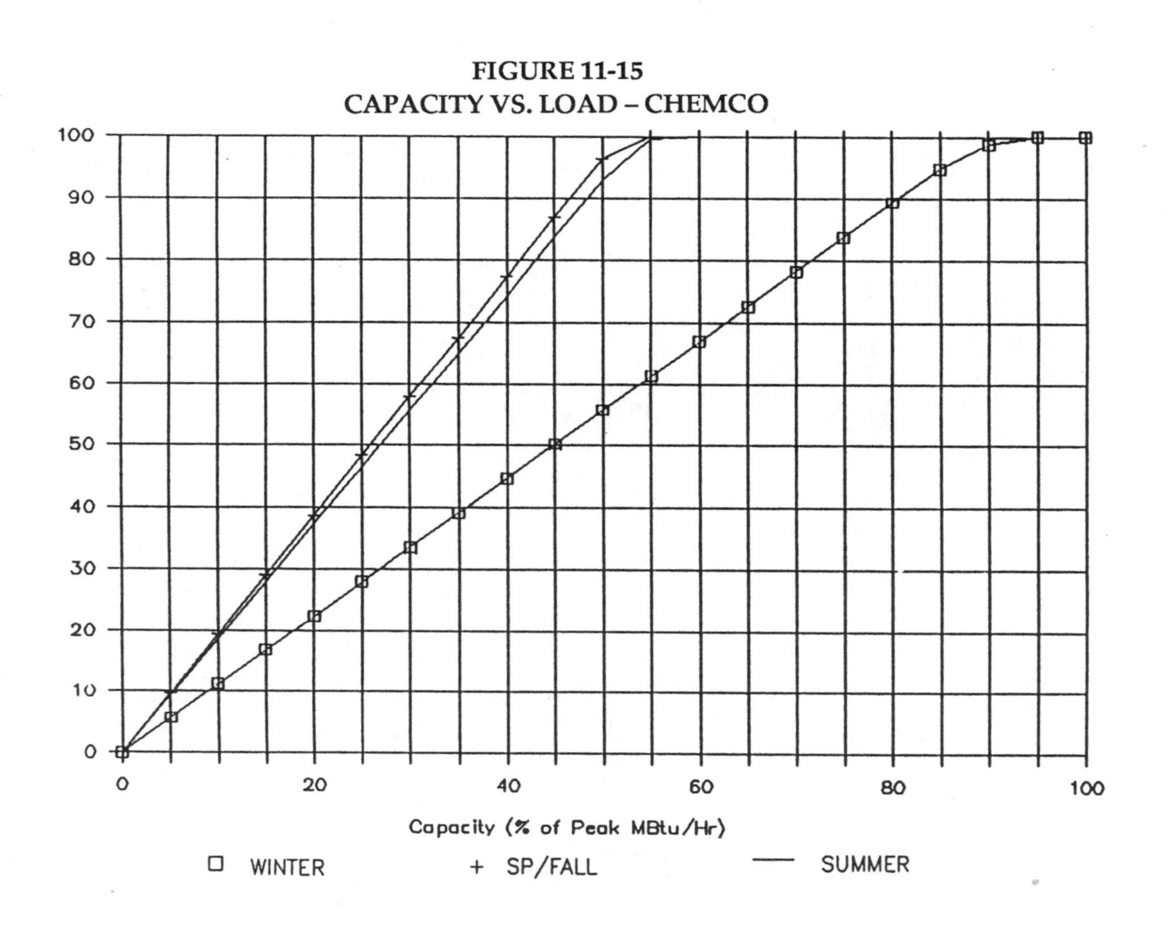

FIGURE 11-16
LARGE USER RATE

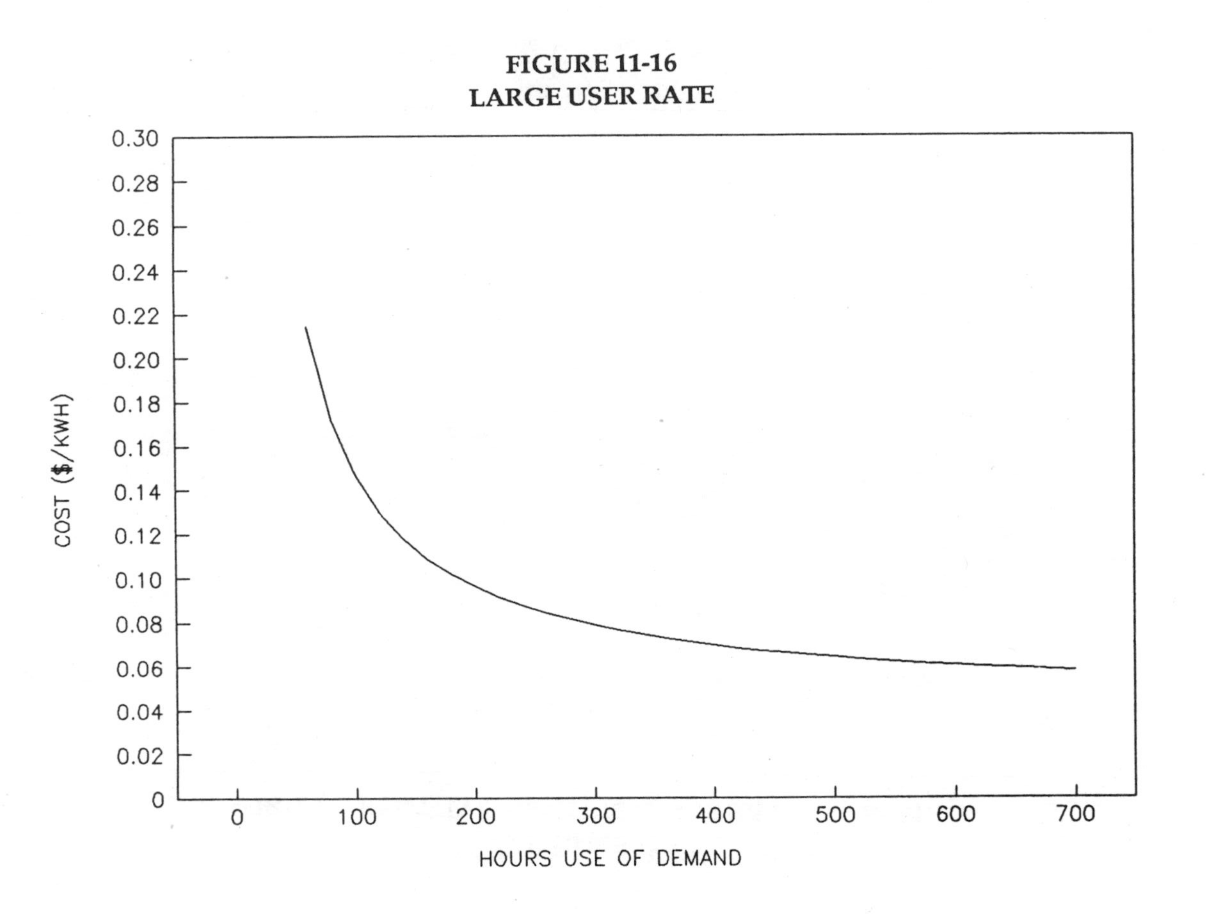

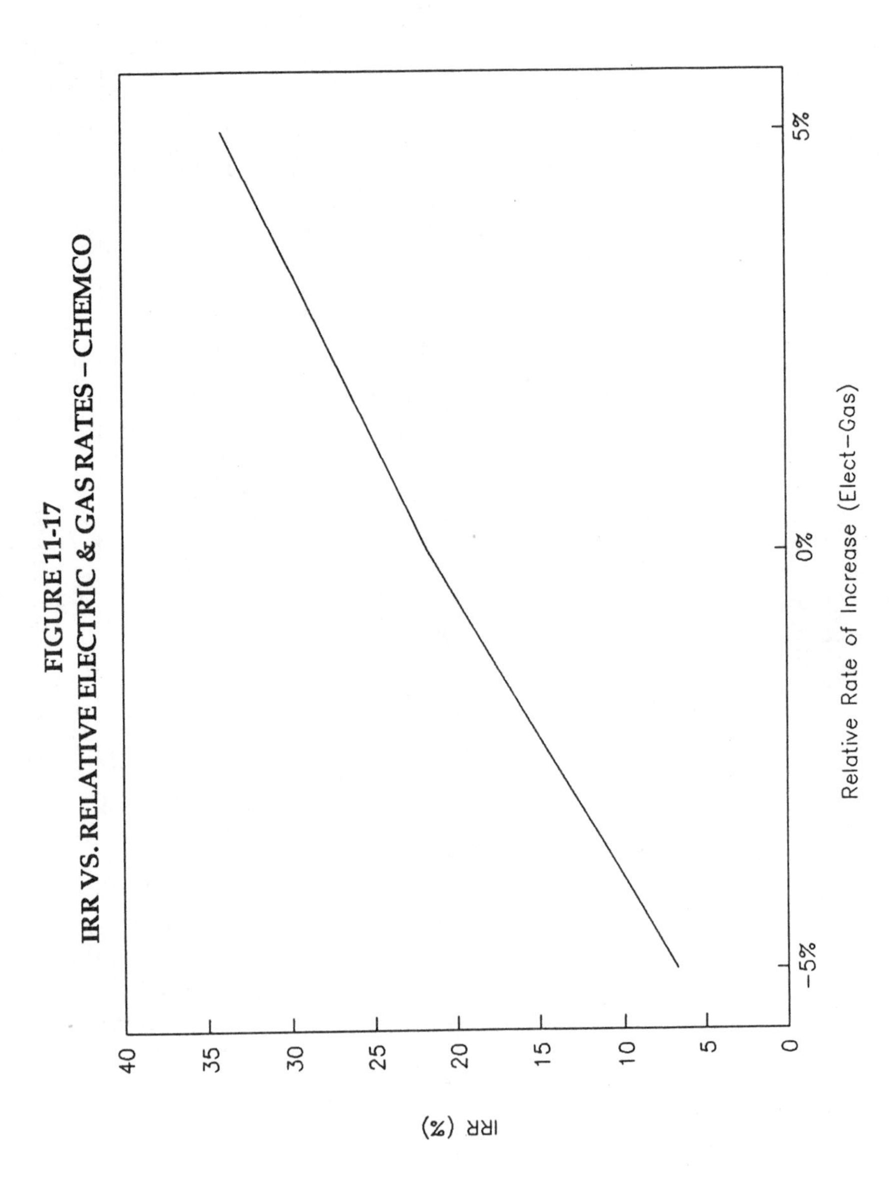

FIGURE 11-17
IRR VS. RELATIVE ELECTRIC & GAS RATES – CHEMCO

FIGURE 11-18
IRR VS. GAS PRICE DIFFERENTIAL – CHEMCO

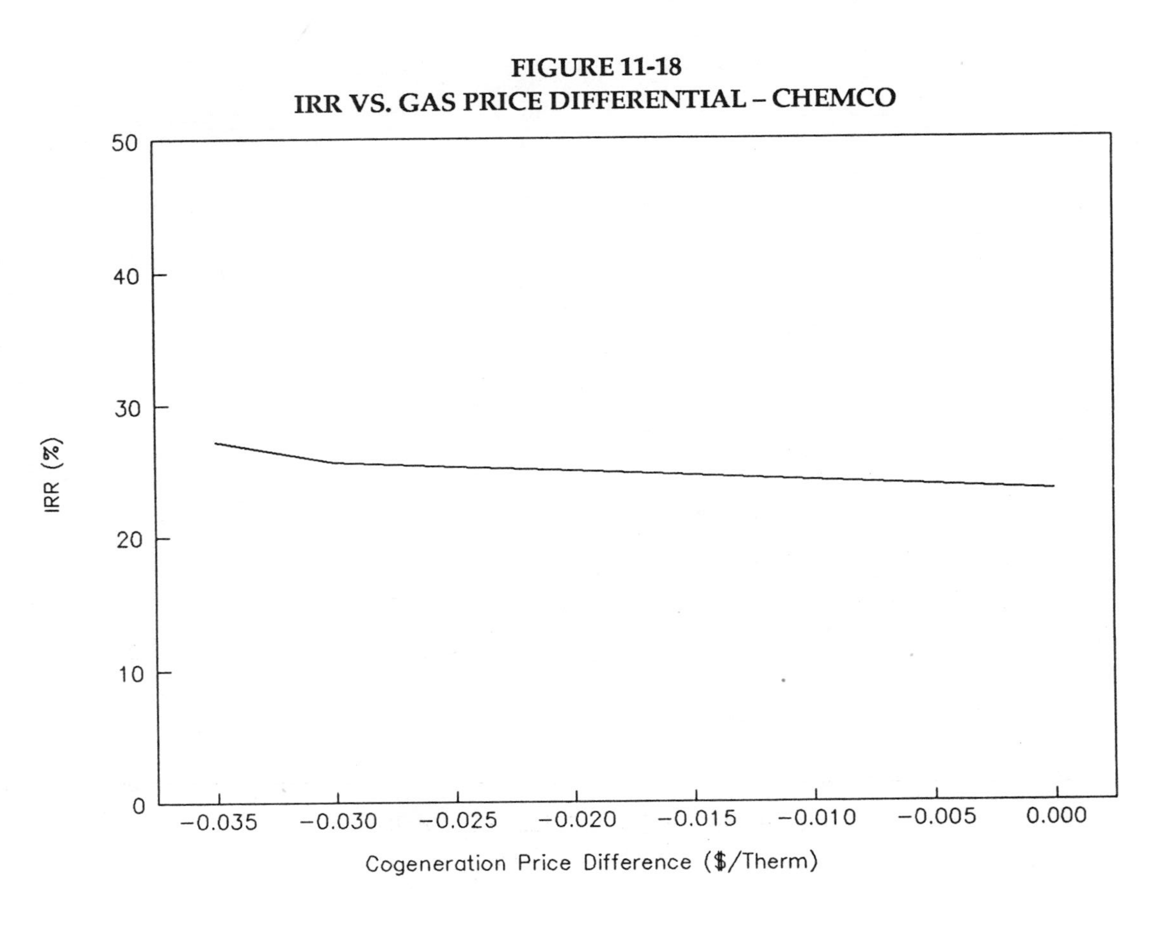

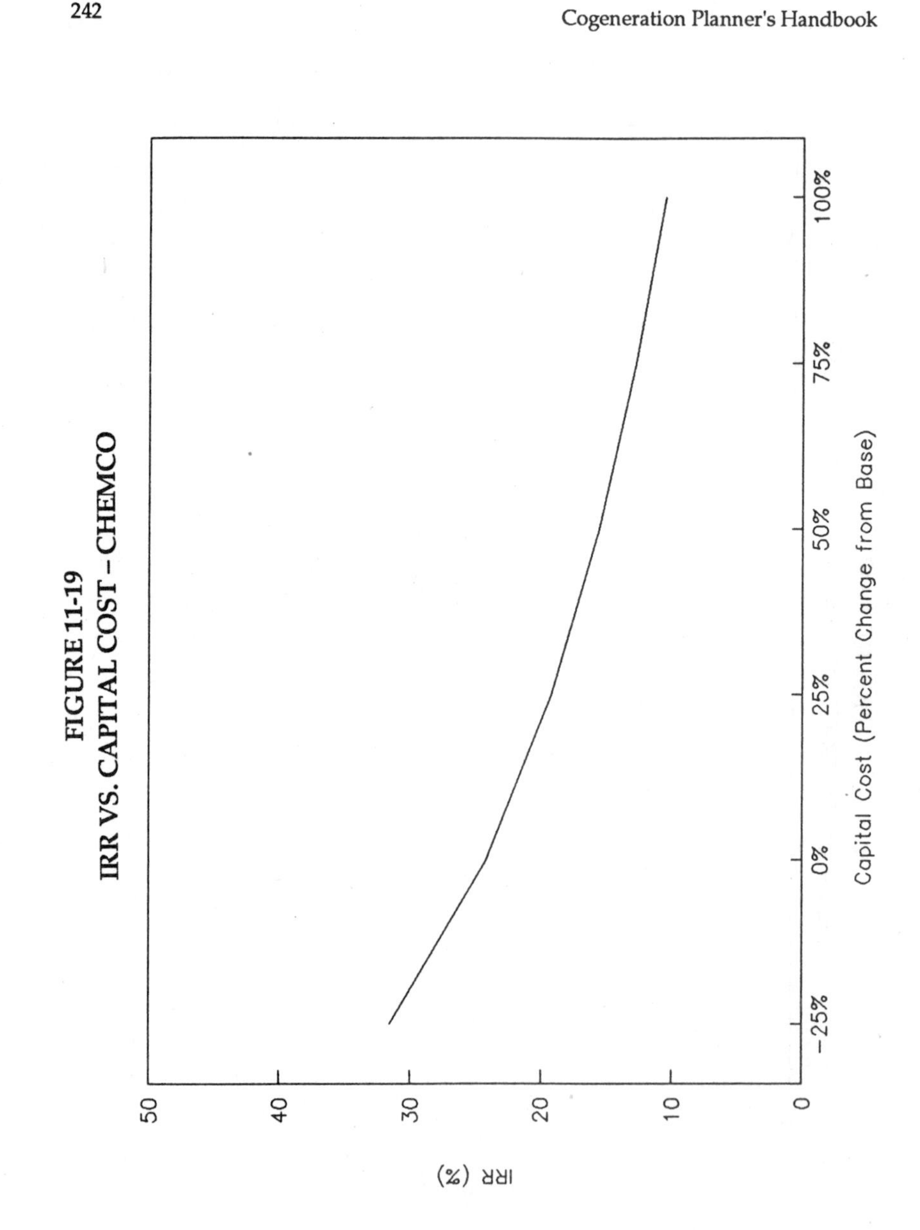

FIGURE 11-19
IRR VS. CAPITAL COST – CHEMCO

TABLE 11-4
CHEMCO ENERGY ANALYSIS – TURBINE

MODE = 1

CAPACITY	7,424 kW	HEAT RATE	13,290 Btu/kWh	ALT HTG FUEL	140,000 Btu/Gal	DUCT BNR CAP	7,700 Btu/kWh
GEN CAP	8,909 kW	HEAT REC	4,600 Btu/kWh	STEAM	1,000,000 Btu/Lb	DUCT BNR EFF	90.0%
STEAM CAP	0 kW	HOUR/YR	8,500	BLR EFFICIENCY	80.0%		
COMP LOSS	70 kW						

MONTH	DAYS	COGEN MIN ENG. CAP (kW)	COGEN AVE OUT (kW)	COGEN ENG. OUT (kWh)	COGEN SYS OUT (kWh)	REC'V HEAT AVAILABLE (THERMS)	TOTAL DISPL'BLE FUEL REQ'D (THERMS)	AVAIL/REQ'D HEAT	USED/AVAIL HEAT	SUPP'L HEAT (THERMS)	DUCT BURNER CAPACITY (THERMS)	DUCT BURNER GAS (THERMS)
JAN	31	8,172	8,369	6,041,784	5,984,817	277,922	1,317,098	26.4%	94.3%	791,499	465,217	516,908
FEB	28	8,107	8,315	5,422,145	5,371,020	249,419	1,174,725	26.5%	93.8%	705,941	417,505	463,895
MAR	31	7,825	8,056	5,815,720	5,760,884	267,523	1,199,137	27.9%	89.2%	720,611	447,810	497,567
APR	30	7,455	7,732	5,401,460	5,350,531	248,467	1,059,905	29.3%	100.0%	599,457	415,912	462,125
MAY	31	7,156	7,447	5,375,811	5,325,123	247,287	896,041	34.5%	100.0%	469,546	413,937	459,930
JUN	30	6,888	7,181	5,017,131	4,969,825	230,788	753,992	38.3%	100.0%	372,406	386,319	413,784
JUL	31	6,784	7,069	5,102,904	5,054,789	234,734	731,736	40.1%	83.8%	388,698	392,924	431,887
AUG	31	6,823	7,102	5,127,343	5,078,998	235,858	704,895	41.8%	91.8%	347,372	394,805	385,969
SEP	30	6,995	7,280	5,086,113	5,038,157	233,961	780,260	37.5%	100.0%	390,247	391,631	433,608
OCT	31	7,320	7,588	5,477,642	5,425,994	251,972	916,206	34.4%	100.0%	480,994	421,778	468,643
NOV	30	7,698	7,909	5,525,628	5,473,528	254,179	1,003,739	31.7%	69.9%	625,389	425,473	472,748
DEC	31	8,031	8,211	5,927,734	5,871,842	272,676	1,179,291	28.9%	86.1%	708,685	456,436	507,151
TOT	365			65,321,416	64,705,508	3,004,785	11,717,026			6,600,845	5,029,749	5,514,215
MAX								41.8%	100.0%			
MIN								26.4%	69.9%			
DATE	02/15											

MONTH	CONV'TL BOILER HEAT (THERMS)	CONV'L BLR GAS USE (THERMS)	SUPP'L BLR USE (THERMS)	GAS FOR POWER (THERMS)	TOTAL BLR & COGEN GAS USE (THERMS)	ST TURBINE CAPACITY (kW)	ST TURBINE ENERGY (kWh)
JAN	326,281	407,852	924,760	802,953	1,727,713	0	0
FEB	288,436	360,545	824,440	720,603	1,545,043	0	0
MAR	272,801	341,001	838,568	772,909	1,611,477	0	0
APR	183,544	229,431	691,555	717,854	1,409,410	0	0
MAY	55,608	69,510	529,441	714,445	1,243,886	0	0
JUN	0	0	413,784	666,777	1,080,561	0	0
JUL	0	0	431,887	678,176	1,110,063	0	0
AUG	0	0	385,969	681,424	1,067,393	0	0
SEP	0	0	433,608	675,944	1,109,552	0	0
OCT	59,215	74,019	542,662	727,979	1,270,640	0	0
NOV	199,916	249,894	722,643	734,356	1,456,999	0	0
DEC	252,249	315,312	822,462	787,796	1,610,258	0	0
TOTALS	1,638,052	2,047,564	7,561,779	8,681,216	16,242,995		0
MAX							
MIN							
DATE	02/15						

TABLE 11-5
CHEMCO COGENERATION SUPPLEMENTAL COST – TURBINE

Ave Elect 0.094 $/kWh
FAC 0.005 $/kWh

MONTH	DAYS	SUPP'TL DEMAND (kW)			ON-PEAK ENERGY (kWh)		OFF-PEAK ENERGY (kWh)	SUPP'TL TOTAL (kWh)	SUPP'TL LD FACT (HOURS)	SUPP'TL COST ($)
JAN	31	441.7			(600,017)		0	0	0	4,617
FEB	28	845.7			(129,420)		0	0	0	8,657
MAR	31	1,044.6			88,916		0	88,916	85	15,842
APR	30	1,145.7			332,669		0	332,669	290	28,127
MAY	31	1,407.5			403,677		0	403,677	287	33,940
JUN	30	1,575.0			367,175		0	367,175	233	33,972
JUL	31	1,874.2			494,011		0	494,011	264	42,672
AUG	31	1,771.6			681,002		0	681,002	384	50,061
SEP	30	1,526.4			406,043		0	406,043	266	35,236
OCT	31	1,326.0			290,006		0	290,006	219	28,010
NOV	30	1,418.1			348,272		0	348,272	246	31,554
DEC	31	1,053.8			41,758		0	41,758	40	13,452
TOTALS	365				2,724,092	0	0	3,453,529	193	326,140
MAX		1,874.2	0.0	0.0					384	
MIN		441.7	0.0	0.0					0	
DATE	02/15									

SUP ST DEM M LB/HR 0 $/MO
SUPP'L PGA 0 $/TH
GAS HTG 2.60 $/MMBtu FAC STEAM 0 $/M LB

MONTH	SUPP'TL GAS USE (THERMS)	SUPP'TL NON-DISPL'B (THERMS)	SUPP'TL DISPL'BLE (THERMS)	SUPP'TL OIL USE (GAL)	SUPP'TL STEAM USE (M LB)	TOTAL SUP. FUEL (MMBTU)	TOTAL SUP. FUEL ($)
JAN	1,094,840	170,080	924,760	0		109,484	284,658
FEB	1,002,633	178,193	824,440	0		100,263	260,685
MAR	1,040,702	202,134	838,568	0		104,070	270,583
APR	826,364	134,809	691,555	0		82,636	214,855
MAY	700,614	171,173	529,441	0		70,061	182,160
JUN	555,833	142,049	413,784	0		55,583	144,517
JUL	563,341	131,454	431,887	0		56,334	146,469
AUG	501,807	115,838	385,969	0		50,181	130,470
SEP	649,878	216,270	433,608	0		64,988	168,968
OCT	760,203	217,541	542,662	0		76,020	197,653
NOV	898,322	175,679	722,643	0		89,832	233,564
DEC	985,211	162,749	822,462	0		98,521	256,155
TOTALS	9,579,750	2,017,971	7,561,779	0	0	957,975	2,490,735
MAX							
MIN							
DATE	02/15						

TABLE 11-6
CHEMCO COGENERATION SYSTEM COSTS – TURBINE

O&M	0.005 $/kWh	GAS POWER	0.2600 $/TH
OIL	0.7 $/GAL	BUYBACK RATE	0.02 $/kWh
DIST'N O&M	0.00 $/M LB	STANDBY RATE	4.00 $/kW

MONTH	DAYS	COST GAS POWER ($)	MAINT COST ($)	LABOR COST ($)	STBY COST ($)	POWER SALES ($)	STANDBY USE CHARGE ($)	OWN & ADM COST ($)	STEAM SYS COST ($)	TOTAL ON-SITE ($)
JAN	31	301,692	29,924	0	29,516	(12,000)	44,224	2,458	0	395,815
FEB	28	273,088	26,855	0	29,516	(2,588)	0	2,458	0	329,329
MAR	31	290,412	28,804	0	29,516	0	0	2,458	0	351,191
APR	30	264,996	26,753	0	29,516	0	0	2,458	0	323,723
MAY	31	258,787	26,626	0	29,516	0	0	2,458	0	317,387
JUN	30	238,462	24,849	0	29,516	0	0	2,458	0	295,286
JUL	31	242,205	25,274	0	29,516	0	0	2,458	0	299,454
AUG	31	240,648	25,395	0	29,516	0	44,224	2,458	0	342,241
SEP	30	245,099	25,191	0	29,516	0	0	2,458	0	302,265
OCT	31	265,319	27,130	0	29,516	0	0	2,458	0	324,423
NOV	30	272,932	27,368	0	29,516	0	0	2,458	0	332,274
DEC	31	292,609	29,359	0	29,516	0	0	2,458	0	353,943
TOTALS	365	3,186,250	323,528	0	354,192	(14,589)	88,448	29,500	0	3,967,329

DATE 02/15

TABLE 11-7
CHEMCO OPERATING COST SUMMARY
TURBINE

CONVENTIONAL SYSTEM	
Purchased Power	$4,006,940
Heating Gas	3,845,799
Fuel Oil	0
Purchased Steam	0
Total	$7,852,739
COGENERATION SYSTEM	
Purchased Power	$ 326,140
Heating Gas	2,490,735
Fuel Oil	0
Purchased Steam	0
Fuel for Power	3,186,250
Maintenance	323,528
Standby Power	442,640
Own & Admin	29,500
Labor Costs	0
Dist'n System	0
Total	$6,798,792
POWER SALES REVENUE	(14,589)
TOTAL OPERATING COST SAVING	$1,068,536

TABLE 11-8

Cash Flow

General Inflation	5.0%			CHEMCO	Turbine	
All Rates Relative To General Inflation					Tax Rates	
Year	1-5	6-10	11 & up		Federal	0.00%
Purchase Power	1.00%				State	0.00%
Htg Gas	2.00%				Local	0.00%
Fuel Oil					Property	0.00%
Purchased Steam						
Fuel For Power					Tax Life	15
O & M					Disc Rate	13.00%
Power Sales					Econ Life	15

PROJECT YEAR	0	1	2	3	4	5	6	7	8	9	10
YEAR	1989	1990	1991	1992	1993	1994	1995	1996	1997	1998	1999
Conventional Cost											
Purchased Power	4,006,940	4,247,357	4,502,198	4,772,330	5,058,670	5,362,190	5,630,299	5,911,814	6,207,405	6,517,775	6,843,664
Heating Gas	3,845,799	4,115,005	4,403,055	4,711,269	5,041,058	5,393,932	5,663,629	5,946,810	6,244,151	6,556,358	6,884,176
Fuel Oil	0	0	0	0	0	0	0	0	0	0	0
Purchased Steam	0	0	0	0	0	0	0	0	0	0	0
Total Conventional	7,852,739	8,362,362	8,905,253	9,483,599	10,099,728	10,756,122	11,293,928	11,858,625	12,451,556	13,074,134	13,727,840
Cogeneration Cost											
Purchased Power	4,006,940	4,247,357	4,502,198	388,438	411,744	436,449	458,271	481,184	505,244	530,506	557,031
Heating Gas	3,845,799	4,115,005	4,403,055	3,051,257	3,264,845	3,493,385	3,668,054	3,851,456	4,044,029	4,246,231	4,458,542
Fuel Oil	0	0	0	0	0	0	0	0	0	0	0
Purchased Steam	0	0	0	0	0	0	0	0	0	0	0
Fuel for Power	0	0	0	3,688,483	3,872,907	4,066,553	4,269,880	4,483,374	4,707,543	4,942,920	5,190,066
Maintenance	0	0	0	374,524	393,250	412,912	433,558	455,236	477,998	501,897	526,992
Standby Power	0	0	0	527,191	558,823	592,352	621,970	653,068	685,722	720,008	756,008
Own & Admin	0	0	0	34,150	35,857	37,650	39,533	41,509	43,585	45,764	48,052
Labor Costs	0	0	0	0	0	0	0	0	0	0	0
Dist'n System	0	0	0	0	0	0	0	0	0	0	0
Total Cogeneration	7,852,739	8,362,362	8,905,253	8,064,043	8,537,426	9,039,300	9,491,265	9,965,829	10,464,120	10,987,326	11,536,692
Power Sales	0	0	0	(16,888)	(17,733)	(18,619)	(19,550)	(20,528)	(21,554)	(22,632)	(23,764)
Net Cogeneration Cost	7,852,739	8,362,362	8,905,253	8,047,155	8,519,694	9,020,681	9,471,715	9,945,301	10,442,566	10,964,694	11,512,929
Total Oper Cost Saving	0	0	0	1,436,445	1,580,034	1,735,441	1,822,213	1,913,324	2,008,990	2,109,439	2,214,911
User Share		0	0	0	0	0	0	0	0	0	0
User payments		0	0	0	0	0	0	0	0	0	0
Depreciation		0	0	439,200	439,200	439,200	439,200	439,200	439,200	439,200	439,200
Owner Taxes		0	0	0	0	0	0	0	0	0	0
Interest		0	0	0	0	0	0	0	0	0	0
Before Tax Income		0	0	997,245	1,140,834	1,296,241	1,383,013	1,474,124	1,569,790	1,670,239	1,775,711
Income Tax		0	0	0	0	0	0	0	0	0	0
After Tax Income		0	0	997,245	1,140,834	1,296,241	1,383,013	1,474,124	1,569,790	1,670,239	1,775,711
Depreciation		0	0	439,200	439,200	439,200	439,200	439,200	439,200	439,200	439,200
Principal		0	0	0	0	0	0	0	0	0	0
Equity	0	2,196,000	4,392,000	0	0	0	0	0	0	0	0
Net to Owner	0	(2,196,000)	(4,392,000)	1,436,445	1,580,034	1,735,441	1,822,213	1,913,324	2,008,990	2,109,439	2,214,911
Present Value	0	(1,943,363)	(3,439,580)	995,528	969,064	941,928	875,243	813,279	755,701	702,200	652,487

TABLE 11-8 (Cont'd)

CHEMCO

Project Cost	
Equipment Cost	2,950,000
Installation Cost	2,450,000
Development Cost	648,000
Contingency	540,000
Int Dur Const	0
Total	6,588,000

Debt			
Equity	100.00%		
Int Rate	10.00%		
Term	15		
Payments	$0		
Shared Savings			
Year	1-5	6-10	11 & up
% To Host	0	0	0

PROJECT YEAR	11	12	13	14	15	16	17	18	19	20	21	22
YEAR	2000	2001	2002	2003	2004	2005	2006	2007	2008	2009	2010	2011
Conventional Cost												
Purchased Power	7,185,847	7,545,140	7,922,397	8,318,516	8,734,442	9,171,164	9,629,722	10,111,209	10,616,769	11,147,607	11,704,988	12,290,237
Heating Gas	7,228,385	7,589,804	7,969,295	8,367,759	8,786,147	9,225,455	9,686,727	10,171,064	10,679,617	11,213,598	11,774,278	12,362,991
Fuel Oil	0	0	0	0	0	0	0	0	0	0	0	0
Purchased Steam	0	0	0	0	0	0	0	0	0	0	0	0
Total Conventional	14,414,232	15,134,944	15,891,691	16,686,276	17,520,589	18,396,619	19,316,450	20,282,272	21,296,386	22,361,205	23,479,265	24,653,229
Cogeneration Cost												
Purchased Power	584,883	614,127	644,833	677,075	710,929	746,475	783,799	822,989	864,138	907,345	952,712	1,000,348
Heating Gas	4,681,469	4,915,543	5,161,320	5,419,386	5,690,355	5,974,873	6,273,617	6,587,298	6,916,663	7,262,496	7,625,620	8,006,901
Fuel Oil	0	0	0	0	0	0	0	0	0	0	0	0
Purchased Steam	0	0	0	0	0	0	0	0	0	0	0	0
Fuel for Power	5,449,569	5,722,048	6,008,150	6,308,558	6,623,986	6,955,185	7,302,944	7,668,091	8,051,496	8,454,071	8,876,774	9,320,613
Maintenance	553,342	581,009	610,059	640,562	672,591	706,220	741,531	778,608	817,538	858,415	901,336	946,402
Standby Power	793,809	833,499	875,174	918,933	964,879	1,013,123	1,063,779	1,116,968	1,172,817	1,231,458	1,293,030	1,357,682
Own & Admin	50,455	52,978	55,627	58,408	61,328	64,395	67,615	70,995	74,545	78,272	82,186	86,295
Labor Costs	0	0	0	0	0	0	0	0	0	0	0	0
Dist'n System	0	0	0	0	0	0	0	0	0	0	0	0
Total Cogeneration	12,113,527	12,719,203	13,355,163	14,022,922	14,724,068	15,460,271	16,233,285	17,044,949	17,897,196	18,792,056	19,731,659	20,718,242
Power Sales	(24,952)	(26,199)	(27,509)	(28,885)	(30,329)	(31,845)	(33,438)	(35,110)	(36,865)	(38,708)	(40,644)	(42,676)
Net Cogeneration Cost	12,088,575	12,693,004	13,327,654	13,994,037	14,693,739	15,428,426	16,199,847	17,009,839	17,860,331	18,753,348	19,691,015	20,675,566
Total Oper Cost Saving	2,325,657	2,441,940	2,564,037	2,692,239	2,826,851	2,968,193	3,116,603	3,272,433	3,436,055	3,607,857	3,788,250	3,977,663
User Share	0	0	0	0	0	0	0	0	0	0	0	0
User payments	0	0	0	0	0	0	0	0	0	0	0	0
Depreciation	439,200	439,200	439,200	439,200	439,200	439,200	439,200	439,200	0	0	0	0
Owner Taxes	0	0	0	0	0	0	0	0	0	0	0	0
Interest	0	0	0	0	0	0	0	0	0	0	0	0
Before Tax Income	1,886,457	2,002,740	2,124,837	2,253,039	2,387,651	2,528,993	2,677,403	2,833,233	3,436,055	3,607,857	3,788,250	3,977,663
Income Tax	0	0	0	0	0	0	0	0	0	0	0	0
After Tax Income	1,886,457	2,002,740	2,124,837	2,253,039	2,387,651	2,528,993	2,677,403	2,833,233	3,436,055	3,607,857	3,788,250	3,977,663
Depreciation	439,200	439,200	439,200	439,200	439,200	439,200	439,200	439,200	0	0	0	0
Principal	0	0	0	0	0	0	0	0	0	0	0	0
Equity	0	0	0	0	0	0	0	0	0	0	0	0
Net to Owner	2,325,657	2,441,940	2,564,037	2,692,239	2,826,851	0	0	0	0	0	0	0
Present Value	606,293	563,370	523,485	486,424	451,987	0	0	0	0	0	0	0

TABLE 11-9
HOSPITAL CONVENTIONAL SYSTEM COST

FAC 0.004 $/kWh   AVE COST 0.070 $/kWh

MONTH	DAYS	ON PEAK DEMAND (kW)			ON PEAK ENERGY (kWh)		OFF PEAK ENERGY (kWh)	TOTAL ENERGY (kWh)	CONV'L LD FACT (HOURS)	COST ($)
JAN	31	2,278.8			361,081		1,137,179	1,498,260	657	99,722
FEB	28	2,309.2			367,707		986,793	1,354,500	587	95,240
MAR	31	2,312.4			384,648		1,018,152	1,402,800	607	97,373
APR	30	2,363.8			384,023		1,132,657	1,516,680	642	102,289
MAY	31	2,501.5			361,801		963,479	1,325,280	530	97,303
JUN	30	2,768.0			424,995		1,260,825	1,685,820	609	115,979
JUL	31	2,969.1			452,168		1,250,272	1,702,440	573	120,521
AUG	31	2,969.0			468,760		1,262,900	1,731,660	583	121,911
SEP	30	2,771.8			444,404		1,328,956	1,773,3[illegible]	640	119,575
OCT	31	2,482.9			375,430		1,028,570	1,404,000	565	100,087
NOV	30	2,318.7			363,012		979,488	1,342,500	579	94,872
DEC	31	2,270.2			341,649		972,891	1,314,540	579	92,611
TOTALS	365				4,729,678	0	13,322,162	18,051,840	596	1,257,483
MAX		2,969.1	ERR	0.0					657	
MIN		2,270.2	ERR	0.0					530	
DATE	02/15									
Revision		1.4								

AVE COST 3.33 $/MMBtu
ALT HTG FUEL 150,000 BTU/GAL
MAX STEAM DEMAND 0 M Lb/HR
BOILER EFFICIENCY 84.0%

MONTH	CONV'L GAS USE (THERMS)	NON-DISPL'B GAS USE (THERMS)	DISPL'BLE GAS USE (THERMS)	FUEL OIL USE (GAL)	STEAM USE (M LB)	TOTAL DISP FUEL (MMBTU)	TOTAL FUEL COST ($)	BOILER EFFIC. (%)
JAN	189,351		189,351		15,657	18,935	63,209	82.7
FEB	178,975		178,975		14,708	17,897	59,831	82.2
MAR	170,507		170,507		14,590	17,051	57,075	85.6
APR	137,299		137,299		11,589	13,730	46,266	84.4
MAY	119,505		119,505		10,242	11,951	40,474	85.7
JUN	215,210		215,210		18,362	21,521	71,626	85.3
JUL	304,282		304,282		25,925	30,428	100,619	85.2
AUG	333,533		333,533		28,345	33,353	110,140	85.0
SEP	328,579		328,579		27,861	32,858	108,528	84.8
OCT	179,178		179,178		15,168	17,918	59,897	84.7
NOV	120,234		120,234		10,253	12,023	40,711	85.3
DEC	123,362		123,362		10,854	12,336	41,729	88.0
TOTALS	2,400,014	0	2,400,014	0	203,554	240,001	800,104	84.9
MAX								88.0
MIN								82.2
DATE	02/15							

FIGURE 11-20
MONTHLY DEMAND – HOSPITAL

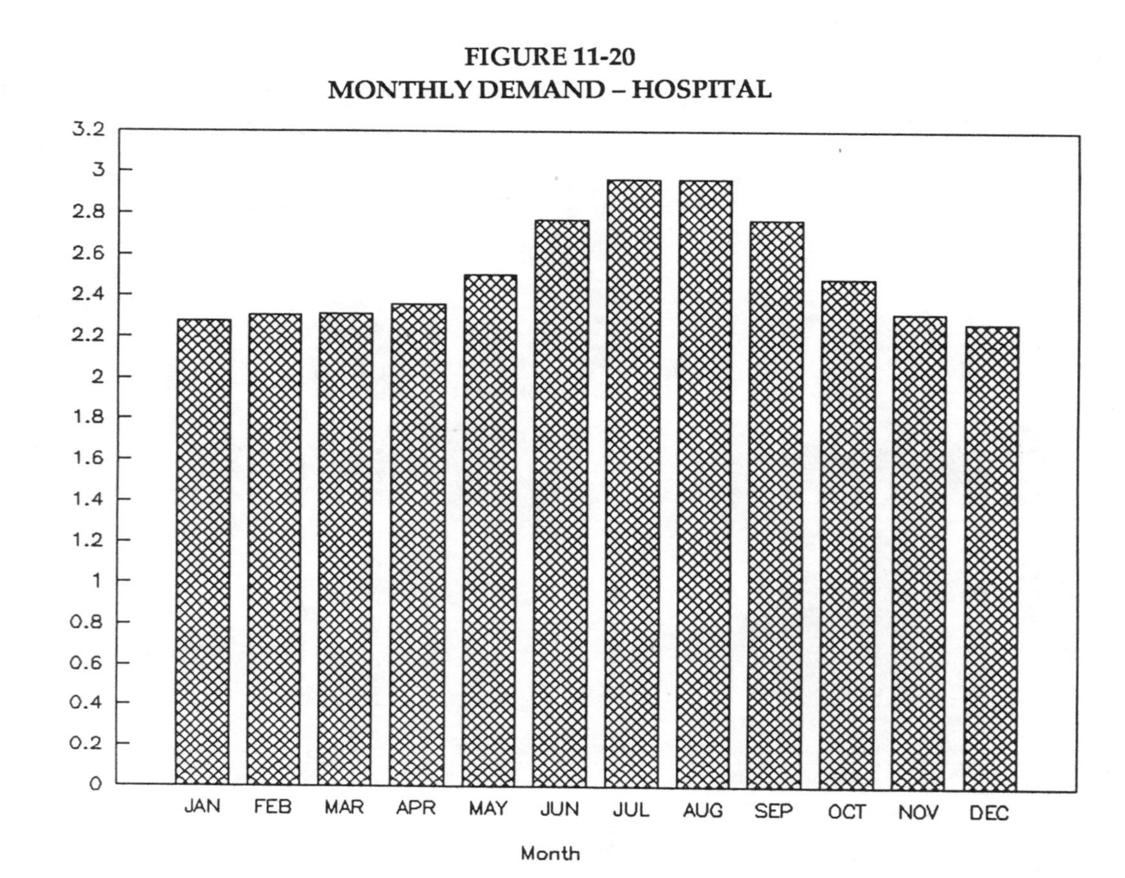

FIGURE 11-21
MONTHLY ELECTRIC USAGE - HOSPITAL

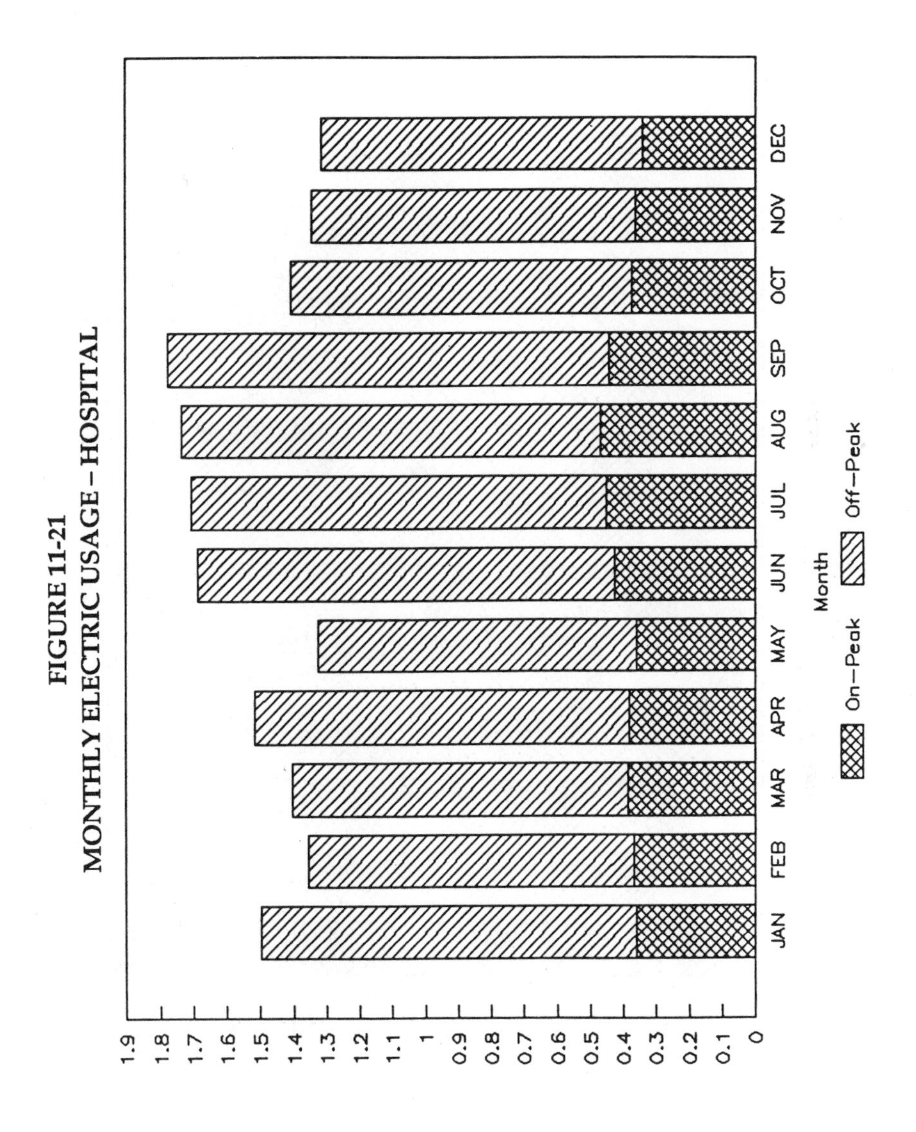

FIGURE 11-22
MONTHLY LOAD FACTORS – HOSPITAL

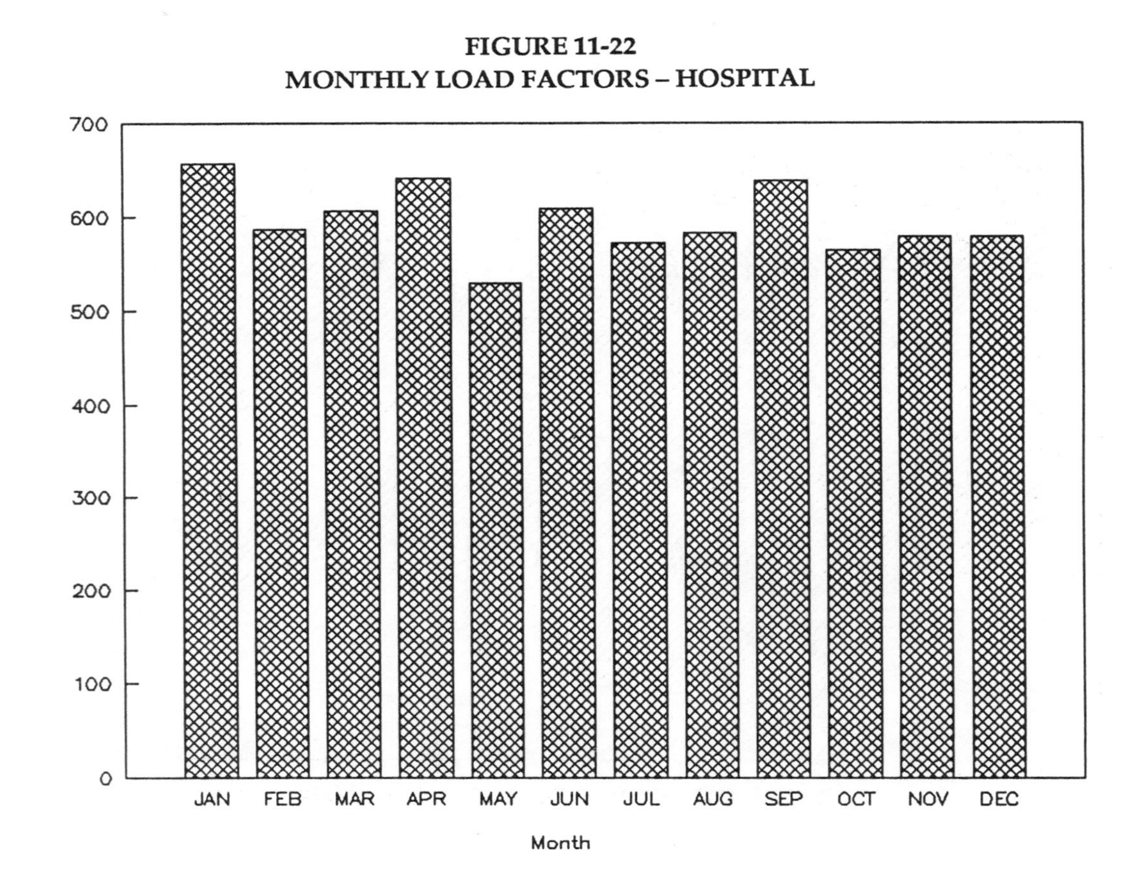

FIGURE 11-23
MONTHLY GAS USAGE – HOSPITAL

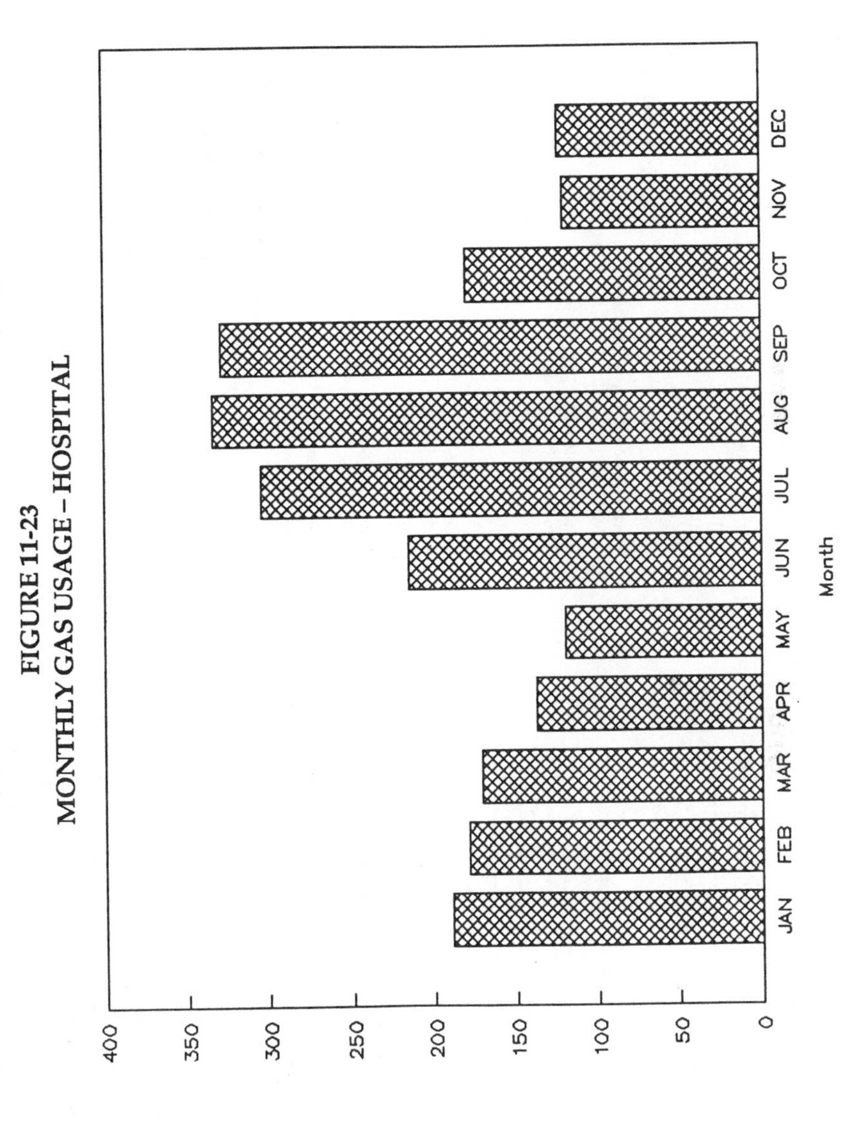

TABLE 11-10
WALKTHROUGH WORKSHEET

SITE: HOSPITAL

ELECTRIC UTILITY: ________

FUEL SUPPLIER: ________

TECHNICAL FEASIBILITY

HVAC System ________

Compatibility with Cogeneration (Table 9–3) ________

A. HVAC retrofit costs (Add to "M") $ 0

B. Minimum Demand 2,270 kW

ECONOMIC FEASIBILITY

C. Average Purchased Power Cost $ .0700 /kWh

D. Standby Charge $ 2.75 /kW/Month

E. Heating Fuel Cost $ 3.33 /MMBtu

F. Cogeneration System Fuel Cost $ 2.89 /MMBtu

G. Lowest Monthly Fuel Requirement 12,000 MMBtu/Month

H. Monthly Electric Output (See Table 9–4 for value of HRF)
 .8 x "G" / HRF = 1,371,000 kWh/Month

I. Maximum Installed Capacity (See Table 9–5 for value of HOURS)
 "H" / HOURS = 1,912 kW

J. System Size (Less than or equal to "B" or "I") 1,910 kW

K. System Unit Cost (See Figure 9–5 for COSTS) $ 1,000 /kW

L. Cogeneration System Cost
 "J" x "K" = $ 1,910,000

M. PROJECT INVESTMENT
 "A" + "L" = $ 1,910,000

N. Annual Electric Output
 12 x "J" x HOURS = 16.43×10^6 kWh/Year

O. Annual Fuel Use (See Table 9–4 for value for FUEL)
 "N" x FUEL = 219,000 MMBtu

P. Annual Cogeneration Fuel Cost
 "F" x "O" = $ 632,000

Q. Annual O & M Cost (See Table 9–6 for value for OM)
 "N" x OM = $ 82,000

R. Annual Standby Costs
 "D" x "J" x 12 = $ 63,000

S. TOTAL OPERATING COSTS
 "P" + "Q" + "R" = $ 777,000

T. Recovered Thermal Energy
 "N" x HRF = 115,000 MMBtu/Year

U. Annual Conventional Fuel Cost $ 800,000

V. Saved Heating Fuel Costs
 "E" x "T" / .8 = $ 479,000

W. Saved Purchased Power Costs
 "C" x "N" = $ 1,150,000

X. TOTAL OPERATING COST REDUCTION
 "V" + "W" = $ 1,629,000

Y. ANNUAL SAVINGS
 "X" – "S" = $ 852,000

Z. SIMPLE PAYBACK
 "M" / "Y" = 2.24 Years

FIGURE 11-24
HOSPITAL

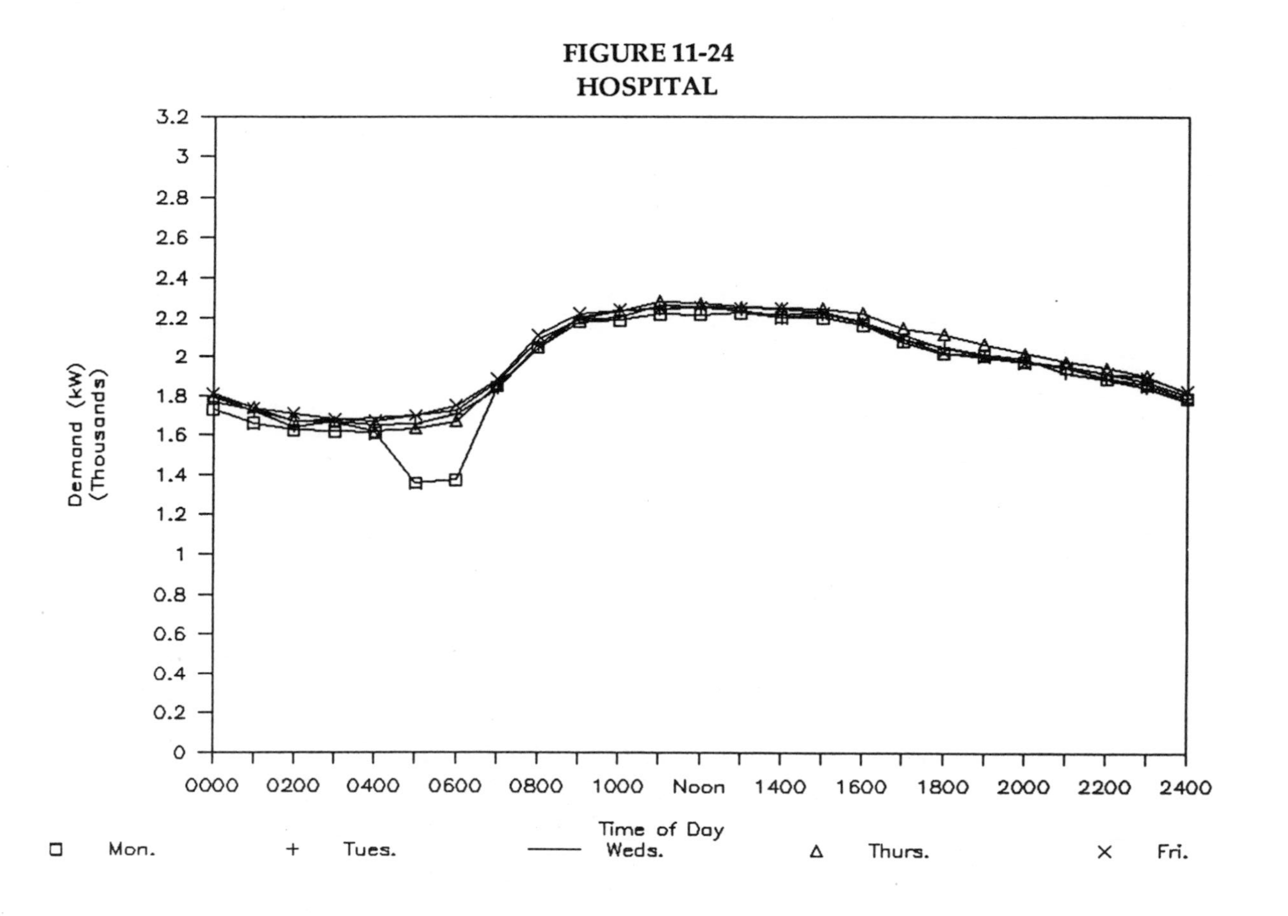

FIGURE 11-25
HOSPITAL

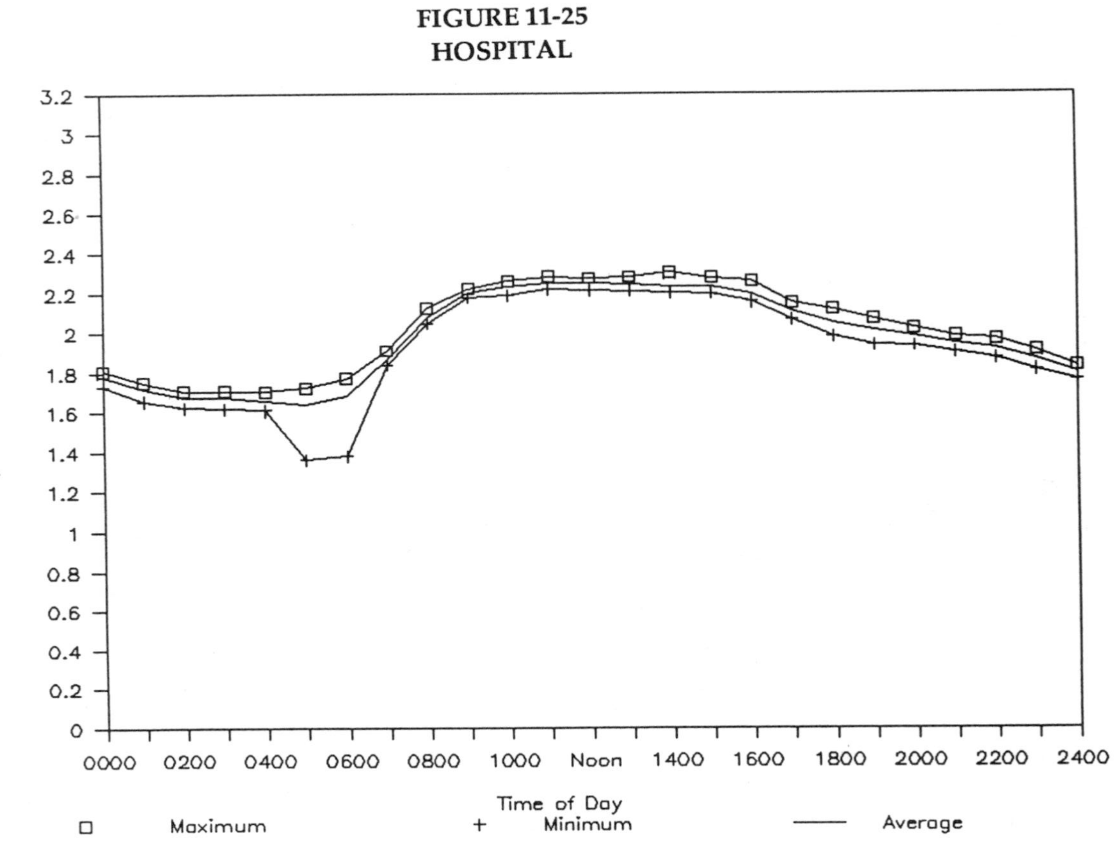

FIGURE 11-26
HOSPITAL

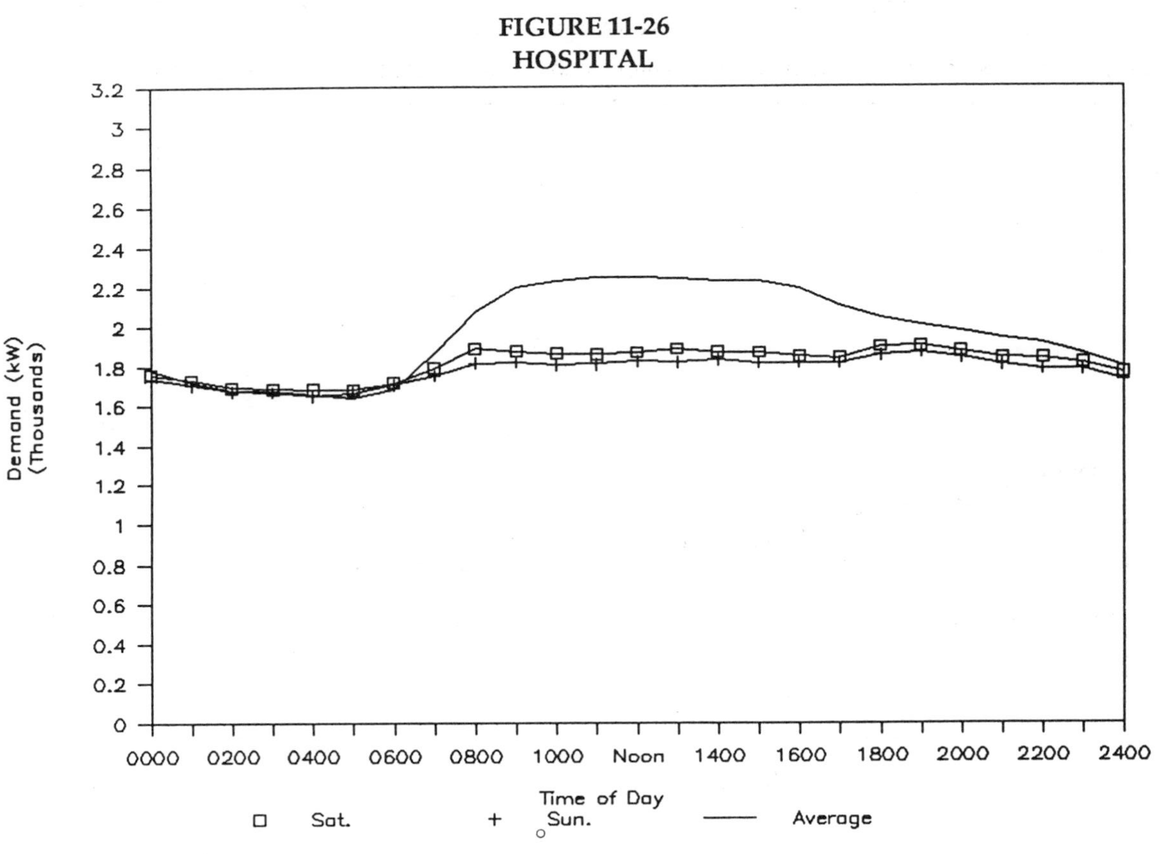

FIGURE 11-27
HOSPITAL – AVERAGE WEEKDAY

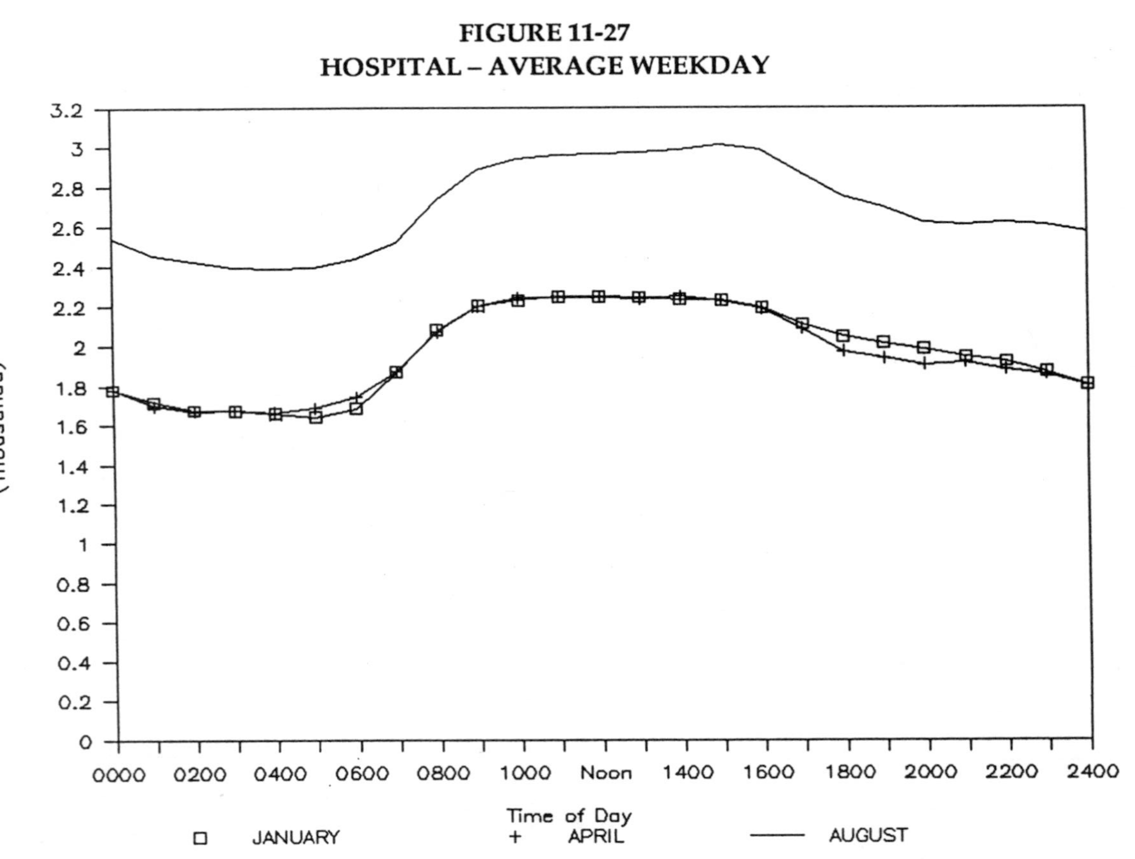

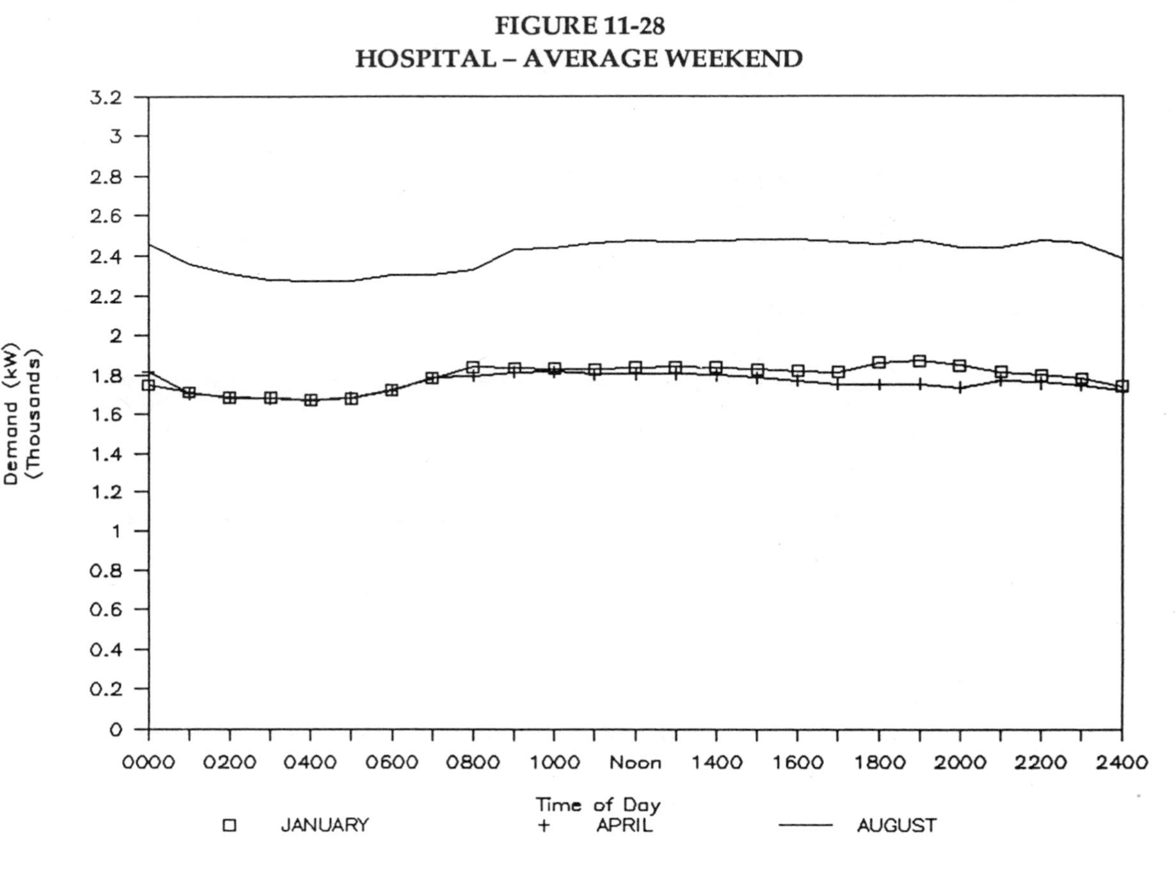

FIGURE 11-28
HOSPITAL - AVERAGE WEEKEND
Demand (kW)
(Thousands)
3.2
3
2.8
2.6
2.4
2.2
2
1.8
1.6
1.4
1.2
1
0.8
0.6
0.4
0.2
0
0000
0200
0400
0600
0800
1000
Noon
1400
1600
1800
2000
2200
2400
Time of Day
JANUARY
+ APRIL
AUGUST

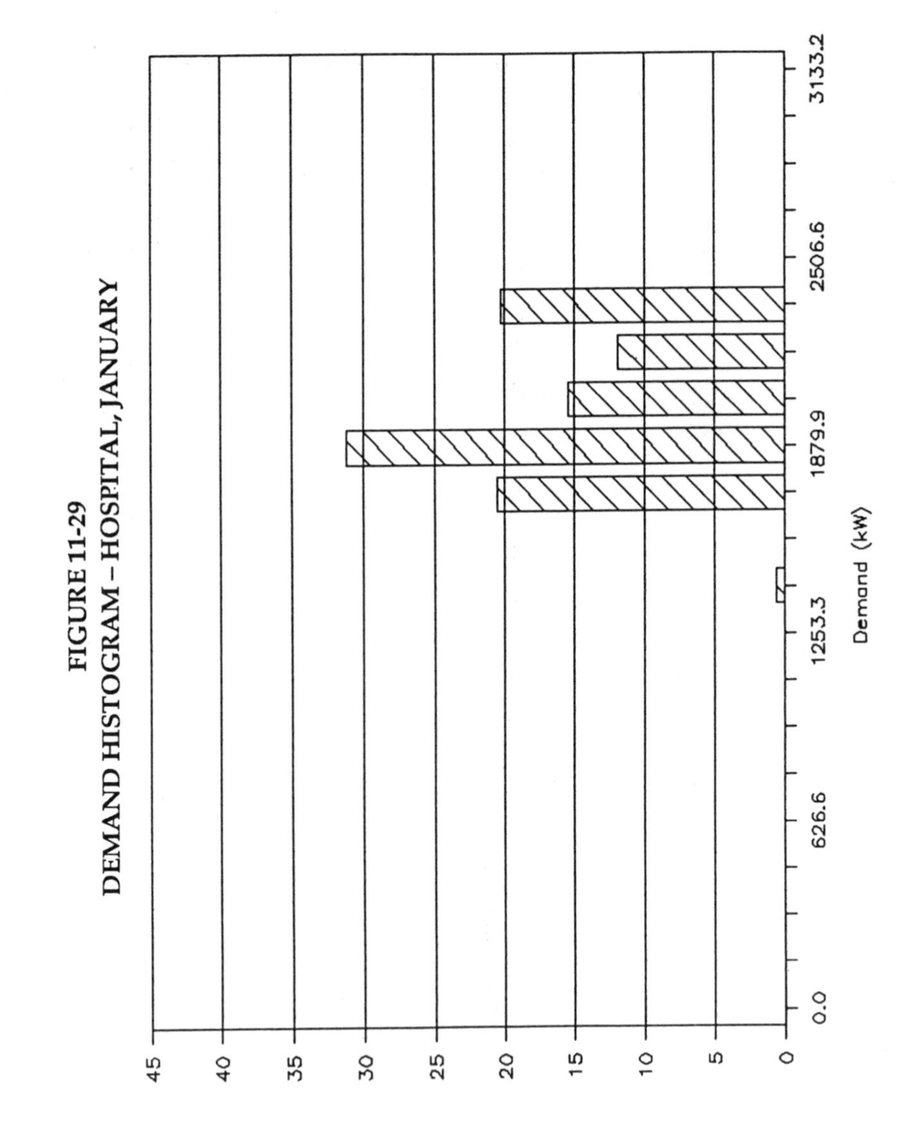

FIGURE 11-29
DEMAND HISTOGRAM – HOSPITAL, JANUARY

FIGURE 11-30
LOAD VS. DURATION – HOSPITAL, JANUARY

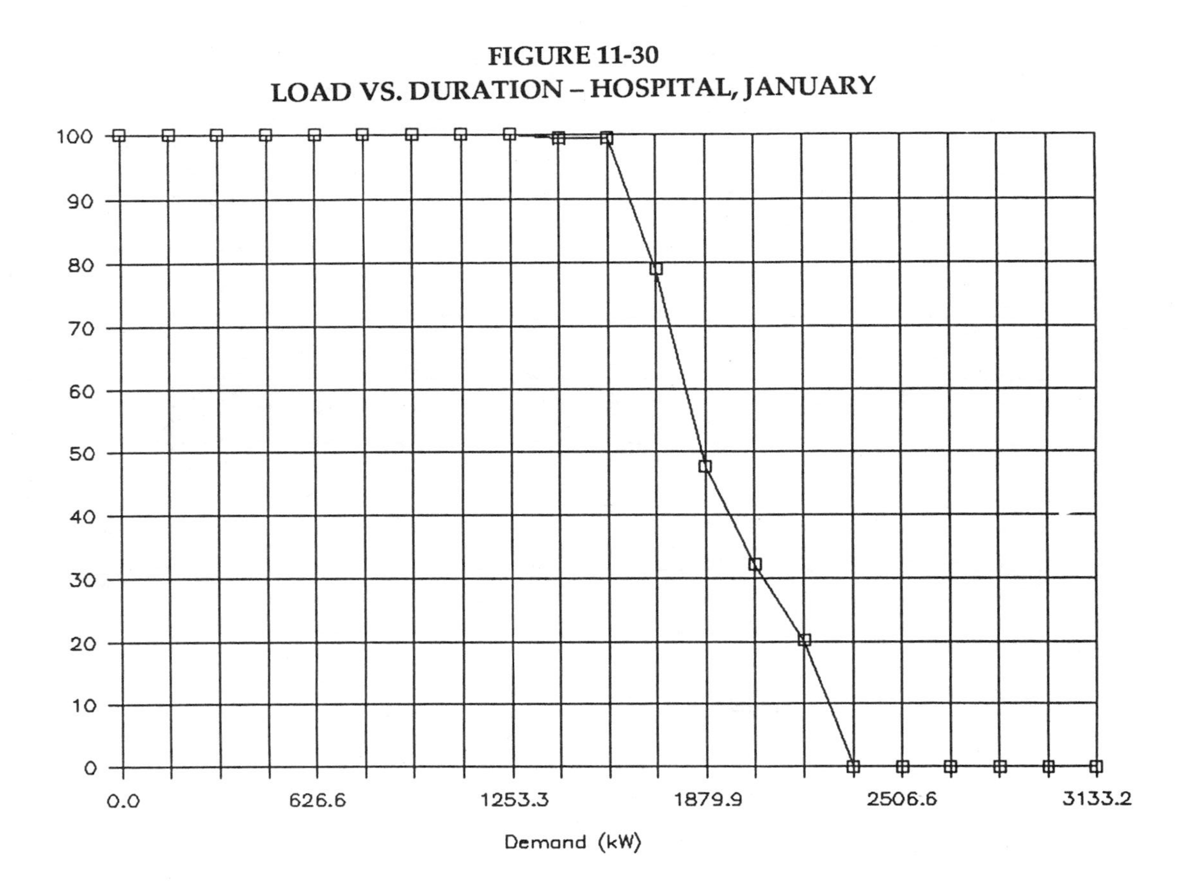

FIGURE 11-31
CAPACITY VS. LOAD – HOSPITAL, JANUARY

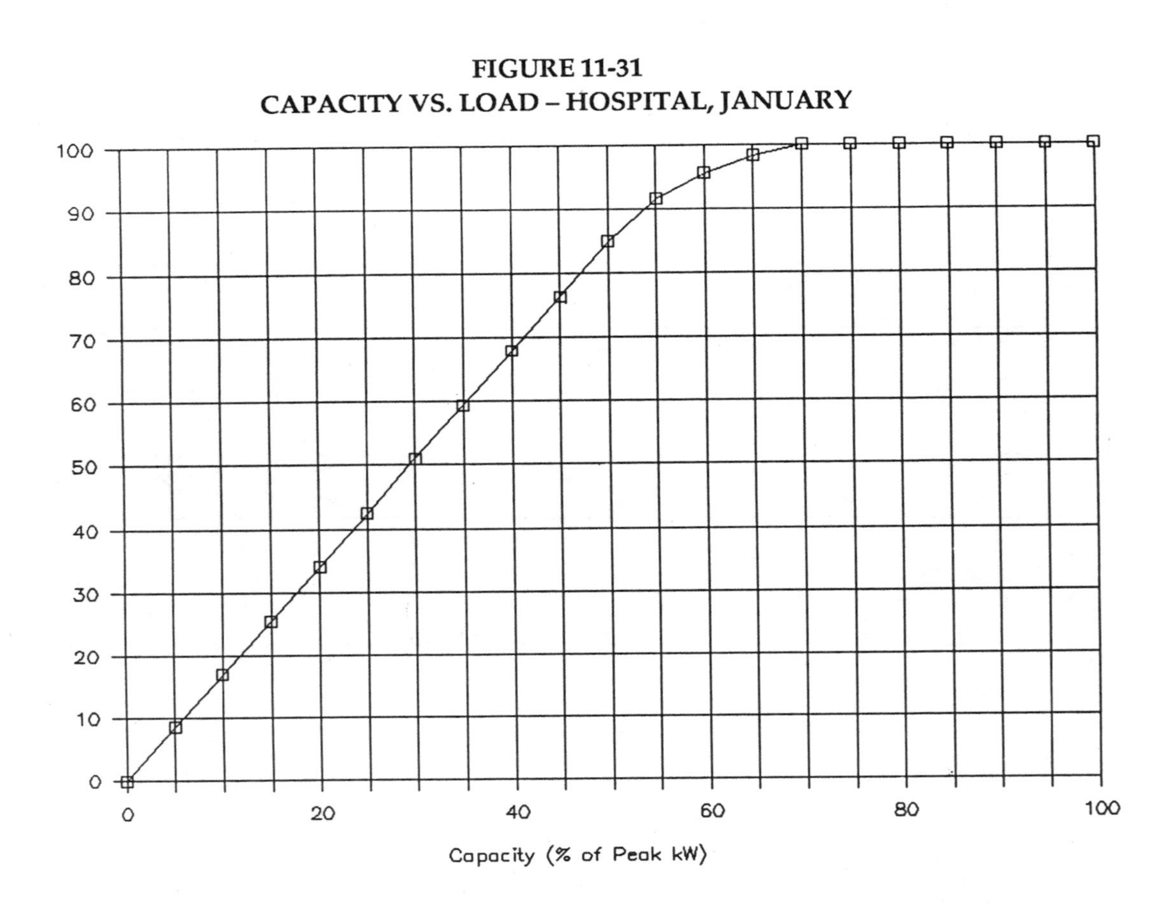

FIGURE 11-32
CAPACITY VS. LOAD - HOSPITAL

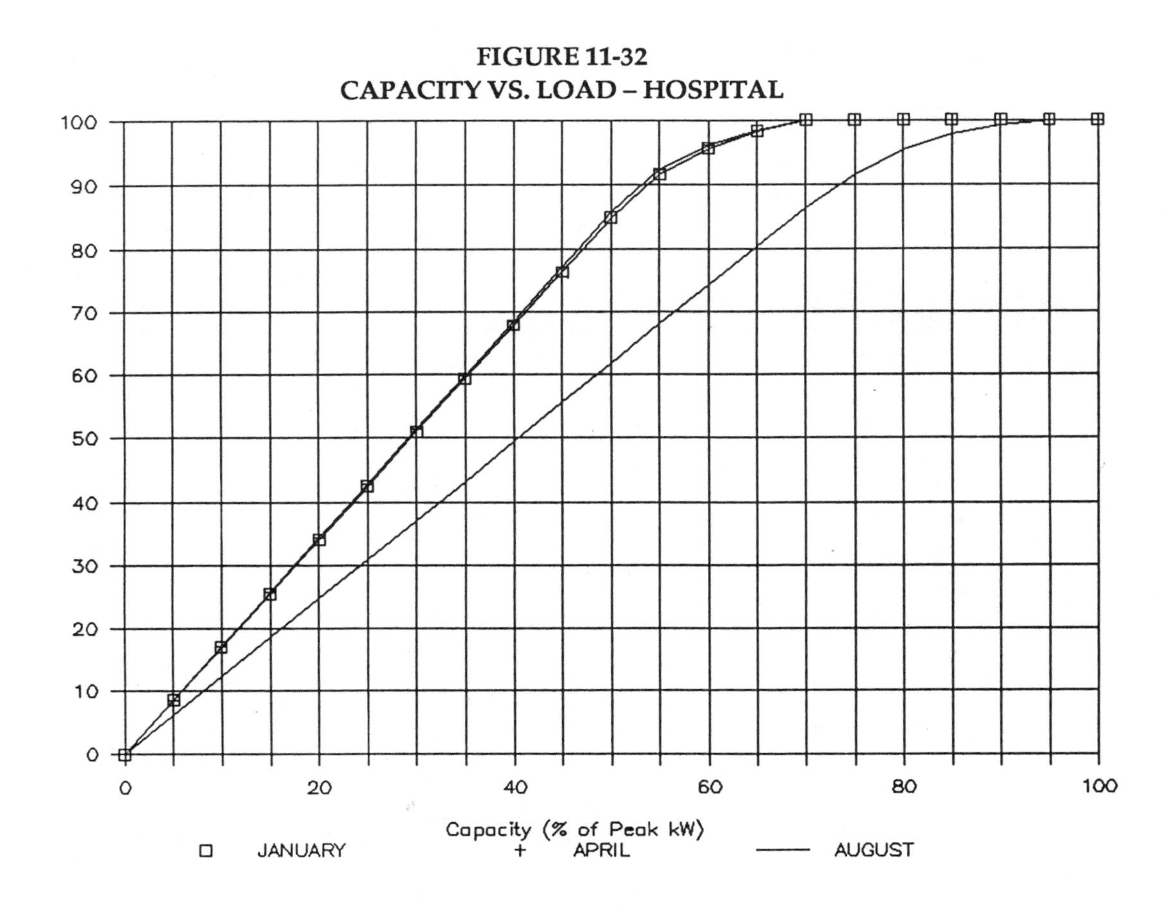

FIGURE 11-33
HOSPITAL – JANUARY

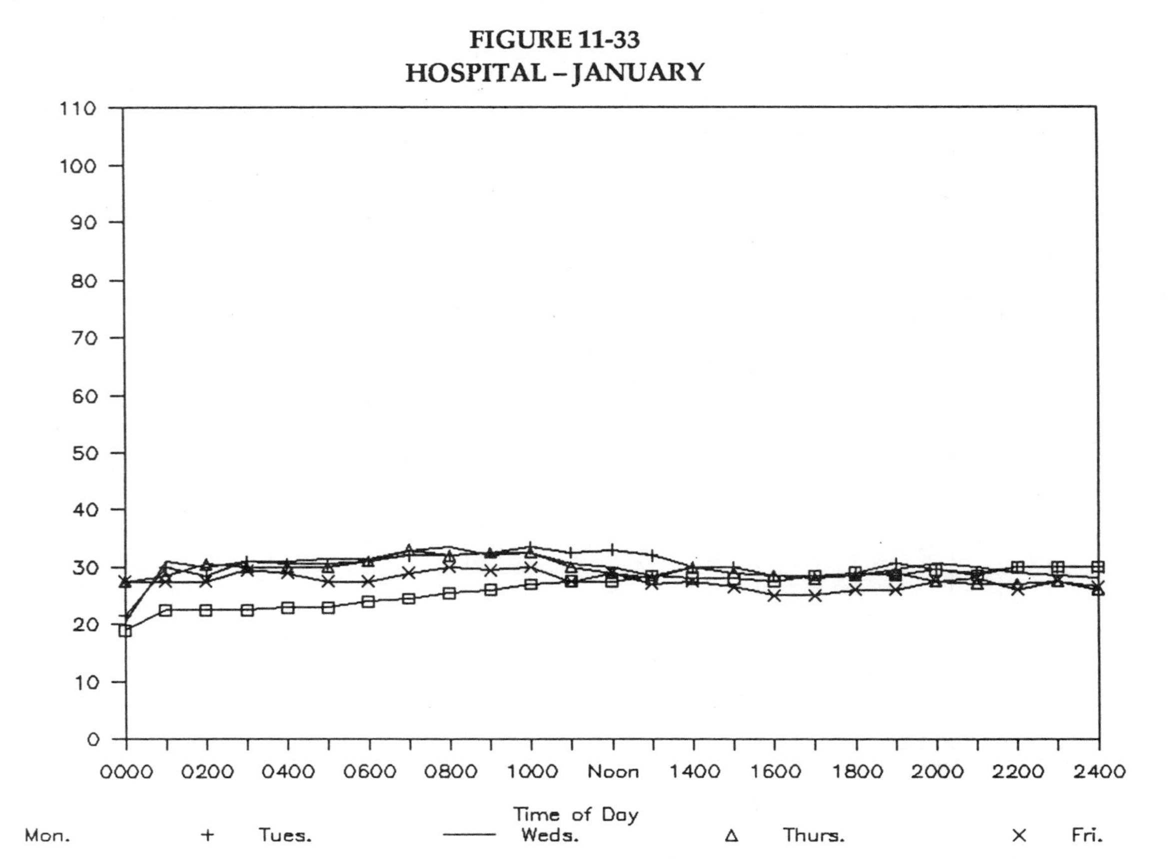

FIGURE 11-34
HOSPITAL

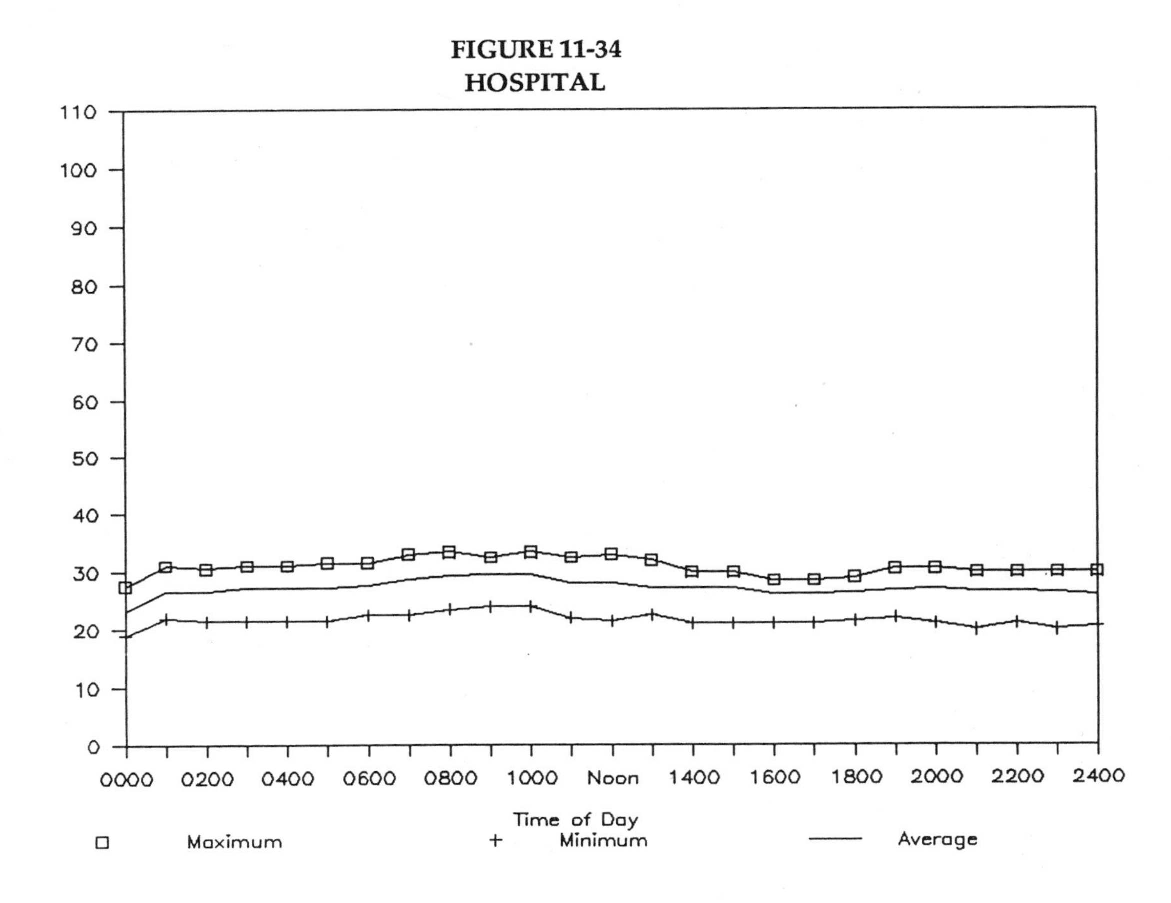

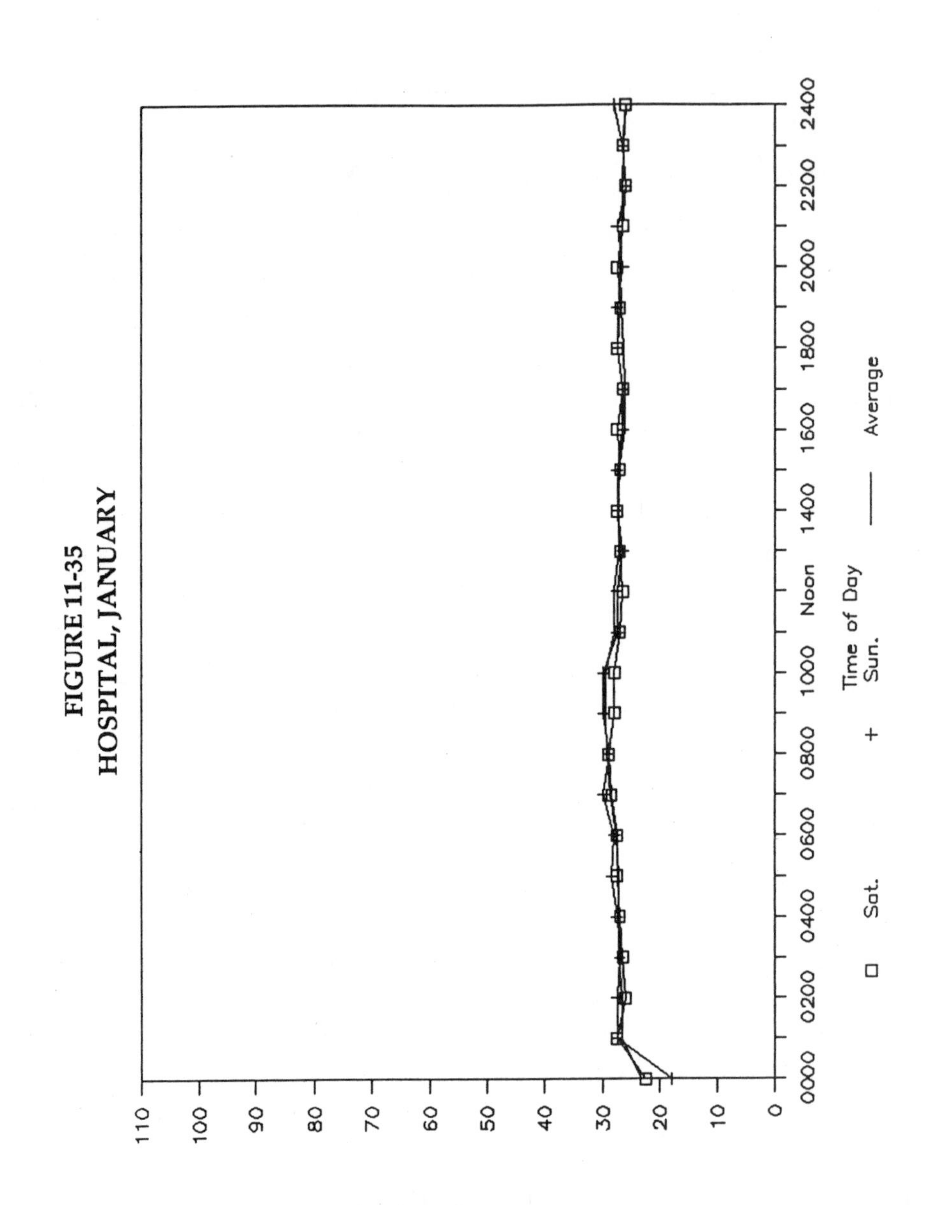

FIGURE 11-35
HOSPITAL, JANUARY

FIGURE 11-36
LOAD VS. DURATION – HOSPITAL, JANUARY

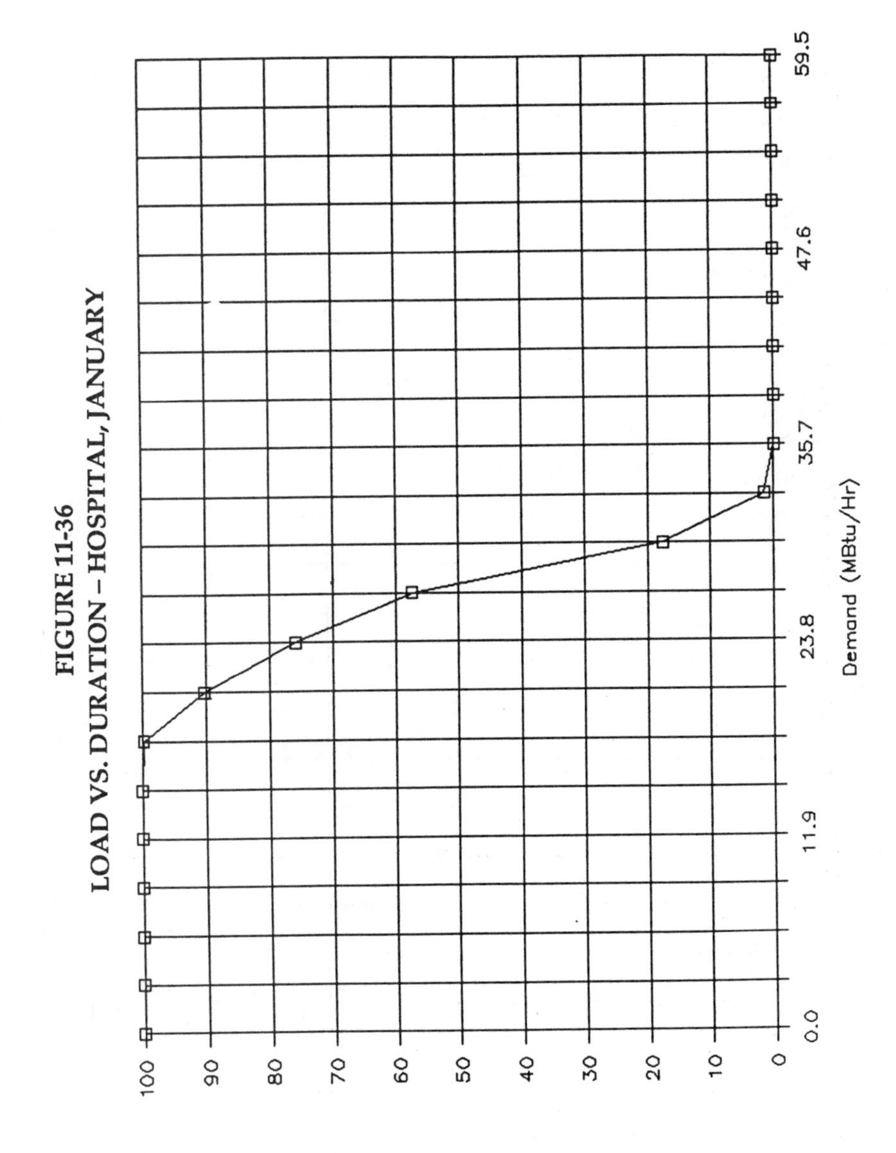

FIGURE 11-37
CAPACITY VS. LOAD – HOSPITAL, JANUARY

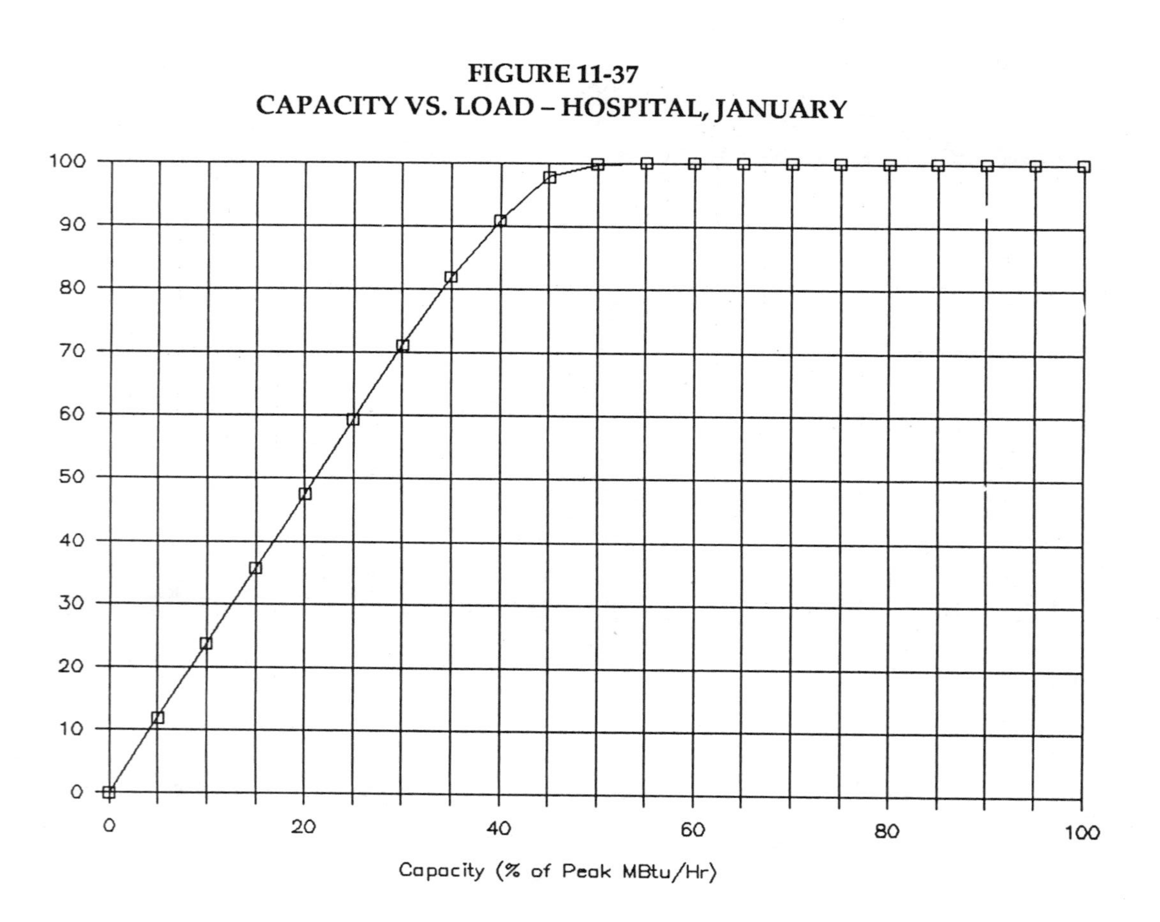

FIGURE 11-38
CAPACITY VS. LOAD – HOSPITAL

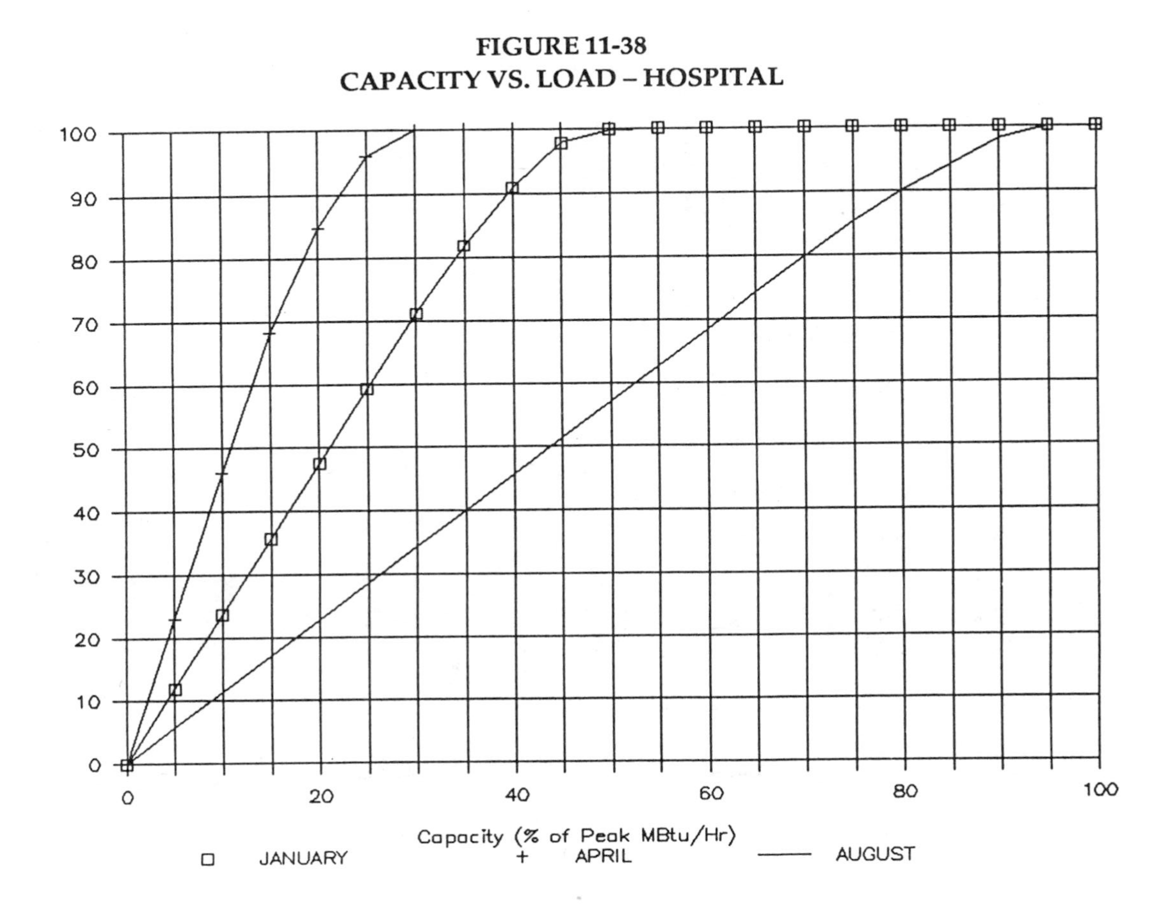

TABLE 11-11
HOSPITAL ENERGY ANALYSIS – TURBINE

CAPACITY	1,020 kW					MODE = 1
GEN CAP	1,200 kW	HEAT RATE	16,430 Btu/kWh	ALT HTG FUEL	150,000 Btu/Gal	
STEAM CAP	0 kW	HEAT REC	7,770 Btu/kWh	STEAM	1,000,000 Btu/Lb	DUCT BNR CAP 8,200 Btu/kWh
COMP LOSS	10 kW	HOUR/YR	8,500	BLR EFFICIENCY	84.0%	DUCT BNR EFF 90.0%

MONTH	DAYS	COGEN MIN ENG. CAP (kW)	COGEN AVE OUT (kW)	COGEN ENG. OUT (kWh)	COGEN SYS OUT (kWh)	REC'V HEAT AVAILABLE (THERMS)	TOTAL DISPL'BLE FUEL REQ'D (THERMS)	AVAIL/REQ'D HEAT	USED/AVAIL HEAT	SUPP'L HEAT (THERMS)	DUCT BURNER CAPACITY (THERMS)	DUCT BURNER GAS (THERMS)
JAN	31	1,121	1,147	828,106	819,987	64,344	189,351	40.5%	100.0%	94,711	67,905	75,450
FEB	28	1,111	1,139	742,380	735,102	57,683	178,975	38.4%	100.0%	92,656	60,875	67,639
MAR	31	1,076	1,106	798,210	790,384	62,021	170,507	43.3%	95.9%	83,740	65,453	72,726
APR	30	1,025	1,062	741,784	734,511	57,637	137,299	50.0%	100.0%	57,695	60,826	64,105
MAY	31	983	1,024	738,933	731,689	57,415	119,505	57.2%	100.0%	42,969	60,593	47,744
JUN	30	949	989	690,904	684,130	53,683	215,210	29.7%	57.8%	149,774	56,654	62,949
JUL	31	934	974	703,110	696,216	54,632	304,282	21.4%	80.2%	211,763	57,655	64,061
AUG	31	940	979	706,718	699,789	54,912	333,533	19.6%	87.5%	232,120	57,951	64,390
SEP	30	965	1,005	701,878	694,997	54,536	328,579	19.8%	86.8%	228,673	57,554	63,949
OCT	31	1,008	1,045	754,655	747,256	58,637	179,178	39.0%	100.0%	91,873	61,882	68,757
NOV	30	1,061	1,090	761,238	753,775	59,148	120,234	58.6%	70.9%	59,050	62,422	65,611
DEC	31	1,104	1,129	814,962	806,972	63,323	123,362	61.1%	68.0%	60,586	66,827	67,318
TOT	365			8,982,878	8,894,810	697,970	2,400,014			1,405,610	736,596	784,698
MAX								61.1%	100.0%			
MIN								19.6%	57.8%			
DATE	02/15											

MONTH	CONV'TL BOILER HEAT (THERMS)	CONV'L BLR GAS USE (THERMS)	SUPP'L GAS USE (THERMS)	GAS FOR POWER (THERMS)	TOTAL GAS USE (THERMS)	ST TURBINE CAPACITY (kW)	ST TURBINE ENERGY (kWh)
JAN	26,806	31,912	107,362	136,058	243,420	0	0
FEB	31,781	37,834	105,473	121,973	227,446	0	0
MAR	18,287	21,770	94,496	131,146	225,642	0	0
APR	0	0	64,105	121,875	185,980	0	0
MAY	0	0	47,744	121,407	169,150	0	0
JUN	93,120	110,857	173,806	113,516	287,322	0	0
JUL	154,108	183,462	247,523	115,521	363,044	0	0
AUG	174,169	207,345	271,734	116,114	387,848	0	0
SEP	171,119	203,713	267,662	115,319	382,980	0	0
OCT	29,991	35,703	104,461	123,990	228,451	0	0
NOV	0	0	65,611	125,071	190,682	0	0
DEC	0	0	67,318	133,898	201,216	0	0
TOTALS	699,381	832,596	1,617,295	1,475,887	3,093,182		0
MAX							
MIN							
DATE	02/15						

TABLE 11-12
HOSPITAL COGENERATION SUPPLEMENTAL COST – TURBINE

Ave Elect 0.077 $/kWh
FAC 0.004 $/kWh

MONTH	DAYS	SUPP'TL ON PEAK (kW)	SUPP'TL MAXIMUM (kW)		ON-PEAK ENERGY (kWh)		OFF-PEAK ENERGY (kWh)	SUPP'TL TOTAL (kWh)	SUPP'TL LD FACT (HOURS)	SUPP'TL COST ($)
JAN	31	1,168.4	1,166.0		164,383		513,890	678,273	581	50,358
FEB	28	1,209.1	1,216.8		191,684		427,714	619,398	509	48,829
MAR	31	1,247.3	1,249.2		195,965		416,451	612,416	490	48,670
APR	30	1,349.2	1,415.8		210,092		572,076	782,169	552	55,369
MAY	31	1,528.3	1,551.5		189,399		404,192	593,591	383	51,211
JUN	30	1,828.7	1,918.0		263,972		737,717	1,001,690	522	72,393
JUL	31	2,044.3	2,090.2		288,345		717,878	1,006,224	481	76,689
AUG	31	2,038.5	2,167.7		303,935		727,936	1,031,871	476	77,835
SEP	30	1,816.2	2,080.8		280,594		797,770	1,078,363	518	75,270
OCT	31	1,485.3	1,549.1		198,707		458,037	656,744	424	52,939
NOV	30	1,268.4	1,317.1		182,961		405,764	588,725	447	47,551
DEC	31	1,176.8	1,243.6		147,956		359,611	507,568	408	43,918
TOTALS	365				2,617,994	0	6,539,036	9,157,030	483	701,033
MAX		2,044.3	2,167.7	0.0					581	
MIN		1,168.4	1,166.0	0.0					383	
DATE	02/15									

SUP ST DEM M LB/HR 0 $/MO
SUPP'L PGA 0 $/TH
GAS HTG 3.37 $/MMBtu FAC STEAM 0 $/M LB

MONTH	SUPP'TL GAS USE (THERMS)	SUPP'TL NON-DISPL'B (THERMS)	SUPP'TL DISPL'BLE (THERMS)	SUPP'TL OIL USE (GAL)	SUPP'TL STEAM USE (M LB)	TOTAL SUP. FUEL (GALLONS)	TOTAL SUP. FUEL ($)
JAN	107,362	0	107,362	0		71,575	36,521
FEB	105,473	0	105,473	0		70,316	35,907
MAR	94,496	0	94,496	0		62,997	32,333
APR	64,105	0	64,105	0		42,737	22,441
MAY	47,744	0	47,744	0		31,829	17,116
JUN	173,806	0	173,806	0		115,871	58,149
JUL	247,523	0	247,523	0		165,015	82,144
AUG	271,734	0	271,734	0		181,156	90,025
SEP	267,662	0	267,662	0		178,441	88,699
OCT	104,461	0	104,461	0		69,641	35,577
NOV	65,611	0	65,611	0		43,741	22,931
DEC	67,318	0	67,318	0		44,878	23,487
TOTALS	1,617,295	0	1,617,295	0	0	1,078,197	545,329
MAX							
MIN							
DATE	02/15						

TABLE 11-13
HOSPITAL COGENERATION SYSTEM COSTS – TURBINE

O&M	0.005 $/kWh	GAS POWER	2.7500 $/MMBtu
OIL	0.5 $/GAL	BUYBACK RATE	0.022 $/kWh
DIST'N O&M	0.00 $/M LB	STANDBY RATE	2.75 $/kW

MONTH	DAYS	COST GAS POWER ($)	MAINT COST ($)	LABOR COST ($)	STBY COST ($)	POWER SALES ($)	TURBINE OIL COST ($)	OWN & ADM COST ($)	STEAM SYS COST ($)	TOTAL ON-SITE ($)
JAN	31	39,287	4,100	0	3,137	0	0	825	0	47,349
FEB	28	35,220	3,676	0	3,137	0	0	825	0	42,857
MAR	31	37,868	3,952	0	3,137	0	0	825	0	45,782
APR	30	35,191	3,673	0	3,137	0	0	825	0	42,826
MAY	31	35,056	3,658	0	3,137	0	0	825	0	42,677
JUN	30	32,778	3,421	0	3,137	0	0	825	0	40,160
JUL	31	33,357	3,481	0	3,137	0	0	825	0	40,800
AUG	31	33,528	3,499	0	3,137	0	0	825	0	40,989
SEP	30	33,298	3,475	0	3,137	0	0	825	0	40,735
OCT	31	35,802	3,736	0	3,137	0	0	825	0	43,500
NOV	30	36,114	3,769	0	3,137	0	0	825	0	43,845
DEC	31	38,663	4,035	0	3,137	0	0	825	0	46,660
TOTALS	365	426,162	44,474	0	37,643	0	0	9,900	0	518,179

DATE 02/15

TABLE 11-14
HOSPITAL
OPERATING COST SUMMARY
Turbine

CONVENTIONAL SYSTEM	
Purchased Power	$1,257,483
Heating Gas	800,104
Fuel Oil	0
Purchased Steam	0
Total	$2,057,587
COGENERATION SYSTEM	
Purchased Power	$ 701,033
Heating Gas	545,329
Fuel Oil	0
Purchased Steam	0
Fuel for Power	426,162
Maintenance	44,474
Standby Power	37,643
Own & Admin	9,900
Labor Costs	0
Dist'n System	0
Total	$1,764,541
POWER SALES REVENUE	0
TOTAL OPERATING COST SAVING	$ 293,046

TABLE 11-15

				HOSPITAL	Turbine		Cash Flow
General Inflation	5.0%						
All Rates Relative To General Inflation					Tax Rates		
Year	1-5	6-10	11 & up		Federal	0.00%	
Purchase Power	1.50%				State	0.00%	
Htg Gas	0.50%				Local	0.00%	
Fuel Oil					Property	0.00%	
Purchased Steam							
Fuel For Power	1.30%				Tax Life	20	
O & M					Disc Rate	9.00%	
Power Sales					Econ Life	20	

PROJECT YEAR	0	1	2	3	4	5	6	7	8	9	10
YEAR	1990	1991	1992	1993	1994	1995	1996	1997	1998	1999	2000
Conventional Cost											
Purchased Power	1,257,483	1,339,219	1,426,268	1,518,976	1,617,709	1,722,860	1,809,003	1,899,453	1,994,426	2,094,147	2,198,855
Heating Gas	800,104	844,110	890,536	939,516	991,189	1,045,705	1,097,990	1,152,889	1,210,534	1,271,060	1,334,613
Fuel Oil	0	0	0	0	0	0	0	0	0	0	0
Purchased Steam	0	0	0	0	0	0	0	0	0	0	0
Total Conventional	2,057,587	2,183,329	2,316,805	2,458,491	2,608,898	2,768,565	2,906,993	3,052,343	3,204,960	3,365,208	3,533,468
Cogeneration Cost											
Purchased Power	1,257,483	1,339,219	795,129	846,812	901,855	960,476	1,008,500	1,058,925	1,111,871	1,167,464	1,225,837
Heating Gas	800,104	844,110	606,965	640,348	675,568	712,724	748,360	785,778	825,067	866,320	909,636
Fuel Oil	0	0	0	0	0	0	0	0	0	0	0
Purchased Steam	0	0	0	0	0	0	0	0	0	0	0
Fuel for Power	0	0	481,550	511,888	544,137	578,417	607,338	637,705	669,590	703,070	738,223
Maintenance	0	0	49,033	51,484	54,058	56,761	59,599	62,579	65,708	68,994	72,444
Standby Power	0	0	42,695	45,470	48,426	51,573	54,152	56,860	59,703	62,688	65,822
Own & Admin	0	0	10,915	11,460	12,034	12,635	13,267	13,930	14,627	15,358	16,126
Labor Costs	0	0	0	0	0	0	0	0	0	0	0
Dist'n System	0	0	0	0	0	0	0	0	0	0	0
Total Cogeneration	2,057,587	2,183,329	1,986,287	2,107,464	2,236,077	2,372,587	2,491,216	2,615,777	2,746,566	2,883,894	3,028,089
Power Sales	0	0	0	0	0	0	0	0	0	0	0
Net Cogeneration Cost	2,057,587	2,183,329	1,986,287	2,107,464	2,236,077	2,372,587	2,491,216	2,615,777	2,746,566	2,883,894	3,028,089
Total Oper Cost Saving	0	0	330,518	351,028	372,821	395,978	415,777	436,565	458,394	481,313	505,379
User Share			0	0	0	0	0	0	0	0	0
User payments			0	0	0	0	0	0	0	0	0
Depreciation			77,165	77,165	77,165	77,165	77,165	77,165	77,165	77,165	77,165
Owner Taxes			0	0	0	0	0	0	0	0	0
Interest			0	0	0	0	0	0	0	0	0
Before Tax Income			253,353	273,863	295,656	318,813	338,612	359,400	381,229	404,148	428,214
Income Tax			0	0	0	0	0	0	0	0	0
After Tax Income			253,353	273,863	295,656	318,813	338,612	359,400	381,229	404,148	428,214
Depreciation			77,165	77,165	77,165	77,165	77,165	77,165	77,165	77,165	77,165
Principal			0	0	0	0	0	0	0	0	0
Equity	514,433	1,028,867	0	0	0	0	0	0	0	0	0
Net to Owner	(514,433)	(1,028,867)	330,518	351,028	372,821	395,978	415,777	436,565	458,394	481,313	505,379
Present Value	(514,433)	(943,914)	278,190	271,058	264,116	257,358	247,914	238,816	230,052	221,610	213,478

TABLE 11-15 (Cont'd)

HOSPITAL

Project Cost	
Equipment Cost	990,000
Installation Cost	275,000
Development Cost	151,800
Contingency	126,500
Int Dur Const	0
Total	1,543,300

Debt			
Equity	100.00%		
Int Rate	9.00%		
Term	0		
Payments	$0		
Shared Savings			
Year	1-5	6-10	11 & up
% To Host	0	0	0

PROJECT YEAR	11	12	13	14	15	16	17	18	19	20	21	22
YEAR	2001	2002	2003	2004	2005	2006	2007	2008	2009	2010	2011	2012
Conventional Cost												
Purchased Power	2,308,797	2,424,237	2,545,449	2,672,722	2,806,358	2,946,676	3,094,009	3,248,710	3,411,145	3,581,703	3,760,788	3,948,827
Heating Gas	1,401,344	1,471,411	1,544,982	1,622,231	1,703,342	1,788,510	1,877,935	1,971,832	2,070,423	2,173,945	2,282,642	2,396,774
Fuel Oil	0	0	0	0	0	0	0	0	0	0	0	0
Purchased Steam	0	0	0	0	0	0	0	0	0	0	0	0
Total Conventional	3,710,141	3,895,649	4,090,431	4,294,953	4,509,700	4,735,185	4,971,944	5,220,542	5,481,569	5,755,647	6,043,430	6,345,601
Cogeneration Cost												
Purchased Power	1,287,129	1,351,486	1,419,060	1,490,013	1,564,514	1,642,739	1,724,876	1,811,120	1,901,676	1,996,760	2,096,598	2,201,428
Heating Gas	955,118	1,002,874	1,053,018	1,105,669	1,160,952	1,219,000	1,279,950	1,343,947	1,411,144	1,481,702	1,555,787	1,633,576
Fuel Oil	0	0	0	0	0	0	0	0	0	0	0	0
Purchased Steam	0	0	0	0	0	0	0	0	0	0	0	0
Fuel for Power	775,135	813,891	854,586	897,315	942,181	989,290	1,038,755	1,090,692	1,145,227	1,202,488	1,262,613	1,325,743
Maintenance	76,066	79,869	83,862	88,056	92,458	97,081	101,935	107,032	112,384	118,003	123,903	130,098
Standby Power	69,113	72,569	76,198	80,007	84,008	88,208	92,619	97,250	102,112	107,218	112,578	118,207
Own & Admin	16,932	17,779	18,668	19,601	20,581	21,610	22,691	23,826	25,017	26,268	27,581	28,960
Labor Costs	0	0	0	0	0	0	0	0	0	0	0	0
Dist'n System	0	0	0	0	0	0	0	0	0	0	0	0
Total Cogeneration	3,179,494	3,338,468	3,505,392	3,680,661	3,864,694	4,057,929	4,260,825	4,473,867	4,697,560	4,932,438	5,179,060	5,438,013
Power Sales	0	0	0	0	0	0	0	0	0	0	0	0
Net Cogeneration Cost	3,179,494	3,338,468	3,505,392	3,680,661	3,864,694	4,057,929	4,260,825	4,473,867	4,697,560	4,932,438	5,179,060	5,438,013
Total Oper Cost Saving	530,648	557,180	585,039	614,291	645,006	677,256	711,119	746,675	784,009	823,209	864,370	907,588
User Share	0	0	0	0	0	0	0	0	0	0	0	0
User payments	0	0	0	0	0	0	0	0	0	0	0	0
Depreciation	77,165	77,165	77,165	77,165	77,165	77,165	77,165	77,165	77,165	77,165	77,165	0
Owner Taxes	0	0	0	0	0	0	0	0	0	0	0	0
Interest	0	0	0	0	0	0	0	0	0	0	0	0
Before Tax Income	453,483	480,015	507,874	537,126	567,841	600,091	633,954	669,510	706,844	746,044	787,205	907,588
Income Tax	0	0	0	0	0	0	0	0	0	0	0	0
After Tax Income	453,483	480,015	507,874	537,126	567,841	600,091	633,954	669,510	706,844	746,044	787,205	907,588
Depreciation	77,165	77,165	77,165	77,165	77,165	77,165	77,165	77,165	77,165	77,165	77,165	0
Principal	0	0	0	0	0	0	0	0	0	0	0	0
Equity	0	0	0	0	0	0	0	0	0	0	0	0
Net to Owner	530,648	557,180	585,039	614,291	645,006	677,256	711,119	746,675	784,009	823,209	864,370	0
Present Value	205,644	198,097	190,827	183,825	177,079	170,580	164,321	158,290	152,482	146,886	141,496	0

FIGURE 11-39
IRR VS. CAPITAL COST – HOSPITAL

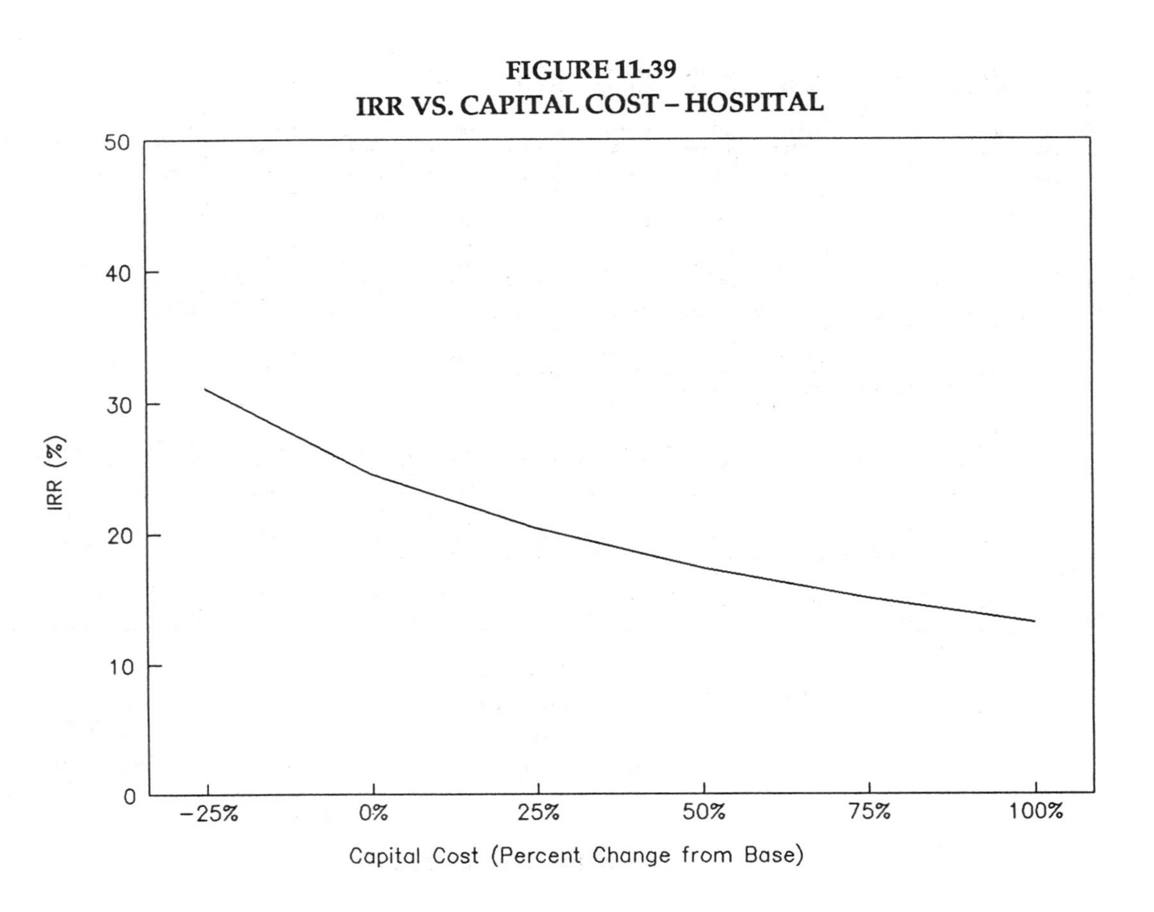

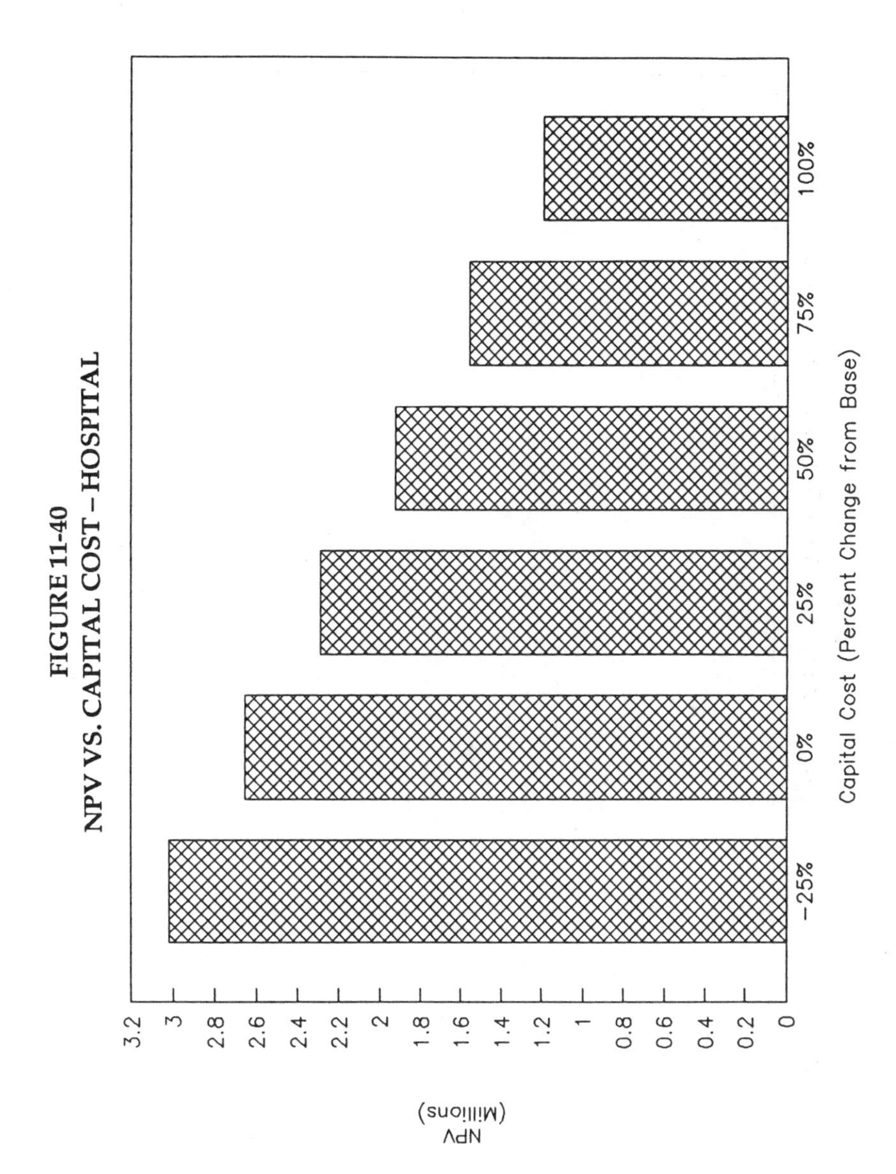

FIGURE 11-40
NPV VS. CAPITAL COST - HOSPITAL

FIGURE 11-41
IRR VS. RELATIVE ELECTRIC & GAS RATES – HOSPITAL

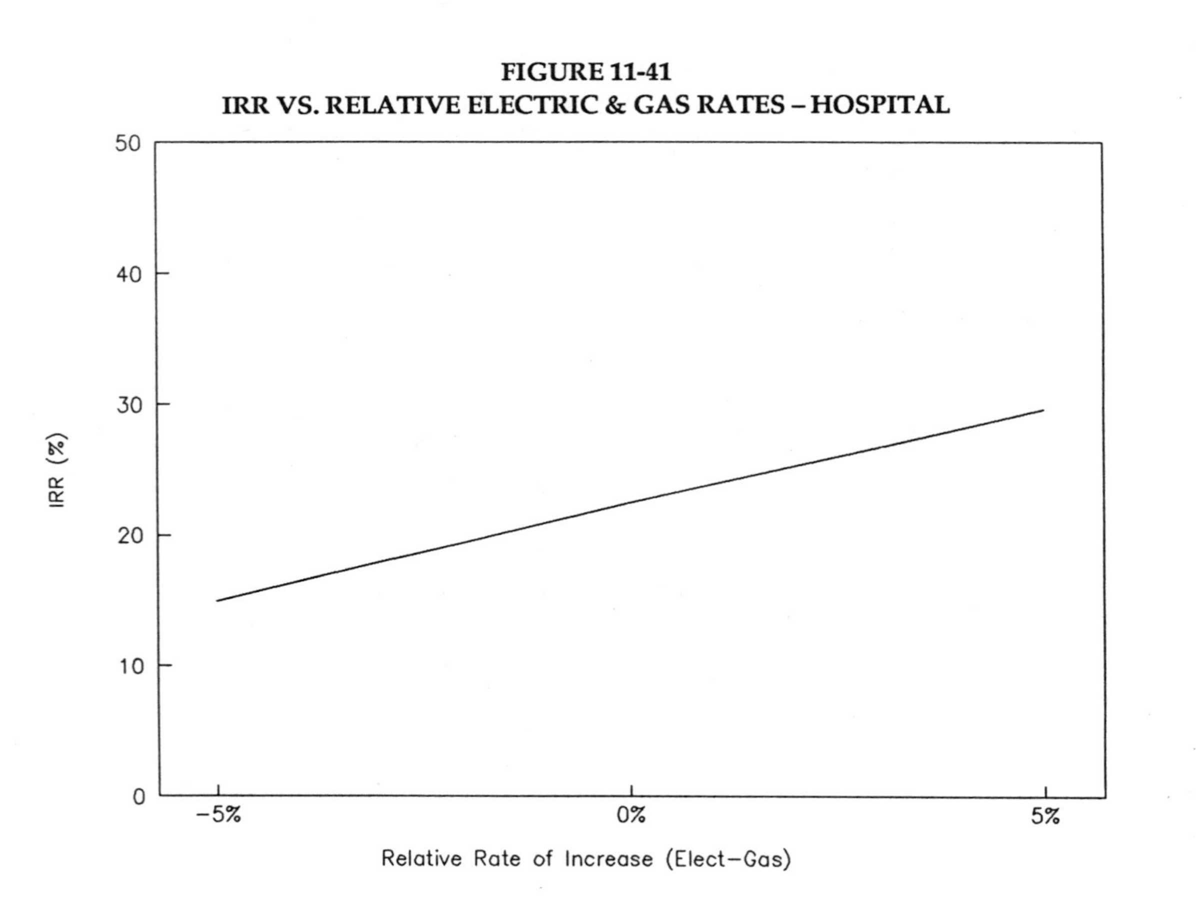

FIGURE 11-42
NPV VS. RELATIVE ELECTRIC & GAS RATES – HOSPITAL

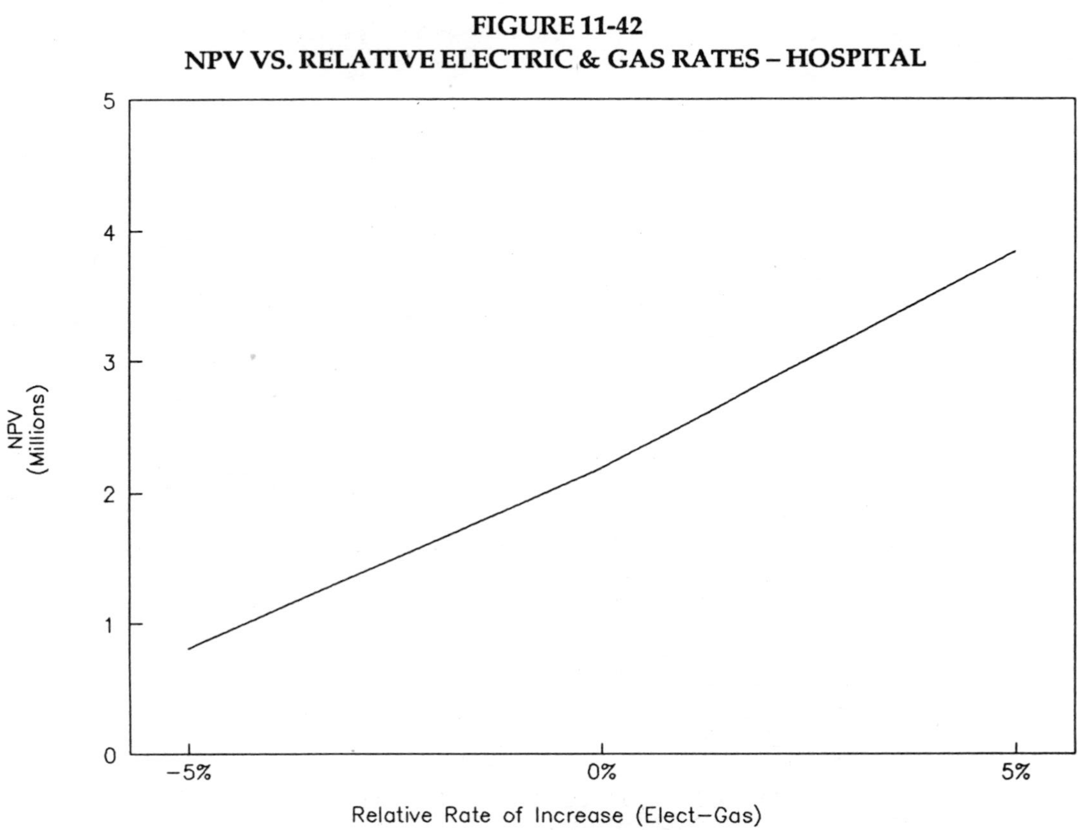

TABLE 11-16
LAUNDRY CONVENTIONAL SYSTEM COST

FAC 0.001 $/kWh AVE COST 0.107 $/kWh

MONTH	DAYS	ON PEAK DEMAND (kW)			BASE ENERGY (kWh)		PROCESS ENERGY (kWh)	TOTAL ENERGY (kWh)	CONV'L LD FACT (HOURS)	COST ($)
JAN	31	0.0			13,064		74,536	87,600	0	9,414
FEB	28	0.0			11,800		67,640	79,440	0	8,538
MAR	31	0.0			13,064		59,176	72,240	0	7,765
APR	30	0.0			12,643		64,637	77,280	0	8,306
MAY	31	0.0			13,064		68,776	81,840	0	8,796
JUN	30	0.0			12,643		78,077	90,720	0	9,749
JUL	31	0.0			13,064		116,776	129,840	0	13,949
AUG	31	0.0			13,064		116,056	129,120	0	13,872
SEP	30	0.0			12,643		86,957	99,600	0	10,703
OCT	31	0.0			13,064		61,096	74,160	0	7,971
NOV	30	0.0			12,643		63,917	76,560	0	8,229
DEC	31	0.0			13,064		66,856	79,920	0	8,590
TOTALS	365				153,821	0	924,499	1,078,320	0	115,883
MAX		0.0	0.0	0.0					0	
MIN		0.0	0.0	0.0					0	
DATE	02/16									
Revision		1.4								

AVE COST 3.23 $/MMBtu
ALT HTG FUEL 150,000 BTU/GAL
MAX STEAM DEMAND 0 M Lb/HR
BOILER EFFICIENCY 80.0%

MONTH	CONV'L GAS USE (THERMS)	NON-DISPL'B GAS USE (THERMS)	DISPL'BLE GAS USE (THERMS)	FUEL OIL USE (GAL)	STEAM USE (M LB)	TOTAL DISP FUEL (MMBTU)	TOTAL FUEL COST ($)	BOILER EFFIC. (%)
JAN	41,369	10,000	31,369			3,137	13,307	80.0
FEB	32,796	10,000	22,796			2,280	10,572	80.0
MAR	38,739	10,000	28,739			2,874	12,468	80.0
APR	28,223	10,000	18,223			1,822	9,113	80.0
MAY	26,318	10,000	16,318			1,632	8,505	80.0
JUN	22,797	10,000	12,797			1,280	7,382	80.0
JUL	21,240	10,000	11,240			1,124	6,886	80.0
AUG	21,759	10,000	11,759			1,176	7,051	80.0
SEP	22,873	10,000	12,873			1,287	7,406	80.0
OCT	25,644	10,000	15,644			1,564	8,290	80.0
NOV	26,483	10,000	16,483			1,648	8,558	80.0
DEC	33,574	10,000	23,574			2,357	10,820	80.0
TOTALS	341,815	120,000	221,815	0	0	22,182	110,359	80.0
MAX								80.0
MIN								80.0
DATE	02/16							

FIGURE 11-43
MONTHLY ELECTRIC USAGE – LAUNDRY

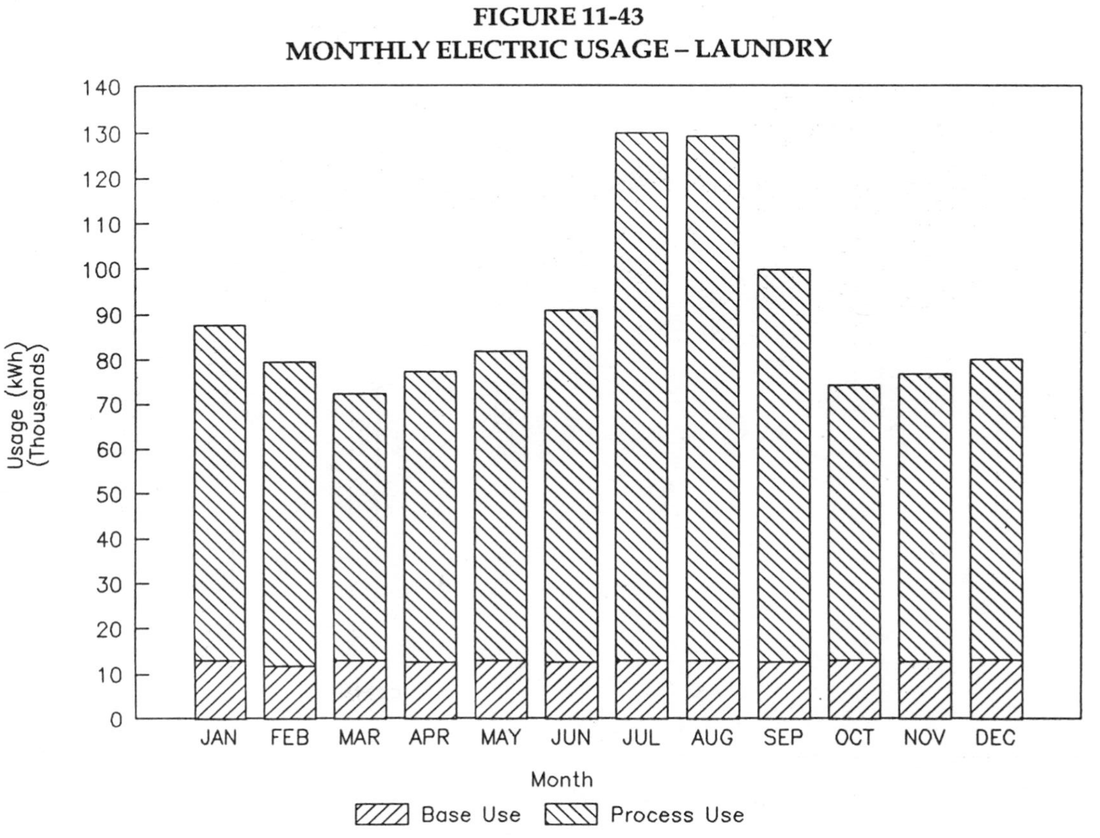

FIGURE 11-44
MONTHLY GAS USAGE – LAUNDRY

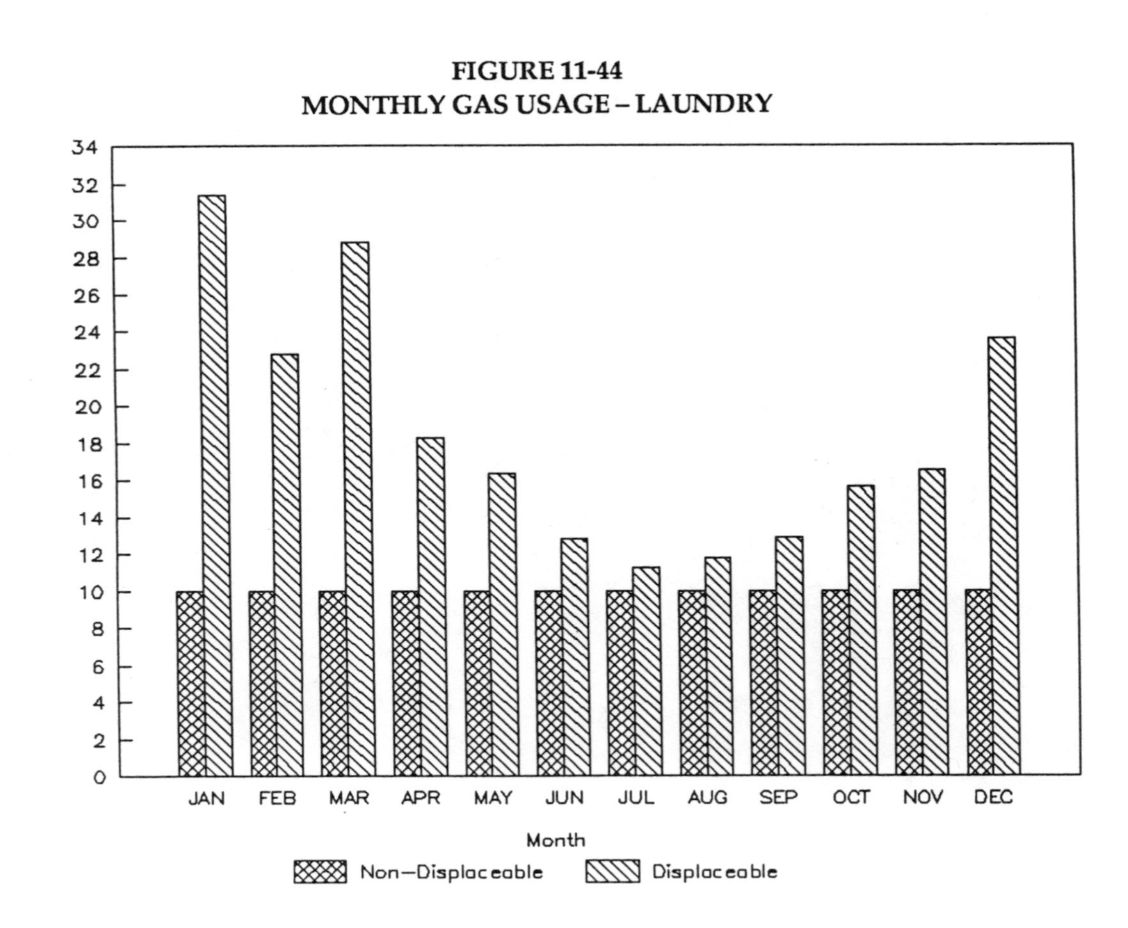

TABLE 11-17
WALKTHROUGH WORKSHEET

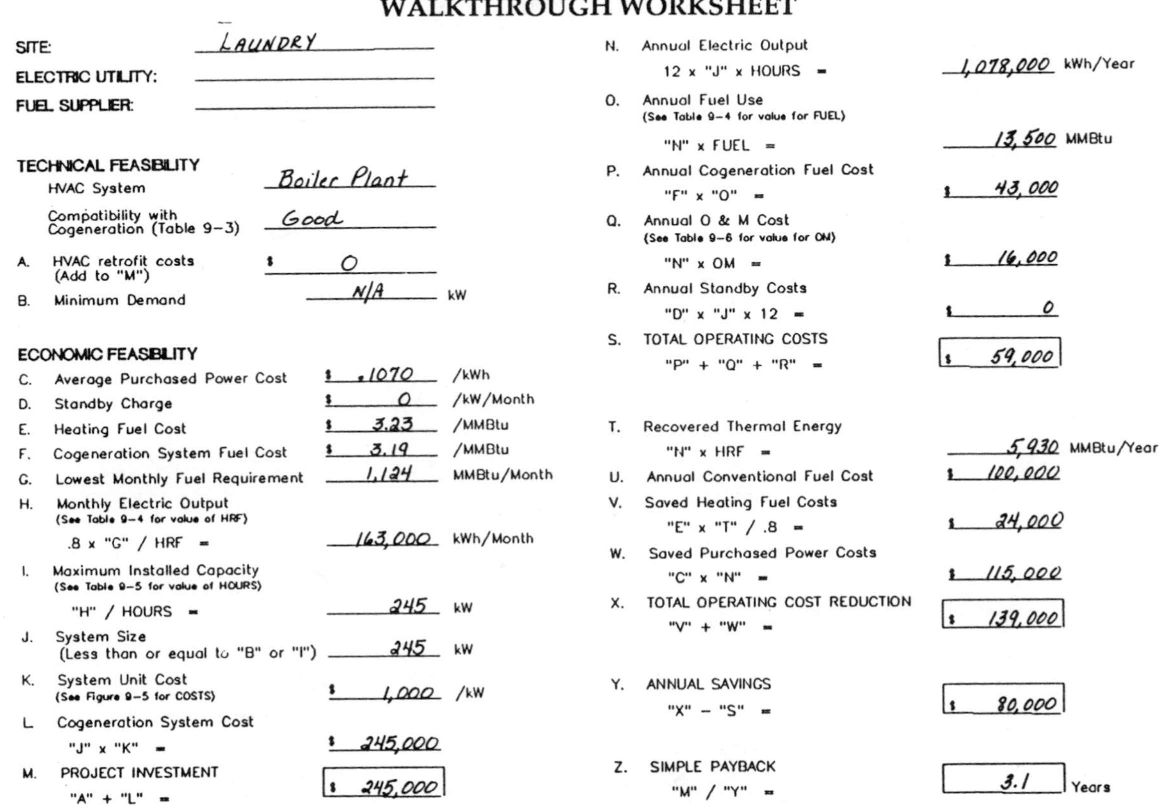

SITE: LAUNDRY
ELECTRIC UTILITY: ______
FUEL SUPPLIER: ______

TECHNICAL FEASIBILITY

HVAC System: Boiler Plant
Compatibility with Cogeneration (Table 9-3): Good

A. HVAC retrofit costs (Add to "M"): $ 0
B. Minimum Demand: N/A kW

ECONOMIC FEASIBILITY

C. Average Purchased Power Cost: $ .1070 /kWh
D. Standby Charge: $ 0 /kW/Month
E. Heating Fuel Cost: $ 3.23 /MMBtu
F. Cogeneration System Fuel Cost: $ 3.19 /MMBtu
G. Lowest Monthly Fuel Requirement: 1,124 MMBtu/Month
H. Monthly Electric Output (See Table 9-4 for value of HRF)
.8 x "G" / HRF = 163,000 kWh/Month
I. Maximum Installed Capacity (See Table 9-5 for value of HOURS)
"H" / HOURS = 245 kW
J. System Size (Less than or equal to "B" or "I") 245 kW
K. System Unit Cost (See Figure 9-5 for COSTS) $ 1,000 /kW
L. Cogeneration System Cost
"J" x "K" = $ 245,000
M. PROJECT INVESTMENT
"A" + "L" = $ 245,000

N. Annual Electric Output
12 x "J" x HOURS = 1,078,000 kWh/Year
O. Annual Fuel Use (See Table 9-4 for value for FUEL)
"N" x FUEL = 13,500 MMBtu
P. Annual Cogeneration Fuel Cost
"F" x "O" = $ 43,000
Q. Annual O & M Cost (See Table 9-6 for value for OM)
"N" x OM = $ 16,000
R. Annual Standby Costs
"D" x "J" x 12 = $ 0
S. TOTAL OPERATING COSTS
"P" + "Q" + "R" = $ 59,000

T. Recovered Thermal Energy
"N" x HRF = 5,930 MMBtu/Year
U. Annual Conventional Fuel Cost $ 100,000
V. Saved Heating Fuel Costs
"E" x "T" / .8 = $ 24,000
W. Saved Purchased Power Costs
"C" x "N" = $ 115,000
X. TOTAL OPERATING COST REDUCTION
"V" + "W" = $ 139,000

Y. ANNUAL SAVINGS
"X" − "S" = $ 80,000

Z. SIMPLE PAYBACK
"M" / "Y" = 3.1 Years

FIGURE 11-45
HOURLY DEMAND – LAUNDRY

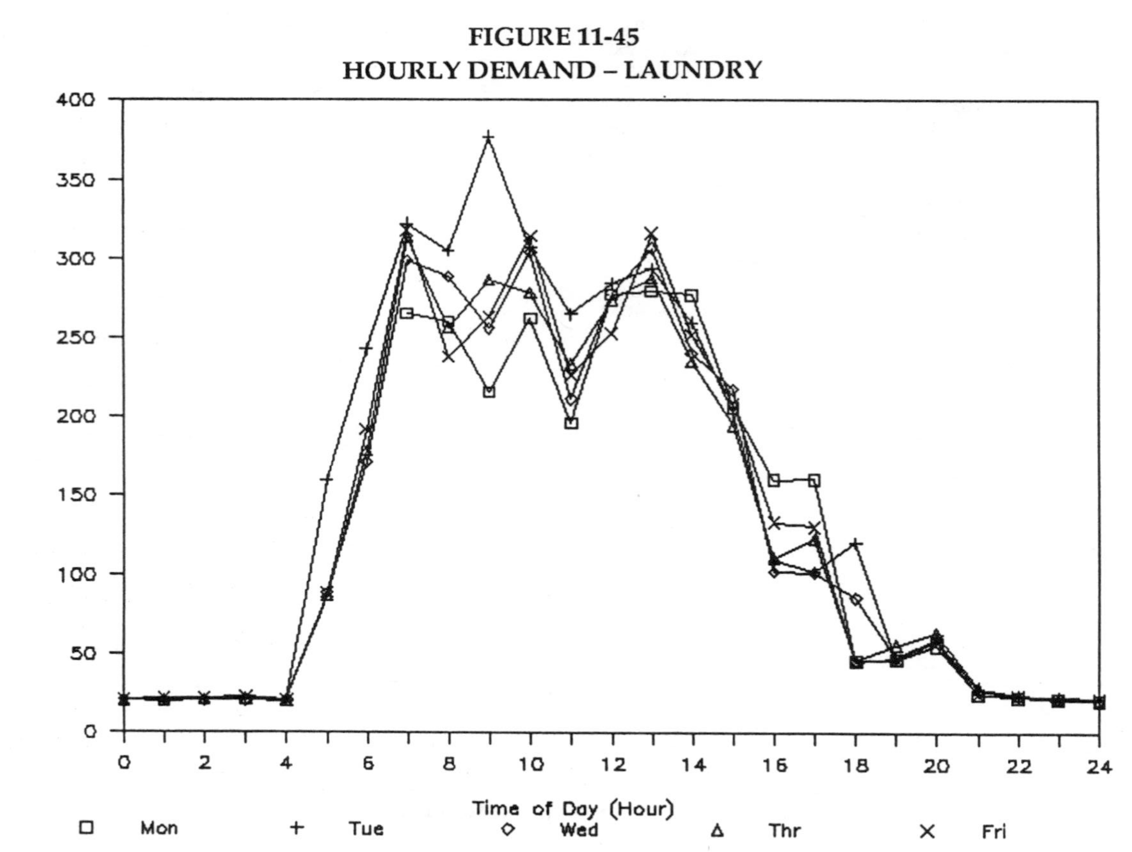

FIGURE 11-46
AVERAGE WEEKDAY HOURLY DEMAND – LAUNDRY

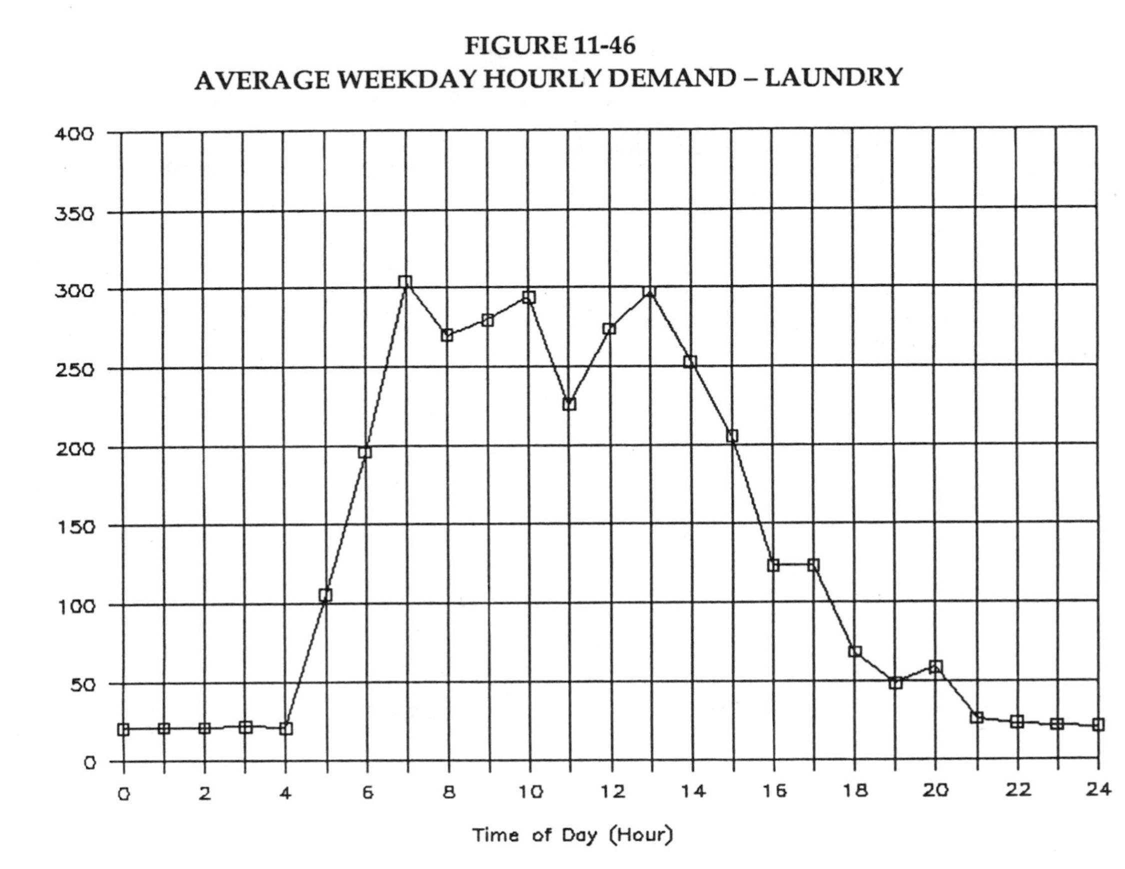

FIGURE 11-47
HOURLY DEMAND – WEEKEND, LAUNDRY

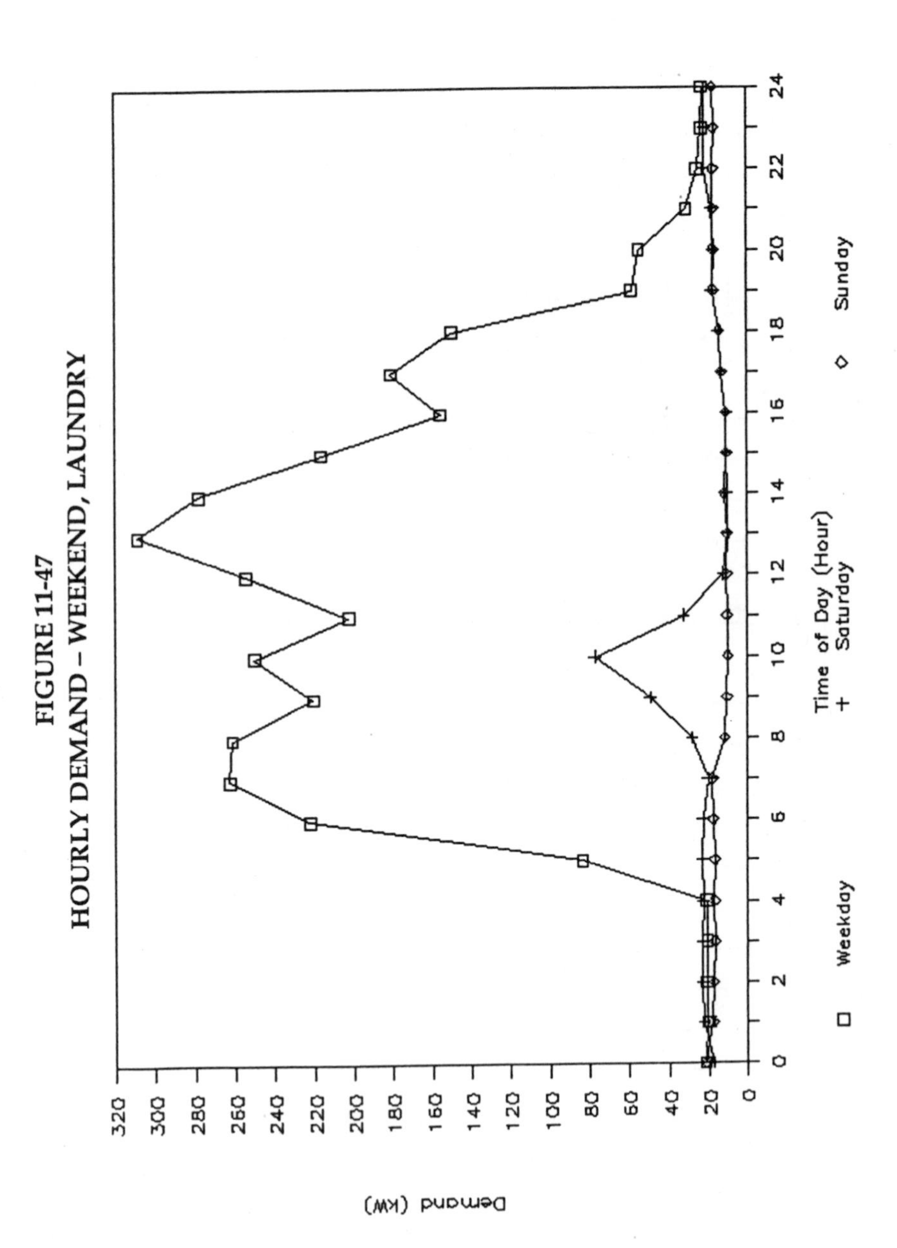

FIGURE 11-48
HOT WATER STEAM REQUIREMENTS – LAUNDRY

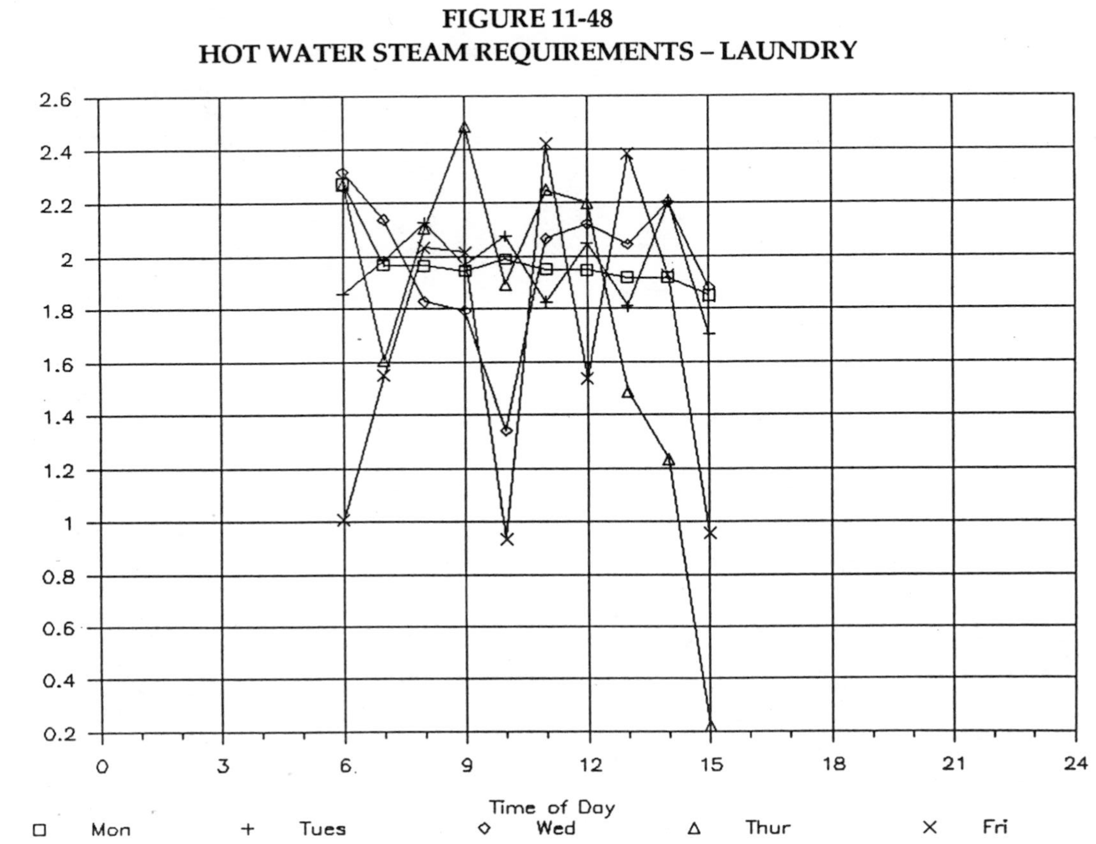

TABLE 11-18
LAUNDRY ENERGY ANALYSIS – RECIP

CAPACITY	325 kW					MODE = 1	
GEN CAP	325 kW	HEAT RATE	13,025 Btu/kWh	ALT HTG FUEL	150,000 Btu/Gal		
STEAM CAP	0 kW	HEAT REC	6,000 Btu/kWh	STEAM	1,000,000 Btu/Lb	DUCT BNR CAP	0 Btu/kWh
COMP LOSS	0 kW	HOUR/YR	8,000	BLR EFFICIENCY	80.0%	DUCT BNR EFF	92.0%

MONTH	DAYS	COGEN MIN ENG. CAP (kW)	COGEN AVE OUT (kW)	COGEN ENG. OUT (kWh)	COGEN SYS OUT (kWh)	REC'V HEAT AVAILABLE (THERMS)	TOTAL DISPL'BLE FUEL REQ'D (THERMS)	AVAIL/REQ'D HEAT	USED/AVAIL HEAT	SUPP'L HEAT (THERMS)	DUCT BURNER CAPACITY (THERMS)	DUCT BURNER GAS (THERMS)
JAN	31	325	325	74,536	74,536	4,472	31,369	17.8%	100.0%	20,623	0	0
FEB	28	325	325	67,640	67,640	4,058	22,796	22.3%	100.0%	14,178	0	0
MAR	31	325	325	59,176	59,176	3,551	28,739	15.4%	100.0%	19,441	0	0
APR	30	325	325	64,637	64,637	3,878	18,223	26.6%	100.0%	10,700	0	0
MAY	31	325	325	68,776	68,776	4,127	16,318	31.6%	100.0%	8,928	0	0
JUN	30	325	325	78,077	78,077	4,685	12,797	45.8%	100.0%	5,553	0	0
JUL	31	325	325	86,357	86,357	5,181	11,240	57.6%	100.0%	3,811	0	0
AUG	31	325	325	86,357	86,357	5,181	11,759	55.1%	100.0%	4,226	0	0
SEP	30	325	325	83,571	83,571	5,014	12,873	48.7%	100.0%	5,284	0	0
OCT	31	325	325	61,096	61,096	3,666	15,644	29.3%	100.0%	8,849	0	0
NOV	30	325	325	63,917	63,917	3,835	16,483	29.1%	100.0%	9,351	0	0
DEC	31	325	325	66,856	66,856	4,011	23,574	21.3%	100.0%	14,848	0	0
TOT	365			860,996	860,996	51,660	221,815			125,792	0	0
MAX								57.6%	100.0%			
MIN								15.4%	100.0%			
DATE	02/16											

MONTH	CONV'TL BOILER HEAT (THERMS)	CONV'L BLR GAS USE (THERMS)	SUPP'L GAS USE (THERMS)	GAS FOR POWER (THERMS)	TOTAL GAS USE (THERMS)	ST TURBINE CAPACITY (kW)	ST TURBINE ENERGY (kWh)
JAN	20,623	25,779	25,779	9,708	45,487	0	0
FEB	14,178	17,723	17,723	8,810	36,533	0	0
MAR	19,441	24,301	24,301	7,708	42,008	0	0
APR	10,700	13,375	13,375	8,419	31,794	0	0
MAY	8,928	11,160	11,160	8,958	30,118	0	0
JUN	5,553	6,941	6,941	10,170	27,111	0	0
JUL	3,811	4,763	4,763	11,248	26,011	0	0
AUG	4,226	5,282	5,282	11,248	26,530	0	0
SEP	5,284	6,605	6,605	10,885	27,490	0	0
OCT	8,849	11,062	11,062	7,958	29,020	0	0
NOV	9,351	11,689	11,689	8,325	30,014	0	0
DEC	14,848	18,560	18,560	8,708	37,268	0	0
TOTALS	125,792	157,240	157,240	112,145	389,385		0
MAX							
MIN							
DATE	02/16						

TABLE 11-19
LAUNDRY COGENERATION SUPPLEMENTAL COST – RECIP

Ave Elect 0.108 $/kWh
FAC 0.001 $/kWh

MONTH	DAYS	SUPP'TL ON PEAK (kW)	SUPP'TL ON PEAK (kVa)	SUPP'TL OFF PEAK (kW)	BASE ENERGY (kWh)		PROCESS ENERGY (kWh)	SUPP'TL TOTAL (kWh)	SUPP'TL LD FACT (HOURS)	SUPP'TL COST ($)
JAN	31	0.0			13,064		0	13,064		1,412
FEB	28	0.0			11,800		0	11,800		1,276
MAR	31	0.0			13,064		0	13,064		1,412
APR	30	0.0			12,643		0	12,643		1,367
MAY	31	0.0			13,064		0	13,064		1,412
JUN	30	0.0			12,643		0	12,643		1,367
JUL	31	0.0			13,064		30,419	43,483		4,678
AUG	31	0.0			13,064		29,699	42,763		4,601
SEP	30	0.0			12,643		3,386	16,029		1,730
OCT	31	0.0			13,064		0	13,064		1,412
NOV	30	0.0			12,643		0	12,643		1,367
DEC	31	0.0			13,064		0	13,064		1,412
TOTALS	365				153,821	0	63,503	217,324	0	23,446
MAX		0.0	0.0	0.0					0	
MIN		0.0	0.0	0.0					0	
DATE	02/16									

SUP ST DEM M LB/HR 0 $/MO
SUPP'L PGA 0 $/TH
GAS HTG 0.3238 $/TH FAC STEAM 0 $/M LB

MONTH	SUPP'TL GAS USE (THERMS)	SUPP'TL NON-DISPL'B (THERMS)	SUPP'TL DISPL'BLE (THERMS)	SUPP'TL OIL USE (GAL)	SUPP'TL STEAM USE (M LB)	TOTAL SUP. FUEL (MMBTU)	TOTAL SUP. FUEL ($)
JAN	35,779	10,000	25,779	0		3,578	11,523
FEB	27,723	10,000	17,723	0		2,772	8,954
MAR	34,301	10,000	24,301	0		3,430	11,052
APR	23,375	10,000	13,375	0		2,338	7,567
MAY	21,160	10,000	11,160	0		2,116	6,860
JUN	16,941	10,000	6,941	0		1,694	5,514
JUL	14,763	10,000	4,763	0		1,476	4,819
AUG	15,282	10,000	5,282	0		1,528	4,985
SEP	16,605	10,000	6,605	0		1,661	5,407
OCT	21,062	10,000	11,062	0		2,106	6,829
NOV	21,689	10,000	11,689	0		2,169	7,029
DEC	28,560	10,000	18,560	0		2,856	9,221
TOTALS	277,240	120,000	157,240	0	0	27,724	89,760
MAX							
MIN							
DATE	02/16						

TABLE 11-20
LAUNDRY COGENERATION SYSTEM COSTS – RECIP

O&M	0.015 $/kWh	GAS POWER	0.3190 $/TH
OIL	0.7 $/GAL	BUYBACK RATE	0.024 $/kWh
DIST'N O&M	0.40 $/M LB	STANDBY RATE	0.00 $/kW

MONTH	DAYS	COST GAS POWER ($)	MAINT COST ($)	LABOR COST ($)	STBY COST ($)	POWER SALES ($)	TURBINE OIL COST ($)	OWN & ADM COST ($)	STEAM SYS COST ($)	TOTAL ON-SITE ($)
JAN	31	3,097	1,118	0	0	0	0	167	0	4,382
FEB	28	2,810	1,015	0	0	0	0	167	0	3,992
MAR	31	2,459	888	0	0	0	0	167	0	3,513
APR	30	2,686	970	0	0	0	0	167	0	3,822
MAY	31	2,858	1,032	0	0	0	0	167	0	4,056
JUN	30	3,244	1,171	0	0	0	0	167	0	4,582
JUL	31	3,588	1,295	0	0	0	0	167	0	5,050
AUG	31	3,588	1,295	0	0	0	0	167	0	5,050
SEP	30	3,472	1,254	0	0	0	0	167	0	4,893
OCT	31	2,539	916	0	0	0	0	167	0	3,622
NOV	30	2,656	959	0	0	0	0	167	0	3,781
DEC	31	2,778	1,003	0	0	0	0	167	0	3,947
TOTALS	365	35,774	12,915	0	0	0	0	2,000	0	50,689

DATE 02/16

TABLE 11-21

LAUNDRY
OPERATING COST SUMMARY
Recip

CONVENTIONAL SYSTEM	
Purchased Power	$ 115,883
Heating Gas	110,359
Fuel Oil	0
Purchased Steam	0
Total	$226,242
COGENERATION SYSTEM	
Purchased Power	$ 23,446
Heating Gas	89,760
Fuel Oil	0
Purchased Steam	0
Fuel for Power	35,774
Maintenance	12,915
Standby Power	0
Own & Admin	2,000
Labor Costs	0
Dist'n System	0
Total	$163,895
POWER SALES REVENUE	0
TOTAL OPERATING COST SAVING	$ 62,347

TABLE 11-22

Cash Flow

General Inflation	5.0%			LAUNDRY	Recip	
All Rates Relative To General Inflation					Tax Rates	
Year	1-5	6-10	11 & up		Federal	0.00%
Purchase Power					State	0.00%
Htg Gas	2.00%				Local	0.00%
Fuel Oil					Property	0.00%
Purchased Steam						
Fuel For Power	2.00%				Tax Life	15
O & M					Disc Rate	9.00%
Power Sales					Econ Life	20

PROJECT YEAR	0	1	2	3	4	5	6	7	8	9	10
YEAR	1990	1991	1992	1993	1994	1995	1996	1997	1998	1999	2000
Conventional Cost											
Purchased Power	115,883	121,677	127,761	134,149	140,856	147,899	155,294	163,059	171,212	179,772	188,761
Heating Gas	110,359	118,084	126,350	135,195	144,658	154,784	162,523	170,650	179,182	188,141	197,548
Fuel Oil	0	0	0	0	0	0	0	0	0	0	0
Purchased Steam	0	0	0	0	0	0	0	0	0	0	0
Total Conventional	226,242	239,761	254,111	269,343	285,514	302,683	317,817	333,708	350,394	367,913	386,309
Cogeneration Cost											
Purchased Power	115,883	24,619	25,849	27,142	28,499	29,924	31,420	32,991	34,641	36,373	38,191
Heating Gas	110,359	96,043	102,766	109,959	117,657	125,893	132,187	138,797	145,736	153,023	160,674
Fuel Oil	0	0	0	0	0	0	0	0	0	0	0
Purchased Steam	0	0	0	0	0	0	0	0	0	0	0
Fuel for Power	0	38,278	40,958	43,825	46,893	50,175	52,684	55,318	58,084	60,988	64,038
Maintenance	0	13,561	14,239	14,951	15,698	16,483	17,307	18,173	19,081	20,035	21,037
Standby Power	0	0	0	0	0	0	0	0	0	0	0
Own & Admin	0	2,100	2,205	2,315	2,431	2,553	2,680	2,814	2,955	3,103	3,258
Labor Costs	0	0	0	0	0	0	0	0	0	0	0
Dist'n System	0	0	0	0	0	0	0	0	0	0	0
Total Cogeneration	226,242	174,600	186,017	198,192	211,177	225,027	236,279	248,093	260,497	273,522	287,198
Power Sales	0	0	0	0	0	0	0	0	0	0	0
Net Cogeneration Cost	226,242	174,600	186,017	198,192	211,177	225,027	236,279	248,093	260,497	273,522	287,198
Total Oper Cost Saving	0	65,161	68,094	71,151	74,337	77,656	81,539	85,616	89,896	94,391	99,111
User Share		0	0	0	0	0	0	0	0	0	0
User payments		0	0	0	0	0	0	0	0	0	0
Depreciation		23,587	23,587	23,587	23,587	23,587	23,587	23,587	23,587	23,587	23,587
Owner Taxes		0	0	0	0	0	0	0	0	0	0
Interest		0	0	0	0	0	0	0	0	0	0
Before Tax Income		41,574	44,507	47,564	50,750	54,069	57,952	62,029	66,310	70,805	75,524
Income Tax		0	0	0	0	0	0	0	0	0	0
After Tax Income		41,574	44,507	47,564	50,750	54,069	57,952	62,029	66,310	70,805	75,524
Depreciation		23,587	23,587	23,587	23,587	23,587	23,587	23,587	23,587	23,587	23,587
Principal		0	0	0	0	0	0	0	0	0	0
Equity	0	330,213	0	0	0	0	0	0	0	0	0
Net to Owner	0	(265,053)	68,094	71,151	74,337	77,656	81,539	85,616	89,896	94,391	99,111
Present Value	0	(243,168)	57,313	54,942	52,662	50,471	48,619	46,835	45,116	43,460	41,865

TABLE 11-22 (Cont'd)

LAUNDRY

Project Cost	
Equipment Cost	200,000
Installation Cost	90,000
Development Cost	34,800
Contingency	29,000
Int Dur Const	0
Total	353,800

Debt			
Equity	100.00%		
Int Rate	10.00%		
Term	5		
Payments	$0		
Shared Savings			
Year	1-5	6-10	11 & up
% To Host	0	0	0

PROJECT YEAR	11	12	13	14	15	16	17	18	19	20	21	22
YEAR	2001	2002	2003	2004	2005	2006	2007	2008	2009	2010	2011	2012
Conventional Cost												
Purchased Power	198,199	208,109	218,514	229,440	240,912	252,958	265,605	278,886	292,830	307,471	322,845	338,987
Heating Gas	207,426	217,797	228,687	240,121	252,127	264,733	277,970	291,869	306,462	321,785	337,874	354,768
Fuel Oil	0	0	0	0	0	0	0	0	0	0	0	0
Purchased Steam	0	0	0	0	0	0	0	0	0	0	0	0
Total Conventional	405,624	425,906	447,201	469,561	493,039	517,691	543,576	570,754	599,292	629,257	660,719	693,755
Cogeneration Cost												
Purchased Power	40,101	42,106	44,211	46,422	48,743	51,180	53,739	56,426	59,247	62,210	65,320	68,586
Heating Gas	168,708	177,143	186,001	195,301	205,066	215,319	226,085	237,389	249,259	261,722	274,808	288,548
Fuel Oil	0	0	0	0	0	0	0	0	0	0	0	0
Purchased Steam	0	0	0	0	0	0	0	0	0	0	0	0
Fuel for Power	67,239	70,601	74,131	77,838	81,730	85,816	90,107	94,613	99,343	104,310	109,526	115,002
Maintenance	22,089	23,193	24,353	25,571	26,849	28,192	29,601	31,081	32,635	34,267	35,981	37,780
Standby Power	0	0	0	0	0	0	0	0	0	0	0	0
Own & Admin	3,421	3,592	3,771	3,960	4,158	4,366	4,584	4,813	5,054	5,307	5,572	5,851
Labor Costs	0	0	0	0	0	0	0	0	0	0	0	0
Dist'n System	0	0	0	0	0	0	0	0	0	0	0	0
Total Cogeneration	301,558	316,636	332,468	349,091	366,546	384,873	404,117	424,323	445,539	467,816	491,206	515,767
Power Sales	0	0	0	0	0	0	0	0	0	0	0	0
Net Cogeneration Cost	301,558	316,636	332,468	349,091	366,546	384,873	404,117	424,323	445,539	467,816	491,206	515,767
Total Oper Cost Saving	104,066	109,270	114,733	120,470	126,493	132,818	139,459	146,432	153,753	161,441	169,513	177,989
User Share	0	0	0	0	0	0	0	0	0	0	0	0
User payments	0	0	0	0	0	0	0	0	0	0	0	0
Depreciation	23,587	23,587	23,587	23,587	23,587	23,587	0	0	0	0	0	0
Owner Taxes	0	0	0	0	0	0	0	0	0	0	0	0
Interest	0	0	0	0	0	0	0	0	0	0	0	0
Before Tax Income	80,480	85,683	91,146	96,883	102,907	109,231	139,459	146,432	153,753	161,441	169,513	177,989
Income Tax	0	0	0	0	0	0	0	0	0	0	0	0
After Tax Income	80,480	85,683	91,146	96,883	102,907	109,231	139,459	146,432	153,753	161,441	169,513	177,989
Depreciation	23,587	23,587	23,587	23,587	23,587	23,587	0	0	0	0	0	0
Principal	0	0	0	0	0	0	0	0	0	0	0	0
Equity	0	0	0	0	0	0	0	0	0	0	0	0
Net to Owner	104,066	109,270	114,733	120,470	126,493	132,818	139,459	146,432	153,753	161,441	0	0
Present Value	40,329	38,849	37,423	36,050	34,727	33,453	32,225	31,043	29,903	28,806	0	0

REFERENCES

American Gas Association, *Cogeneration Feasibility Analysis*, June 1986.

American Gas Association, *Natural Gas Prime Movers*, January 1987.

American Gas Association, *A.G.A. Marketing Manual Natural Gas Cooling*, 1986.

American Public Power Association, *Cogeneration and Small Power Production*, Guidelines for Public Power Systems, November 1980.

American Society of Heating, *ASHRAE Handbook, 1988 Equipment Volume*, Engine and Turbine Drives, Chapter 32, 1983.

American Society of Heating, *ASHRAE Handbook, 1987 HVAC Systems and Applications Volume*, Cogeneration Systems, Chapter 8, 1987.

Babcock and Wilcox, *Steam/Its Generation and Use*, Thirty-ninth edition, 1978.

Industrial Press, Inc., *Gas Engineers Handbook*, 1965.

Institute of Electrical and Electronic Engineers, Inc., *IEEE Guide for Interfacing Dispersed Storage and Generation Facilities with Electric Utility Systems*, prepared for the IEEE Standards Coordinating Committee on Dispersed Storage and Generation, 1988.

International District Heating Association, *District Heating Handbook*, Fourth Edition, 1983.

Orlando, J.A., *Cogeneration Analysis and Development Handbook in North Carolina*, prepared for and distributed by Energy Division, N.C. Department of Commerce, June 1987.

Orlando, J.A., *Cogeneration Technology Handbook*, Government Institutes, Inc., July 1985.

Orlando, J.A., M.J. Zimmer, *Assessment of Factory-Assembled Packaged Cogeneration Systems*, prepared for Cogeneration Technical Services, Inc., GKCO, Inc., Wickwire, Gavin & Gibbs, May 1986.

Reason, John, "Making Interconnection Work," *Power*, June 1982.

Waukesha Power Systems, *Waukesha Cogeneration Handbook*, Second Edition, 1986.

FUEL LHV/HHV VALUES

FUEL	TYPICAL (BTU/CF)	LHV/HHV
Dry Natural Gas	900	0.90
Butane	3,010	0.90
Propane	2,320	0.91
Ethane	1,620	0.91
Carbon Monoxide	320	1.00
Hydrogen	270	0.84
Methane	910	0.90

FUEL HEATING VALUES

FUEL	UNIT	HEATING VALUE (BTU/UNIT)
#2 Oil	Gallons	140,000
#6 Oil	Gallons	150,000
Coal	Ton	25,000,000
Steam	Thousand Pounds	1,000,000

EQUIVALENTS

1 Horsepower = 746 watts
1 kWhr = 3413 Btu
1 therm = 100,000 Btu
12,000 Btu = 1 Ton (air conditioning)

DEFINITIONS

Ambient Temperature: The temperature of the air surrounding the equipment.

Automatic Synchronizer: This device, in its simplest form, is a magnetic-type control relay which will automatically close the generator switch when the conditions for paralleling are satisfied.

Automatic Transfer Switch: This switch is a double-throw, electrically operated switch which will, on a given signal, open one set of contacts and throw over to the second set of contacts. As normally used in hospitals, television and radio stations and other applications where automatic emergency power is used, the switch automatically transfers a load from a normal source of electrical power to an emergency source on failure of the normal.

Availability: The ratio of the time the unit is capable of being in use to the total time.

Avoided Cost: The decremental cost for the electric utility to generate or purchase electricity that is avoided through the purchase of power from a cogeneration facility.

Back-up Power: Electric energy which may be required during an unscheduled outage of the facility's engine generator set. Frequently referred to as standby power.

Baseload: The minimum electric or thermal load which is supplied continuously over a period of time.

Bottoming-Cycle: A cogeneration facility in which the energy input to the system is first applied to a thermal process, and the rejected heat emerging from the process is then used for power production.

Capability: The maximum load which a generating unit can carry under specified conditions for a given period of time, without exceeding approved limits of temperature and stress.

Capacity: The load for which a generating unit is rated. This capacity is applicable only under conditions specified for the rating.

Capacity Credits: The value incorporated into the utility's rate for purchasing energy, based upon the savings due to the reduction or postponement of new generation capacity resulting from the purchase of power.

Capacity Factor: The ratio of the actual annual plant electricity output to the rated plant output.

Circuit Breaker: A special switch used to protect electrical circuits. It is generally designed to open or break the circuit when some abnormal condition occurs.

Combustor: The mechanical component of the gas turbine in which fuel is burned to increase the temperature of the working medium.

Continuous Standby: The rating at which a generator set may be operated for the duration of a power outage. No overload capacity is guaranteed.

Demand: The rate at which electric energy is delivered at a given instant or averaged over any designated period of time (usually one hour or less).

Peak Annual Demand – The greatest of all demands which occurred during a prescribed demand interval in a calendar year.

Billing Demand – The demand upon which billing to a customer is based, as specified in a rate schedule or contract. It may be based on the contract year, a contract minimum, or a previous maximum, and does not necessarily equal the demand actually measured during the billing period.

Coincident Demand – The sum of two or more demands which occur in the same demand interval.

Instantaneous Peak Demand – The maximum demand at the instant of greatest load.

Demand Charge: The specified charge for electrical capacity on the basis of the billing demand.

Dual Fuel System: An engine that can switch back and forth from one fuel (e.g., gas) to another (e.g., oil) or which can simultaneously burn both fuels with no or minimal downtime.

Efficiency: The ratio of the amount of useful output energy divided by the input energy.

Energized System: A system under load (supplying energy to load) or carrying rated voltage and frequency, but not supplying load.

Energy Charge: That portion of the billed charge for electric service based upon the electric energy (kilowatt hours) supplied.

Excitation: The power required to energize the magnetic field of generators.

Exhaust Heat Recovery: The process of extracting heat from the working medium leaving a prime mover and transferring it to a second fluid stream or to a product.

Frequency: The number of cycles per second the current alternates.

Grid: The system of interconnected transmission lines, substations and generating plants of one or more utilities.

Grid Interconnection: The intertie of a cogeneration plant to an electric utility's system to allow electricity flow in either or both directions.

Harmonics: Waveforms whose frequencies are multiples of the fundamental (60 Hz) wave. The combination of harmonies and fundamental waves causes a nonsinusoidal wave.

Heat Rate: A measure of generating station thermal efficiency, generally expressed in Btus (per net kilowatt hour).

Heating Value: The energy content in a fuel which is available as useful heat.

Induction Generator: A nonsynchronous AC generator identical in construction with an AC motor driven above synchronous speed.

Intercooler: A heat exchanger located between two compressor stages to reduce the air temperature entering the high-pressure compressor stage.

Internal Rate of Return: Discount rate at which the present value of an investment is equal to the investment.

Interruptible Power: Electric energy supplied by an electric utility subject to interruption by the electric utility under specified conditions.

Inverter: A device for converting direct current into alternating current.

Kilowatt: Power is the rate of doing work. Electric power is expressed in watts or kilowatts.

Kilowatt Hour: The basic unit of electric energy, equal to one kilowatt of power supplied for one hour (kWh).

Load Following: Operation of equipment to match production to demand.

Magnetizing Current: The current required to magnetize the iron core of a magnetic current in a generator.

Maintenance Power: Electric energy which may be required during scheduled outages of the cogenerator.

National Electrical Code: The National Electrical Code is a volume of standard electrical rules prepared by the National Fire Protection Association.

National Electrical Manufacturers Assn.: An organization of electrical manufacturers set up to provide information pertaining to certain types of electrical equipment.

Network: A system of transmission or distribution lines so cross-connected and operated as to permit multiple power supply to any principle point on it.

Off-Peak: Time periods when power demands are below average. For electric utilities, generally nights and weekends; for gas utilities, summer months.

Payback Period: The time required to completely recover the original capital investment.

Power: The time rate of generating, transferring or using electric energy usually expressed in kilowatts.

Apparent – A quantity of power proportional to the mathematical product of the volts and amperes of a circuit generally designated in kilovoltamperes (kVa).

	Reactive – The portion of "apparent power" that does not do work. It is commercially measured in kilovars.
	Real – The energy or work-producing part of "apparent power." It is commercially measured in kilovars.
Power Factor:	Power factor is the ratio of real power measured in kilowatts to apparent power measured in kilovoltamperes for any given load and time.
Prime Power:	The rating at which a generator may be operated continuously as a sole source of power, with intermittent overloads.
Qualified Facility:	A cogeneration facility which has been granted a "qualified" status by the FERC. To obtain the"qualified" status, a facility must meet the ownership requirement (i.e., less than 50% electric utility ownership), operating, and efficiency standards.
Rate Schedule:	Price list showing how the utility will bill a class of customers.
Regenerator:	A heat exchanger designed to transfer heat from a turbine's exhaust gases to the compressed air stream.
Reheater:	A combustor located between two turbine stages which increases the temperature of the working fluid.
Service Area:	Territory in which a utility system is required or has the right to supply electric service to ultimate customers.

Simple Cycle Turbine: A turbine in which the working medium passes successively through the compressor, combustor, and turbine.

Single-Shaft Gas Turbine: A turbine in which all the rotating components are mechanically coupled together on a single shaft.

Small Power-Producing Facility: As defined in PURPA, a facility that produces energy solely by the use, as a primary energy source, or biomass, waste, renewable resources, or any combination thereof; and has a power production capacity that, together with any other facilities located at the same site, is not greater than 80 megawatts.

Standby Capacity: The capacity that is designed to be used when part or all of the prime source of power is interrupted.

Supplemental Thermal: The heat required when recovered engine heat is insufficient to meet thermal demands.

Supplemental Firing: The injection of fuel into the recovered heat stream (such as turbine exhaust) to raise the energy content of the stream.

Supplemental Power: Electric energy routinely supplied by an electric utility in addition to that which the facility generates itself.

Synchronous Generator:	A machine that generates an alternating voltage when its armature of field is rotated.
Topping Cycle:	A cogeneration facility in which the energy input to the facility is first used to produce useful power, with the heat recovered from power production then used for other purposes.
Total Energy Systems:	The name previously used to refer to a form of cogeneration in which all electrical and thermal energy needs were met by on-site systems.
Utilization Factor:	The ratio of the maximum demand of a system (or part of a system) to its rated capacity.
Voltage Flicker:	Term commonly used to describe a significant fluctuation of voltage.

DATA AND WORKSHEETS FOR "WALKTHROUGH"

The determination of whether a particular cogeneration application is technically and economically feasible will require many steps, each relying on more detailed data and providing more comprehensive analyses. One starting point in the process is the "Walkthrough" during which a preliminary assessment of cogeneration viability can be made.

This appendix provides data and worksheets which can be used in a walkthrough. It is necessary to emphasize that these procedures rely on typical engine characteristics and average costs for purchased power and fuel. They also assume that all recovered heat will be usefully applied. More detailed analyses using incremental costs, actual equipment characteristics and actual electrical and thermal load profiles are likely to produce less favorable results. This procedure is only intended to provide the information required to determine whether the more comprehensive analysis is economically justified. It should not be used as a basis for a final cogeneration investment decision.

FIGURE 1
COGENERATION SYSTEM INSTALLED COSTS

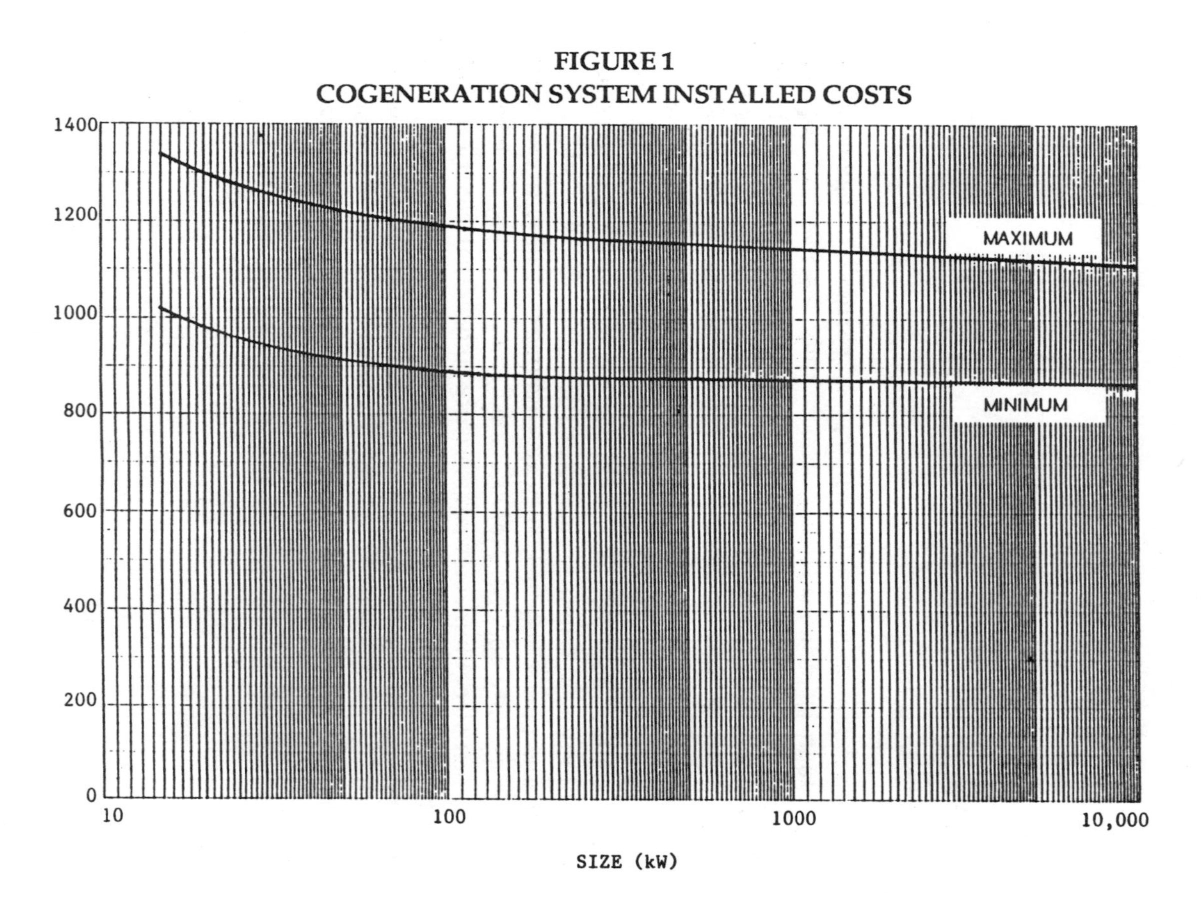

TABLE 1
BUILDING HVAC – COGENERATION COMPATIBILITY

SYSTEM	FEASIBILITY
Central heating and cooling plant	Good match
Two-pipe heating or cooling system	Good match
Water source heat pump	Good match
Radiant hot water panels	Good match
Gas or electric furnaces – both interior and rooftop	Possible, depending on costs
Hot water baseboard	Possible, depending on costs
Unit heaters – steam	Possible, depending on costs
Unit heaters – hot water	Possible, depending on costs
Air handling or built-up air conditioning system with heating coils and electric cooling	Possible, depending on costs
Electric baseboard	May require modifications
Unitary air source heat pumps	May require modifications
Unit heaters – gas or electric	May require modifications
Unitary electric heating-cooling units	May require modifications
Radiant panel, electric heating	May require modifications

TABLE 2
O & M COSTS – "OM"

PRIME MOVER	COST ($/kWh)
Reciprocating Engine	.0150
Gas Turbine	.0050

TABLE 3
ENGINE CHARACTERISTICS – "FUEL" AND "HRF"

PRIME MOVER	FUEL (MMBTU/KWH)	HRF (MMBTU/KWH)
Reciprocating Engine		
Less than 100 kW	.0130	.0060
100 kW to 500 kW	.0125	.0055
More than 500 kW	.0110	.0050
Gas Turbine		
Less than 1,000 kW	.0167	.0095
1,000 kW to 5,000 kW	.0133	.0070
5,000 kW to 15,000 kW	.0111	.0050
Larger than 15,000 kW	.0100	.0045

TABLE 4
ENGINE OPERATING HOURS – "HOURS"

PRIME MOVER	HOURS (HOURS/MO)
Reciprocating Engine	667
Gas Turbine	717

TABLE 5
WALKTHROUGH WORKSHEET

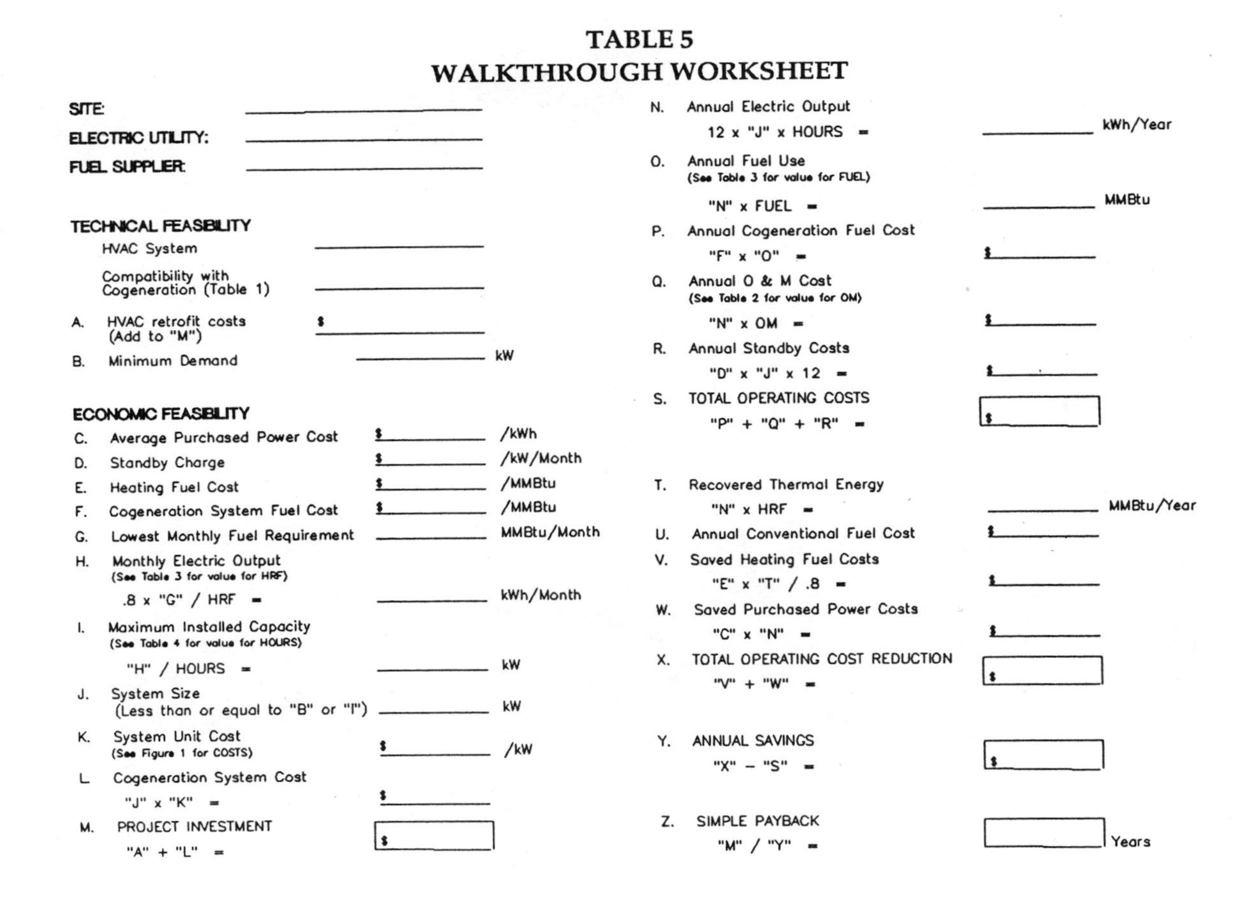

SITE: ____________

ELECTRIC UTILITY: ____________

FUEL SUPPLIER: ____________

TECHNICAL FEASIBILITY

HVAC System ____________

Compatibility with Cogeneration (Table 1) ____________

A. HVAC retrofit costs (Add to "M") $ ____________

B. Minimum Demand ____________ kW

ECONOMIC FEASIBILITY

C. Average Purchased Power Cost $ ____________ /kWh

D. Standby Charge $ ____________ /kW/Month

E. Heating Fuel Cost $ ____________ /MMBtu

F. Cogeneration System Fuel Cost $ ____________ /MMBtu

G. Lowest Monthly Fuel Requirement ____________ MMBtu/Month

H. Monthly Electric Output
(See Table 3 for value for HRF)
.8 x "G" / HRF = ____________ kWh/Month

I. Maximum Installed Capacity
(See Table 4 for value for HOURS)
"H" / HOURS = ____________ kW

J. System Size
(Less than or equal to "B" or "I") ____________ kW

K. System Unit Cost
(See Figure 1 for COSTS) $ ____________ /kW

L. Cogeneration System Cost
"J" x "K" = $ ____________

M. PROJECT INVESTMENT
"A" + "L" = [$ ____________]

N. Annual Electric Output
12 x "J" x HOURS = ____________ kWh/Year

O. Annual Fuel Use
(See Table 3 for value for FUEL)
"N" x FUEL = ____________ MMBtu

P. Annual Cogeneration Fuel Cost
"F" x "O" = $ ____________

Q. Annual O & M Cost
(See Table 2 for value for OM)
"N" x OM = $ ____________

R. Annual Standby Costs
"D" x "J" x 12 = $ ____________

S. TOTAL OPERATING COSTS
"P" + "Q" + "R" = [$ ____________]

T. Recovered Thermal Energy
"N" x HRF = ____________ MMBtu/Year

U. Annual Conventional Fuel Cost $ ____________

V. Saved Heating Fuel Costs
"E" x "T" / .8 = $ ____________

W. Saved Purchased Power Costs
"C" x "N" = $ ____________

X. TOTAL OPERATING COST REDUCTION
"V" + "W" = [$ ____________]

Y. ANNUAL SAVINGS
"X" − "S" = [$ ____________]

Z. SIMPLE PAYBACK
"M" / "Y" = [____________] Years

INDEX